Food Poisoning and Food Hygiene

Food Poisoning and Food Hygiene

Betty C. Hobbs

OStJ, DSc, PhD, FRCPath, DipBact, FRSH, FIBiol, FIFST, FAIFST
Formerly Director, Food Hygiene Laboratory, Central Public Health Laboratory, London

and

Diane Roberts

PhD, FIBiol, Deputy Director, Food Hygiene Laboratory, Central Public Health Laboratory, London

Sixth edition

A member of the Hodder Headline Group
LONDON • SYDNEY • AUCKLAND

First published in Great Britain 1953
Sixth edition 1993
Fourth impression 1997 by Arnold,
a member of the Hodder Headline group,
338 Euston Road, London NW1 3BH

British Library Cataloguing in Publication Data
Hobbs, Betty C
Food Poisoning and Food Hygiene. – 6Rev.ed
I. Title II. Roberts, Diane
363.1927

ISBN 0 340 53740 X

Typeset in 11/12 Times by
Anneset, Weston-super-Mare, Somerset
Printed and bound in Great Britain by
St Edmundsbury Press Ltd, Bury St Edmunds, Suffolk and
J W Arrowsmith Ltd, Bristol

Contents

Preface to sixth edition

The aim of the sixth edition is the same as that of earlier editions, to bring facts about the causes of food poisoning and other food-borne diseases to those concerned with food microbiology and food hygiene, in order to improve preventive measures. The intention has always been to present the facts as simply as possible so that non-technical readers will be interested as well as those trained in the many disciplines required. The book is not intended to be a detailed microbiological teaching manual covering all aspects of the subject, but rather seeks to involve those who need a practical and general knowledge of the relevant organisms which affect food in relation to the consumer. Emphasis is given to the main aspects of hygiene necessary for the production, preparation, sale and service of safe and palatable food.

Since the publication of the fifth edition there has been increasing concern for the safety of food, particularly with regard to the infection of hens' eggs with organisms of the *Salmonella* group and the contamination of various foods with *Listeria monocytogenes*. As a result the Committee on the Microbiological Safety of Food was formed under the chairmanship of Sir Mark Richmond. A large number of recommendations were made, many of which were accepted by the Government. Food law has been strengthened by the introduction of the Food Safety Act, 1990 and subsequent legislation. Some of the chapters in this new edition have been extended and revised to take into account these developments and the difficulties to be overcome by food microbiologists and the food industry during this time. In particular, Chapter 17, *Legislation*, has been completely revised to include many new regulations in the UK since the last edition, and many Directives issued by the European Community relating to food safety.

Chapters on *Cleaning methods* and *Sterilization and disinfection* have been combined to provide a new chapter *Cleaning and disinfection*. The chapter on *Food premises and equipment* has also been extensively revised. The remaining structure of the book remains substantially the same except for the inclusion of Appendix B, *Travelling and camping*, in the chapter on

Education and the short proforma for food-poisoning outbreaks (formerly Appendix C) in *Epidemiology*.

Appendix D, *Bibliography and further reading* has been extended and organized into sections to allow the reader easier access to information, and is now Appendix B.

Acknowledgements

The facts assembled in this book have come from the experience of many workers in the field of food-borne disease and food hygiene. We acknowledge them with gratitude and give our thanks to many helpers and critics. The late Sir Graham S. Wilson, the late Professor R. Cruickshank and the late Lieutenant-Colonel H.J. Bensted are remembered particularly for their inspiration and advice with the groundwork of the book; Sir Graham Wilson gave many helpful suggestions for the improvement of the fourth edition.

We thank Miss Wendy Spence and Mr Malcolm Kane from J. Sainsbury plc, London for writing the chapters on retail and factory hygiene; their experience in these fields is gratefully acknowledged.

We are grateful to Mrs Isobel Maurer and Dr Henrietta Schefferle for the original chapters on *Sterilization and disinfection* and *Legislation* respectively. Mr Peter Hoffman, Division of Hospital Infection, Central Public Health Laboratory, Colindale, has revised and combined the chapters on *Cleaning methods* and *Sterilization and disinfection* into a new chapter, *Cleaning and disinfection*. *Legislation* has also been fully revised by Mr Michael Jacob, formerly Chief Environmental Health Officer, Health Aspects of Environment and Food Division, Department of Health: we are grateful to Mr Jacob and to Mr Hoffman who also contributed many helpful suggestions for the fifth edition when his experience of Indian practice in a large hospital was useful.

Dr Robert Barrell, Public Health Laboratory, Manchester, provided the comments on Africa in the chapter on *Food hygiene in the tropics*, and we thank him. Dr Betty Cowan, OBE, supplemented the tropical data from her experience in rural and urban districts of India.

The advice of Dr Terry Roberts, formerly at the Food Research Institute, Bristol, and Dr Stephen Palmer, CDSC Welsh Unit, Cardiff, reviewed early typescripts of *Microbiological specifications* and *Epidemiology* respectively and we thank them. For *Epidemiology* in the sixth edition, Dr Beverley Booth, Christian Medical College and Hospital, Ludhiana, Punjab, gave corrective advice.

The chapter on *Premises and equipment* originally composed together with the late Mr L. Kluth, amended later by Mr J. Simpson and by Mr

R. Jowett of the Polytechnic of the South Bank, has now been considerably revised by Mr R. Griffin, Principal Environmental Health Officer of the Department of Health, we are very grateful to him. Mr J.R. Jones, Department of Environmental Health and Science, The College of North-East London, has continued to provide up-to-date information on pesticides and we thank him; the *Control of infestation* chapter was originally checked at the Pest Infestation Control Laboratory of the Ministry of Agriculture Fisheries and Food and also by Rentokil Ltd.

Mr R.A. Sprenger, Wakefield Metropolitan District Council, provided information on current food-hygiene courses given in the chapter on *Education*. Dr J.C. Kelsey's Appendix B, giving notes on travelling and camping has been combined with the chapter on *Education*, and the proforma for the investigation of food poisoning, Appendix C, has been moved to *Epidemiology*.

All amendments suggested by Dr Robert Charles, now of the Department of Social Security, were included in the fifth edition and remain in the text, we are grateful to him for the thorough review.

Miss J. Branson, formerly Principal Lecturer in Applied Science at the Ealing College of Further Education, gave many helpful suggestions for the revision of *Education* for the fifth edition.

Appendix A, giving brief lecture notes has been revised, although the order remains substantially the same. They were originally written by Miss Catherine F. Scott, retired School Meals Organizer for the County of North Yorkshire, and the late Miss Irene J. Martin, Domestic Science Organizer for the County of North Yorkshire. The notes in their original form appeared in *Hygienic food handling* (St. John Ambulance Association). They are useful for those giving short courses of instruction to non-technical groups concerned with the study of food-hygiene measures. We are grateful to those who made the notes accessible.

The essential statistics were provided by the PHLS Communicable Disease Surveillance Centre, Colindale, and we thank those who made the data available.

We thank our publisher, Edward Arnold (Publishers) Ltd, whose encouragement and optimism, throughout every edition, including the sixth, gave us the inspiration and the strength to complete the tasks.

We will always remember with gratitude Miss Nancy Cockman, whose cooperation and shared responsibility with earlier editions were invaluable.

There are others who have helped by typing almost illegible scripts, Mrs Joan Hook, Friends of Ludhiana, Mrs Thompson, Office Equipment, Dartmouth, Mr Verghese of the Christian Medical College and Hospital, Ludhiana, Punjab. Miss Betty Whyte, former librarian of the PHLS Central Public Health Laboratory, Colindale, read and corrected the proofs of the fifth edition, and we are grateful to her for the exceptional proficiency with language and references.

Many of the illustrations designed to give better understanding of outbreak strategy are based on those drawn by Mr W. Clifford, now retired, and several photographs were taken by Mr. J. Gibson from the Central Public Health Laboratory, Colindale. The new photographs in chapter

14 were prepared by Mr Peter Hoffman with the help of Tricia Jones and John Dunne. Some of the illustrations in Chapter 6 were reproduced from *Hygienic food handling* by kind permission of the St John Ambulance Association, two were originally drawn by the late Miss Ena Still.

We thank Dr Richard J. Gilbert, Director of the Food Hygiene Laboratory, Colindale, and co-author of the fourth edition, for suggestions with regard to the arrangement of some of the chapters for the fifth edition.

Many organizations have given information and photographs; they are listed as follows:- Hobart Manufacturing Co. Ltd; Clenaglass Electric Washer Ltd, Farnborough; Quantas of Sydney; Imperial Machine Co., Rickmansworth, Herts; Crypto Peerless Ltd; Hendon Precision Engineering Co. Ltd; J. Sainsbury plc; Marks & Spencer plc; Moorwood-Vulcan Ltd; Stott-Benham Ltd, Oldham; Dr Crawford Dow, University of Warwick.

We are grateful to those who contributed to earlier editions: Bowater-Scott Corporation Ltd, London; British Railways Board; Combined Laundry Group, London; Dawson Bros Ltd, Woodford Green, Essex; Euk Catering Machinery Ltd, Oldham; Express Dairy Co. Ltd, London; The Gas Council; Hoover Ltd, Greenford, Middlesex; Mr A.C. Horne, Chief Health Inspector, Hemel Hempstead; Jeyes' Sanitary Compounds Co. Ltd, Chigwell, Essex; King Edward's Hospital Fund for London; Mac-Fisheries Ltd, London; Moreton Engineering Co, Ltd, Moreton-in-Wirral, Cheshire; The Chief Constable, York and North East Yorkshire Police; Pressarts Ltd, Leicester; Quiz Electrics Ltd, Teddington, Middlesex; Staines Kitchen Equipment Co. Ltd, London; James Stott & Co. (Engineers) Ltd, Oldham; Waitrose, Food Group of the John Lewis Partnership; T. Wall & Sons (Ice-cream) Ltd, London.

This edition is dedicated to all those who would make food safe from the dangers of unwanted microorganisms.

Abbreviations

a_w	Water activity
BS	British Standard
CAP	Controlled Atmosphere Packing
CIP	Cleaning in place
CCDC	Consultant in Communicable Disease Control
CCP	Critical Control Point
CDC	Centers for Disease Control
CDR	Communicable Disease Report
CDS	Communicable Diseases (Scotland) Unit
CDSC	Communicable Disease Surveillance Centre
DH	Department of Health
DHSS	Department of Health and Social Security
DSS	Department of Social Security
EAEC	Enteroadhesive *Escherichia coli*
EC	European Community
EHEC	Enterohaemorrhagic *Escherichia coli*
EHO	Environmental Health Officer
EIEC	Enteroinvasive *Escherichia coli*
EPEC	Enteropathogenic *Escherichia coli*
ETEC	Enterotoxigenic *Escherichia coli*
FAO	Food and Agriculture Organization
GCP	Good Commercial Practice
GMP	Good Manufacturing Practice
HACCP	Hazard Analysis Critical Control Point
HAZOP	Hazard and Operability Studies
HTST	High Temperature Short Time
IAMS	International Association of Microbiological Societies
ICMSF	International Commission on Microbiological Specifications for Foods
IEHO	Institution of Environmental Health Officers
ISO	International Organization for Standardization
kGy	kiloGray (unit of radiation measurement)

LT	Heat Labile Toxin
MAFF	Ministry of Agriculture, Fisheries and Food
Mo(s)EH	Medical Officer(s) of Environmental Health
NaDCC	Sodium dichloroisocyanurate
NBGE	Non-bacterial gastroenteritis
NCV	Non-cholera vibrio
NHS	National Health Service
OPCS	Office of Population Censuses and Surveys
PHLS	Public Health Laboratory Service
PT	Phage Type
PVC	Polyvinyl chloride
PVP	Polyvinylpyrrolidone
QC	Quality Control
QAC	Quaternary ammonium compound
RIPHH	Royal Institute of Public Health and Hygiene
RSH	Royal Society of Health
SO	Scottish Office
ST	Heat Stable Toxin
SVS	State Veterinary Service
UHT	Ultra Heat Treatment
UV	Ultra Violet
VT	Verotoxin
VTEC	Verotoxigenic *Escherichia coli*
WHO	World Health Organization
WO	Welsh Office

Part 1

Food poisoning and food-borne infection

1

Introduction

Historical perspective

Food hygiene is a subject of wide scope. It aims to study methods for the production, preparation and presentation of food which is safe and of good keeping quality. It covers not only the proper handling of every variety of foodstuff and drink, and all the utensils and apparatus used in their preparation, service, and consumption, but also the care and treatment of foods known to be contaminated with food-poisoning bacteria which have originated from the animal host supplying the food.

Food should be nourishing and attractive. It must be visibly clean and it must also be free from noxious materials. These harmful substances may be poisonous chemicals and even chemicals harmless in small amounts, but damaging in large quantities. They may enter the food accidentally during growth, cultivation, or preparation, accumulate in the food during storage in metal containers, form in the food through interaction of chemical components, or they may be concentrated from the natural components of the food. Microorganisms (germs) may be introduced directly from infected food animals or from workers, other foods, or the environment during the preparation of foods. Poisonous substances may be produced by the growth of bacteria and moulds in food.

This book has been written for those engaged in food handling, with the purpose of explaining simply, in so far as our knowledge permits, the nature of these various dangers, how they arise, and how some of them can be prevented.

Noxious substances in food give rise to illness called food poisoning or gastroenteritis, which is usually characterized by vomiting and/or diarrhoea, and various abdominal disturbances. Food poisoning is no new disease, it has been recognized throughout the ages. Centuries ago the laws of the Israelites contained detailed information on foods to be eaten and foods to be abhorred, as well as on methods of preparation and the cleanliness of hands.

About 2000 BC, the book of Leviticus records that Moses made laws to protect his people against infectious disease. Hands were to be washed after killing sacrificial animals and before eating. Laws were given about edible animal life of all kinds. Prohibited animals included swine, now known to be frequent excreters of salmonellae as well as reservoirs of parasites, such as *Trichinella spiralis* and *Taenia solium* (tape worm), and small creatures such as mice, rats, lizards, snails and snakes, all known to harbour salmonellae. Of the inhabitants of the water, only those with fins and scales could be eaten, which eliminated sea mammals, shellfish and crustaceans. Among the birds there is a long list of prohibited species known to be scavengers, including vultures, eagles, seagulls and herons. Quails were edible, but when the Israelites disobeyed instructions not to store or gather more than a day's supply, and dried them all about the camp, they were struck with a 'deadly plague'. Flying insects with four legs were forbidden, but those that jump, locusts, crickets and grasshoppers could be eaten. Even the dead bodies of the prohibited creatures were said to defile those that touched them.

The accounts of food poisoning recorded in ancient history have been associated generally with chemical poisons, mostly given with intent to kill. Nowadays, a small proportion of food poisoning, probably less than 1% in the UK, but a greater percentage in Eastern countries, is due to chemical substances. Heavy metals such as lead, zinc, copper, cadmium, mercury and arsenic may reach food, mainly from containers, in quantities beyond the permitted limits. Powders such as nicotinic acid and sodium fluoride may be mistaken for flour. Flour or oils may be placed in barrels previously used for insecticides, herbicides or fungicides. Toxic waste from factories is frequently discharged into rivers and streams killing fish and sometimes causing illness in the local population.

The organic life in drifting plankton includes protozoa, and when mussels feed on the marine dinoflagellate, *Gonyaulax tamarensis* (a member of the Pyrrhophycophyta) they can cause paralytic shellfish poisoning when eaten raw. Saxitoxin is one of the toxins involved. Some fish, mostly in warm waters, can cause ciguatera poisoning, which is a well-recognized and serious illness in tropical and subtropical regions such as the Caribbean Sea waters and the tropical Pacific. Fish, in particular the larger and older specimens, acquire ciguatera toxin by eating smaller herbivorous fish feeding on the dinoflagellates that produce the toxin. The dinoflagellate protozoa occur in certain coral areas at the bottom of the sea. Symptoms include nausea, vomiting and diarrhoea, which may be severe, and in many cases may begin immediately or within a few hours of ingesting the fish. There are neuro-sensory symptoms also, in particular paraesthesia (numbness and tingling) and dyaesthesia (sensation of burning on contact with cold). Severe cases result in shock and convulsions, paralysis and death. The illness is rare in temperate countries, but cases have occurred when fish such as speckled moray eel, amber jack and barracuda are consumed abroad or brought back for consumption at home. Importation of potentially ciguatoxic fish is discouraged in the UK.

The red algae (Rhodophycophyta) produce carrageenans of two types, high and low molecular weight; those of high molecular weight are used as food additives. There is evidence that the low molecular weight carrageenans are toxic and can induce physiological effects, such as inflammation and immunosuppression.

The blue-green algae, *Microcystis* (Cyanobacteria), are commonly found in fresh and brackish water; growth is stimulated by favourable climatic conditions and increased levels of nitrate and phosphate in the water. Blooms and scums may form and although not always harmful, blue-green algae can release toxic substances that have been responsible for the death of animals and sickness in humans.

Naturally-occurring plant life such as toadstools, hemlock and deadly nightshade can cause illness, and even fatalities. Red kidney beans have also caused food poisoning: outbreaks reported in the UK followed the consumption of the raw or incompletely cooked red beans, *Phaseolus vulgaris*. Symptoms include nausea and vomiting followed by diarrhoea and sometimes abdominal pain after an incubation period of 1–7 hours. The toxic factor is a naturally occurring haemagglutinin (lectin) in the bean. It can be destroyed by adequate cooking, such as boiling well-soaked beans for 10 minutes. Apricot kernels contain a glucoside, amygdalin, which can be hydrolysed to release cyanide. Children in Turkey have died from cyanide poisoning after consuming wild apricot kernels. An anti-cancer compound, also extracted from apricot kernels, has caused fatalities.

An unusually high concentration of solanine in the skin or green shoots of potatoes is occasionally reported as a cause of food poisoning both in the UK and overseas. An epidemic of a newly described disease, the 'toxic oil syndrome', affected some 20 000 people and caused more than 350 deaths in Spain after the ingestion of food oil purchased from itinerant vendors in 1981 and 1982. The industrial rapeseed oil from France, denatured by the addition of aniline, was 'refined' in Spain before distribution. Before sale, the oil was further diluted with many other oils of plant and animal origin. Epidemiological evidence linked the disease with oils containing rapeseed oil and anilides not removed by treatment. Symptoms included fever, respiratory distress, nausea and vomiting, eruptions, general discomfort, headache, abdominal pain and myalgia. Cereals may be infected with the fungus *Claviceps purpurans* which forms a poison responsible for ergotism, and *Aspergillus flavus* which produces aflatoxin, as well as other mould products. Acute vomiting usually occurs within a few minutes to half an hour of ingestion; the incubation period of most bacterial and viral food poisoning is longer.

The ptomaines, alkaloids, are basic chemical substances formed by the breakdown or digestion of putrefying tissues; they were previously thought to be poisons formed in tainted foods, because food poisoning was assumed to be associated with taint. It is now recognized that foods heavily contaminated with food-poisoning germs may be normal in appearance, odour and flavour, although our knowledge of bacterial food poisoning dates back no further than the latter part of the nineteenth century. Microbiology is the science that deals with the study of microorganisms, including bacteria, fungi and moulds, viruses and parasites. Bacteriology is so-called from the Greek word 'bactron', rod, because the first germs seen through a microscope were tiny straight rods. Virology is the study of viruses, the smallest known microorganisms. Mycology is the study of fungi and moulds. Parasitology covers organisms of different orders which parasitize man and animals. In the wider field the subject embraces genetics, the study of genes and heredity and their effect on the structure and functioning of living organisms. Immunology can also be considered to be part of the same science with the study of factors involved in the response of a host to a specific challenge with a foreign agent. Immunology has now developed into a medical speciality in its own right.

Bacteria, variously called germs, microbes, microorganisms or simply organisms, were first seen and described in 1675 by A. van Leeuwenhoek, a linen draper in the town of Delft in Holland. He made lenses and magnifying apparatus and mounted lenses together to form a primitive microscope which revealed tiny animalcules, as he called them, in a drop of pond water placed on a slide. He described their size as one thousand times smaller than the eye of a louse. Scrapings from his teeth revealed similar objects. His drawings leave no doubt that these were the first bacteria to be described.

The importance of his observations was not appreciated for nearly 200 years until in 1859 Louis Pasteur, the great French chemist and bacteriologist, demonstrated the role of bacteria in fermentation processes. He developed methods of growing bacteria necessary for detailed studies.

Pasteur investigated the silkworm plague which threatened to ruin the silk trade in France. He showed that the disease was caused by a bacterial infection of the worm, and suggested measures for control. Pasteur investigated many diseases of man and animals, including rabies, and proved that bacteria and viruses were the cause of many of them; he is famous for making the vaccine that controls rabies. Another aspect of his work is of special interest in the study of food hygiene. He showed that the old theory of spontaneous generation, that is, life arising from the inanimate, was false. Thus, if food were sterilized by heat, living bacteria would not reappear unless introduced from the hands, for example, or from some other contaminated material.

At about the same time, Robert Koch, working in Germany, proved that anthrax, tuberculosis and cholera were caused by bacteria and he also devised methods of growing the germs. In Europe, America, Japan and other parts of the world enthusiastic microbiologists established the causative microbes of gonorrhoea, erysipelas, diphtheria, typhoid fever, dysentery, plague, gangrene, boils, tetanus, scarlet fever and other diseases. Thus, after thousands of years the cause of infection was revealed and the door opened for studies on the relationship between bacteria and disease in man and animals. These studies began to show the way in which bacteria spread and invade the human and animal body, and methods of prevention and cure were investigated.

Joseph Lister applied the theory and practice of Pasteur to surgery and found that wounds became septic by the action of bacteria. He introduced the use of antiseptics and disinfectants that would kill bacteria; there was a steady reduction in wound sepsis.

Before 1850, the sanitary conditions in Britain were poor, though from 1840 onwards the 'Great Sanitary Awakening' had begun. Edwin Chadwick, a lawyer, belonged to a family with a strong belief in personal cleanliness, and in 1842 his *Report on the Sanitary Conditions of the*

Labouring Population of Great Britain was published. The report included an analysis of the most important causes of death, 'consumption', typhus and scarlet fever, their distribution and with emphasis on the squalor, inadequate sewage disposal and contaminated sources of water. There was greater awareness that environment influenced the physical and the mental well-being of the individual. The connection between dirty conditions and disease was gradually understood and measures were taken to control the disposal of sewage and the purity of water supplies. In 1859, Florence Nightingale published *Notes on Nursing: What it is and what it is not*, which stressed the importance of hygiene in the home, and scrupulous cleanliness in caring for the young, the sick and the elderly. She gave guidance on diets, storage of food and milk products and prevention of the spread of infection. The book greatly influenced the control of infection in those days and is relevant even today.

In 1854, John Snow recognized that drinking water could spread cholera. William Budd in 1856 concluded that typhoid fever was spread by milk or water polluted by the excreta of infected persons. Prince Albert died of typhoid fever in 1861. John Snow and William Budd were amongst the earliest epidemiologists searching for the sources and chains of infection, even before the discovery of responsible organisms and the means to grow them. In 1874, an outbreak of typhoid fever in the Swiss town of Lauren was traced to polluted water with the result that water supplies and sewage systems were redesigned.

The chlorination of drinking water in Britain was initiated by Alexander Houston in 1905 during a typhoid epidemic in Lincoln, and this development has helped to abolish water-borne diseases in the UK and other countries. Towards the end of the nineteenth century the danger of infection from milk was recognized, and in cities such as London the heat treatment of milk by pasteurization began. Pasteurization kills many bacteria in the milk, including those that are harmful. Incidents of tuberculous infection from raw milk are no longer seen in the UK because of pasteurization and also because the infection of cows has been eliminated. The eradication of brucellosis in cattle has similarly reduced the incidence of undulant fever. Food poisoning or food-borne infection from organisms such as the salmonellae and *Campylobacter* still occur from 'untreated' milk and from imperfectly heat-treated milk (see Chapter 17 for Milk Regulations, 1989). Scottish law (1983) requires the heat treatment of all milk before distribution. Milk-borne and water-borne outbreaks are described in Chapter 4.

There have been many advances in the treatment of the common food and drink commodities to increase safety. Water is purified by sedimentation, filtration and chlorination; milk is heated, cooled quickly and carefully packed in clean bottles or cartons; ice-cream mix is heated, cooled quickly and stored cold until frozen; liquid whole egg mix is heated, cooled quickly and frozen. Many foods are preserved by heat, cold, dehydration, irradiation or chemicals before they reach the kitchen, but many are received in the raw state.

Raw foods can bring food-poisoning organisms into kitchens and processed foods may be contaminated by them directly or by transfer of the organisms via hands, surfaces and equipment. The food handler can reduce the rate of growth of bacteria in food by careful attention to speed of preparation and limiting exposure to kitchen temperature, quick cooling and cold storage (Chapter 8). When raw foods are known to have a consistently high rate of contamination with salmonellae, from animal sources for example, efforts should be made to reduce risks in production. Food hygiene education will help food handlers to be aware of dangerous practices and the precautionary measures necessary.

Food-poisoning bacteria were first described by Gaertner in 1888. They were isolated from the organs of a man who had died during an outbreak of gastroenteritis affecting 59 persons in Germany; similar bacteria were found in the meat served to the victims and also throughout the relevant carcass of beef. The bacteria were later named *Salmonella*. At about the same time, the ptomaine theory of food poisoning was disproved by volunteers who consumed ptomaines (alkaloids), extracted from putrid food, without ill effects. Under the influence of W. G. Savage in the UK and E. O. Jordan in the USA food poisoning came to be associated with specific bacterial contamination.

In 1896, E. van Ermengem in Belgium described *Clostridium botulinum*, the organism responsible for botulism. The toxin from this organism, formed in certain imperfectly preserved foods, affects the nervous system – often fatally. Formerly the toxic disease was not uncommon, when recognized, and it is still reported in many European countries. However, where good commercial practice exists and home preservation of non-acid foods such as meat, fish and vegetables is discouraged (except by blanching and freezing), only exceptional circumstances will lead to outbreaks (see p. 114 and p. 115).

In the years 1909 to 1923 many of the bacteria now known to be responsible for a large proportion of food-poisoning incidents were grouped together under the generic name *Salmonella* in honour of Dr. E. Salmon, who isolated the first member, the hog cholera bacillus, in 1885.

From 1914 onwards the staphylococci became associated with a violent toxic form of reaction. Certain strains while growing in cooked food produce a toxin which, although without visible or flavour effect, gives rise to the rapid onset of vomiting.

From 1945 to 1953, a fourth major cause of diarrhoea, the anaerobic sporing bacillus *Clostridium perfringens*, was recognized as an agent of food poisoning with diarrhoea as the predominant symptom. This organism resembles *C. botulinum*, but the mechanism of reaction is different, and the illness is usually relatively mild with less disastrous results.

From time to time contaminant bacteria in food are shown to be agents of food poisoning, such as *Bacillus cereus* and other aerobic sporing bacilli and also *Escherichia coli*. They may be present in the raw food ingredients and allowed to accumulate on surfaces or equipment; they may survive preparation of the food or they may recontaminate the food after cooking. In all cases the safety of the food depends on keeping the numbers low.

The *Campylobacter* are now a cause for concern as agents of human gastroenteritis. Contaminated milk and water are the usual vehicles of infection. Poultry, frequently contaminated, are also implicated.

Clean food is free from visible dirt and bacterial spoilage. The aim of food hygiene is the production of food which is both safe and clean. Four main factors are important:

(i) the initial safety of raw animal products before entry into the food industry, shops, hotels, restaurants, canteens and home kitchens;
(ii) the hygiene and care of those handling food during production and service;
(iii) the conditions of storage;
(iv) the general design and cleanliness of kitchens and equipment.

A public health bacteriological laboratory service was instituted for England and Wales in 1939. It supplemented the work of the public and private analysts as an emergency service during World War II, and later it was established as the Public Health Laboratory Service (PHLS). The increase in the number of laboratories provided facilities for the detailed investigation of food poisoning and other food-borne disease. Records for food-poisoning outbreaks and incidents were initiated in 1950 when reports to the PHLS and the Ministry of Health were combined, and in 1968 notification of outbreaks of food-borne disease was made statutory. (See Chapter 17). Teaching on food hygiene became popular and requests for lectures and demonstrations came from local authorities, the food industry, catering services and groups from public and private organizations. In 1955, legislation on food hygiene was passed. (See Chapter 17).

The popularity of communal feeding has gradually increased over the years, especially since World War II when food was scarce and the limited resources had to be shared. Some kitchens were not designed for large-scale catering and the staff were liable to be overwhelmed under cramped conditions and with limited equipment. It seems that the incidence of food poisoning began to rise then, although the lack of earlier records may give a wrong impression. Eating houses in earlier days served mostly freshly cooked hot meals; usually there were large cold cellars in the basement. Over the years made-up meat dishes and precooked foods have become ever more popular and 'convenience' foods have proved helpful for the increasing numbers of working housewives. The importation of foods including meats, egg products, coconut and dehydrated feeding meals necessary for animal consumption, increased in post-war years. Investigations indicated the hazards of many imports found to be contaminated with intestinal pathogenic bacteria and, in particular, salmonellae. Factory farming developed for economic reasons, and although it increases production, there are inherent risks of infection in livestock and consequent spread to the human population.

Current situation

In spite of extensive teaching on food hygiene for the prevention of food-borne disease, the incidence of outbreaks and sporadic cases continues to increase (Table 1.1).

In 1990 in England and Wales, food poisoning from all causes, from all types of Salmonella and from *Salm. enteritidis* increased compared with 1989. Of the outbreaks attributed to *Salm. enteritidis* phage type 4 eggs were still a predominant vehicle of infection. Surveys of eggs, both home-produced from retail shops and imported, showed a variable proportion to be contaminated with *Salm. enteritidis* phage type 4. The sampling rate for imported eggs from all countries is usually 60 eggs per 360 000 or lorry load. In 1990, salmonellae were isolated from 48% of chicken reared in the UK with *Salm. enteritidis* phage type 4 the most common type, found in 21% of the birds. The rates of isolation have, so far, been higher in imported chicken, more than 80% positive for salmonellae. The common serotypes were *Salm. virchow* and *Salm. enteritidis*. In the UK, stringent measures were taken to prevent the spread of infection in breeding and laying flocks. Whole flocks were destroyed if one hen was found to be excreting *Salm. typhimurium* or *Salm. enteritidis* and many thousands of birds were destroyed – 3% of all laying flocks in Britain up to the end of 1989. Compulsory slaughter of laying flocks infected with *Salm. typhimurium* ceased in 1991, but continued for *Salm. enteritidis* infection. Poultry farms must now register laying and breeding flocks of more than 100 and 25 birds respectively; birds and flocks will be subject to compulsory testing for salmonellae. Resolute control of poultry feeds, both imported and home-produced, and also of breeding stock together with irradiation of poultry products would restrain the incidence of salmonellosis arising from poultry and eggs.

To combat the situation in 1975, two orders were introduced in the UK, the Zoonoses Order was passed in 1975 and the Protein Processing Order in 1981; both were revised in 1989. Under the Zoonoses Order the State Veterinary Service (SVS) must be notified when *Salmonella* or *Brucella* organisms are isolated from samples taken from farm animals, carcasses, products or the farm environment. Also powers under the Order apply strict control to the production of processed animal protein and animal feed stuffs. The second order, the Protein Processing Order requires that

Table 1.1 Recorded cases of bacterial food poisoning (England and Wales, 1970–1991)

Year	1970	1971	1972	1973	1974
Number of cases	8634	8079	6020	8574	8591
Year	1975	1976	1977	1978	1979
Number of cases	11 943	11 000	9204	10 590	11 881
Year	1980	1981	1982	1983	1984
Number of cases	10 856	10 665	12 684	15 168	15 312
Year	1985	1986	1987	1988	1989*
Number of cases	12 837	15 214	18 573	24 941	21 925
Year	1990*	1991*			
Number of cases	23 532	20 155			

Data from the Public Health Laboratory Service.
* all laboratory reports.

isolations of salmonellae from animal protein and feed stuffs be notified to the SVS. In addition, there are strict controls for imported fish or animal products under the Importation of Processed Animal Protein Order, 1981. The movement of animal protein contaminated with *Salmonella* can be prohibited by the SVS. Fuller details on legislation are given in Chapter 17.

The Campylobacter, vibrio-like organisms, are now regularly reported as agents of food-borne disease. Milk and water are the usual vehicles. In the UK and USA the incidence of gastroenteritis due to *Campylobacter jejuni* may equal or exceed that of salmonellosis, although sporadic cases are more common than outbreaks, and the organism is frequently found in or on poultry carcasses. *Campylobacter jejuni* requires special conditions for growth, which explains the difficulties experienced in tracing its source, means of spread and survival in food.

More attention is required to ensure ample storage facilities at low temperatures and to provide methods to cool food quickly. The Food Hygiene (Amendment) Regulations, 1990, give directions for storage temperatures for retailed high-risk foods (see Chapter 17). An influx of salmonellae on raw meat or poultry and the spread of the organisms within a kitchen environment may result in an unusually high rate of excretion amongst food handlers. The possibility of diarrhoeal cases amongst the staff may increase the spread of infection still more. An outbreak of salmonellosis in a hospital with many deaths amongst geriatric patients resulted from long storage of cooked meats at warm atmospheric temperature; the high rate of infection indicated that an unusual amount of contamination had reached the kitchen in or on a food product, probably the raw meat. In hospital situations the virulence of organisms may be enhanced by rapid passage from person to person, particularly between patients with low immunity, so that small numbers of organisms (low dosage) initiate illness.

The suggested involvement of foods cooked in microwave ovens in outbreaks and cases of salmonellosis due to *Salm. enteritidis* phage type 4 has given rise to concern about the efficiency of the process. Also a growing awareness of the dangers from *L. monocytogenes* has also focussed attention on microwave cooking. There are no reported cases that directly link illness with the use of microwave ovens. It seems unlikely that the hazards are any greater than with traditional methods of cooking. The after-care of food is the most important factor.

The incidence of food poisoning caused by *C. perfringens* is little changed and there have been serious outbreaks in hospitals. The reasons are clear, but it is difficult to persuade architects and food handlers to provide rapid cooling facilities before refrigeration for cooked poultry, especially large turkeys, and meat. With the growing popularity of turkey meat, walk-in cold stores are necessary. The time and temperature of storage for foods after cooking and before eating is of prime importance in all types of food poisoning, but especially for the sporing organisms able to survive cooking. Lack of care in storage can change non-infective or non-toxic numbers of organisms into massive clinical doses. Cooks and other kitchen personnel need to understand the two basic dangers: (a) keeping cooked food at warm or ambient temperatures, and (b) the recontamination of foods after cooking by hands, surfaces and equipment. The serological identification of *C. perfringens* and the rapid diagnosis of enterotoxin in stools are useful epidemiological tools.

Bacillus cereus also survives cooking by means of spores and the organism appears to have a natural habitat in cereals; outbreaks from toxin in rice are not unusual in Britain. Again, growth and toxin production are encouraged by warm or cool storage and the rapid warm-through and quick-fry for 'take away' service do not render the food safe. The serological typing scheme is useful; not all types appear able to produce toxin.

The fatal botulism from the toxin of *C. botulinum* has, in recent years, been rarely seen in Great Britain, except from imported goods, for example, the canned salmon incident described on p. 114. Careful industrial methods of canning and curing when strictly observed are proof against the survival and outgrowth of spores. Any change in formulation without consideration may have disastrous results, for example, the hazelnut yoghurt outbreak, see p. 115. The 'floppy' baby syndrome is believed to be due to the presence of *C. botulinum* in the intestine, but the origin of the organism is still in doubt; honey is suggested as one source, but the organism has not been traced back to hives and bees. *C. botulinum* spores have been isolated from water tanks and soil in the vicinity of houses where the 'floppy' baby syndrome has been observed.

The vibrios are agents of infection in many parts of the world. Cholera and cholera-like illness from non-cholera vibrios (non-01 *Vibrio cholerae*) as well as infections due to *Vibrio parahaemolyticus* are recognized not only in the East, but also in Europe, Australia and the USA. Oysters and mussels fished from polluted waters and eaten without depuration (cleaning) are potential sources of illness in seafood meals; shrimps and prawns may also be infected.

Aeromonas hydrophila should not be neglected as an agent of food poisoning. It needs careful identification to avoid confusion with the vibrios. Incidents have been described in India and by Swedish workers at a Paediatric Clinic investigating diarrhoeal disease in Ethiopian children. Suggestions as to the possible vehicles are not reported.

Yersinia enterocolitica is the subject of investigation with regard to its role in food-borne disease, especially because it can grow in foods held at refrigerator temperatures. *Yersinia enterocolitica* is reported frequently from Belgium where raw ground pork is fed to very young children as a weaning food; pork figures prominently as a food vehicle. A few reports of outbreaks and incidents come from the Scandinavian countries, Norway, Sweden and Denmark and also from Finland, the Netherlands and Yugoslavia. Unusual conditions for isolation may account for the lack of importance elsewhere as an agent of food-borne infection. Eating habits may be an important factor also. *Y. enterocolitica* can grow in foods during refrigeration and should be considered a hazard in long-term storage of chilled food.

Listeriosis following infection with *Listeria monocytogenes* has increased significantly for the foetus, newborn baby and adult; abortion, meningitis and food-borne infection are reported. The association of listeriosis with the consumption of soft mould-ripened cheese prompted national warnings (USA and UK) to pregnant women and others in susceptible groups to avoid all varieties of such cheeses. In 1990 and 1991 the number of cases in the UK dropped significantly.

Environmental contamination by *Listeria* requires efforts in cleanliness to improve the general hygiene of factories preparing food including machinery and equipment. *Listeria* as well as *Y. enterocolitica* grows slowly in food at refrigerator temperatures.

Escherichia coli is well known as an intestinal pathogen; for babies the acute diarrhoea and dehydration bring a high fatality rate. Some serotypes are associated with diarrhoea in adults, and many food-borne outbreaks are reported. 'Traveller's diarrhoea' is commonly attributed to *E. coli* in food and water; the strains responsible are foreign to the travellers, whereas the local inhabitants are immune to them. A simple test is required to establish infectivity and toxigenicity amongst the many *E. coli* inhabitants of the intestine. Many are harmless, but there can be interaction between *E. coli* and salmonellae, for example, and the transfer of genetic material including virulence and resistance factors.

The approximate number of organisms in food is significant not only in relation to spoilage, but also to food poisoning. Numerical specifications or guidelines can help to improve the microbiological condition of foods so that spoilage and health hazards are reduced or eliminated. Standards applied on a strictly legislative basis and which may be too detailed are unlikely to be enforced, and are thus of little or no value as a compulsory measure.

The Hazard Analysis Critical Control Point (HACCP) approach has brought a fresh perspective to microbiological quality and safety (see Chapters 12 and 19). It has a role in the design of food premises and all functions related to food production.

The International Commission on Microbiological Specifications for Foods (ICMSF) continues annual meetings. The Commission's fifth book provides compiled data of characteristics in foods for all the known agents of food poisoning and other food-borne disease, *Characteristics of food-borne agents of disease*.

Changes made to traditional foods, such as reducing fat, sugar, salt and preservative content, may fail to take account of microbiological hazards. It is dangerous to abandon traditional methods of preparing, processing and preserving food in favour of new untested methods. An increase in vacuum-packed food and ready-cooked chilled convenience foods replacing frozen food may well have adverse microbiological consequences.

Careful legislation is necessary to prevent accidental contamination of food with pesticide residues used both in the home and for agriculture. Such accidents are rare in Britain, but they are reported in the Middle and Far East.

More information is required on the danger from mycotoxins (fungal poisons) for man and animals. There is much natural and purposeful contamination of food with moulds of various types, but little is known about the pathogenicity of the vast majority of species. There are many reports on aflatoxin from *Aspergillus flavus*, the toxin from the fusaria and some other fungi. There is more to learn about other mould products and the level of toxins in various cereals in daily use.

The food-borne parasitic infections are mostly important in tropical countries. They includes amoebiasis, giardiasis, and cryptosporidiosis from the protozoans *Entamoeba histolytica, Giardia lamblia* and *Cryptosporidium*; also the tape worm infestations (*Taenia solium* and *Taenia saginata*) and trichinosis (*Trichinella spiralis*). Little is known about the dose of organisms required for infection. Preventative recommendations advise against the consumption of unwashed and uncooked salad vegetables, undercooked or raw meat and untreated water. The precise mode of transfer, hand to food or directly from water is not known. The survival times outside the human and animal body host are significant. As with viruses, complex methods of isolation hinder investigation.

Toxoplasmosis caused by a small protozoan *Toxoplasma gondii* may be congenital or acquired by the ingestion of toxoplasma cysts in raw or undercooked meat of animals or birds. Also, oocysts shed in the faeces of cats and dogs may be accidentally ingested. The organism may pass through the placenta of the mother to the foetus in early pregnancy.

Each newly documented agent of food-borne disease needs careful investigation with regard to source, spread, growth rate, tolerance to external factors such as heat, cold and dehydration, and the mechanism of reaction. All data must be adapted to fit new items of manufactured foods.

Food-borne virus infection is difficult to trace and follow through. Improved techniques with concentration methods and electron microscopy have demonstrated virus particles in stool samples from patients with gastroenteritis when bacterial agents have not been found, for example, some shellfish outbreaks. Epidemiological investigations are hindered by the lack of evidence for small numbers of similar particles in foods. The life cycle of the virus suggests that viral particles, made up of genetic material and a protein capsid, may lie inert in food and water. Inside certain living cells the protein capsid may be removed, enabling the viral genetic material to participate actively in cell function. It seems that the free particles and viral shapes seen by electron microscopy may be active or passive. Poliomyelitis and viral hepatitis are thought to be transmitted by infected water and food. Shellfish harvested in areas polluted by human or animal sewage may harbour the viruses infecting persons and animals on shore. A virus requires living tissue for growth and is unable to multiply in food, but living shellfish may still contain large numbers of viral particles when swallowed. Viral infections from food and water are still difficult to trace satisfactorily. Various immunological techniques are now available, such as the ELISA (enzyme linked immunosorbent assay) which can identify rotavirus, for example, in the stools of children.

Epidemiology is an expanding science; for food poisoning and other food-borne disease it includes a study of each microbiological agent in relation to reservoirs and food vehicles of infection and intoxication as well as ways of spread. Investigations are not complete without knowledge of the behaviour of organisms in the various food vehicles. The information required is extensive, but necessary to comprehend preventive measures leading to control.

The chapters in Part I indicate the stages of present day information. There is a brief and simple description of the general characteristics of microorganisms, the agents concerned in food poisoning and other food-borne disease, the common food vehicles in which the organisms are consumed, the behaviour or ecology of the organisms in foods and the environment, the rationale of epidemiology and examples of outbreaks. A chapter on spoilage and preservation has been included. Statistics are relevant to the commonest agents and food vehicles in relation to numbers of outbreaks and cases; they must be collated and studied in order to focus attention on the faults in food production, processing and home preparation which need correction.

In Part II the chapters include the disciplines which must be studied before preventive measures can be effective. Consideration is given to care of the hands and other points of personal hygiene, to storage preparation and cooking methods, retail sale and factory practice. Coverage for the kitchen includes cleaning methods, design of premises and equipment, sterilization and disinfection. Microbiological specifications, legislation and education are necessary corollaries to control the incidence of disease from food. There is also a chapter on food hygiene in the tropics.

Suggestions are given for short courses on food poisoning and food hygiene; the facts may not be altogether comprehensive, but they will help to promote a better understanding of the subject. Teaching methods can be centred around actual outbreaks in the community and the home. Paths of spread from the original source to the food concerned should be clearly outlined. It is important also that outstanding features during the subsequent preparation of the food as eaten be highlighted to indicate the faults leading to gross contamination with the agent responsible. Pictorial charts are helpful. New legislative measures have been introduced and former legislation strengthened.

2

Elementary bacteriology

The size, shape and habits of bacteria

It is difficult to understand the chain of circumstances which must precede an outbreak of disease caused by contaminated food, and which must be broken by those handling food, without knowing something about the organisms responsible.

The characteristics of bacteria in relation to their source and reservoir and a knowledge of their ability to survive and grow in food are important factors in the control of both food poisoning and food spoilage. Prevention is much concerned with the destruction of bacteria and with the inhibition of growth.

Bacteria are minute, single-celled organisms of variable shape and activity. Along with the viruses, algae, fungi or moulds, and the lichens, they are classified as the lowest forms of plant life. Bacteria are everywhere – in soil, water, dust, silage, compost and in air. There are thousands of different types and many perform useful functions. Some turn decaying vegetable matter into manure; others, within the human or animal body, assist in the development of certain vitamins essential to health; some can be harnessed for fermentation processes such as in the production of beer or wine and the manufacture of cheese; and others are used to produce antibiotics for the cure of disease. Only a very small proportion of the total bacterial population are dangerous because they can cause disease in man and animals.

Bacteria are so small that individually they cannot be seen without a microscope. They may be as small as 1/2000 mm and clusters of a thousand or more are only just visible to the naked eye. Fifty thousand placed side by side may measure barely an inch (25 mm). Viruses are still smaller. Seen under the high-powered lens of a microscope with a magnification of 500 to 1000 times (Fig. 2.1), the bacteria which may be spread by food are round or rod shaped and appear in various forms according to the type of organism. Some are mobile in fluids and swim about by means of hair-like processes which arise singly or in clusters from one end, both ends, or from all around the cells. Some possess capsules, outer mucinous coats, which

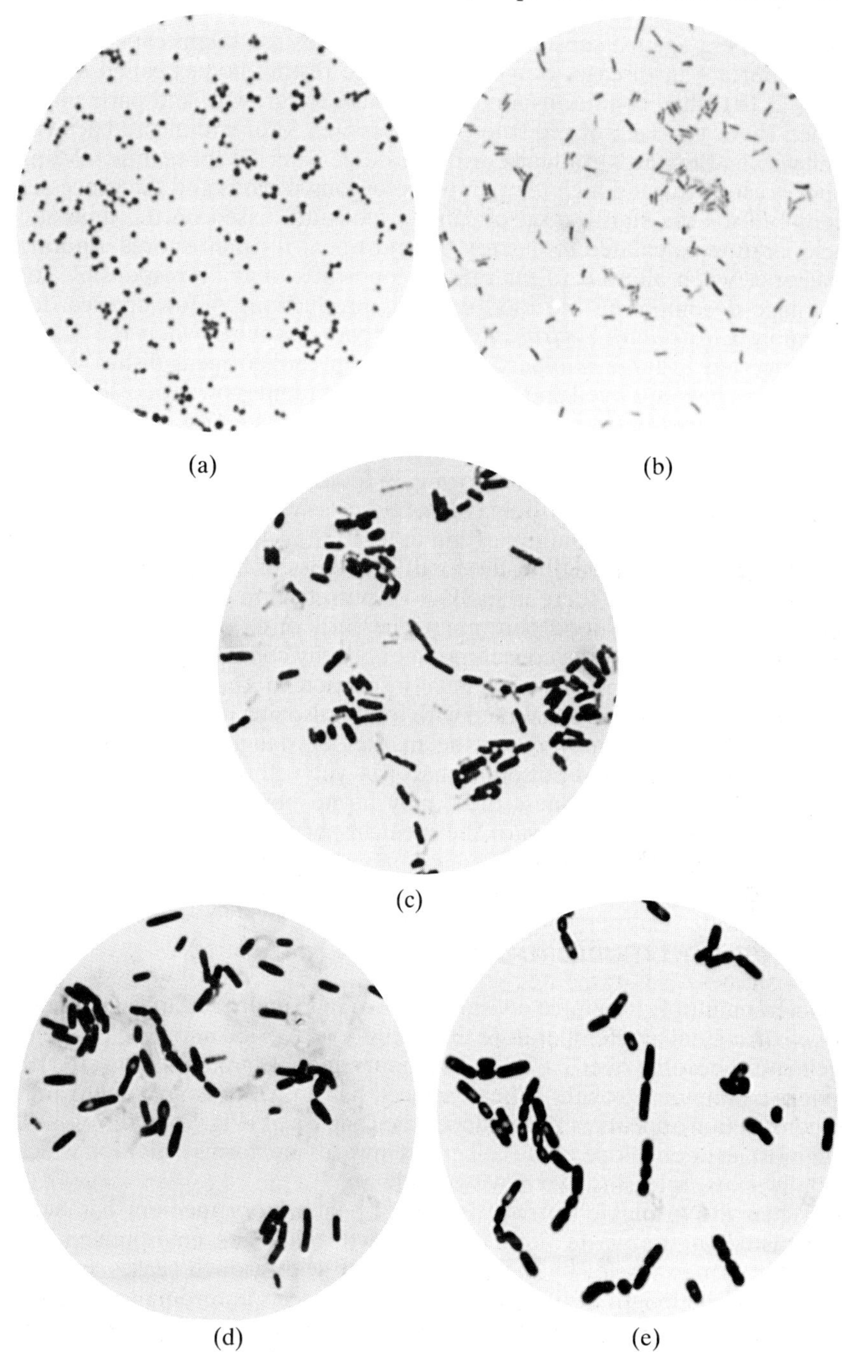

Fig. 2.1 Food-poisoning bacteria: (a) *Staphylococcus aureus* (b) *Salmonella* (c) *Clostridium perfringens* (d) *Clostridium botulinum*, showing a few spores (e) *Bacillus cereus* – showing spores

protect them against substances which might destroy them; capsules may be important in disease. Some can produce resting bodies called spores (Fig. 2.1d) when conditions are unfavourable for growth, and particularly when there is a lack of moisture. These spores form within the bacterial cell which afterwards gradually disintegrates leaving the spore intact. Many spores can withstand high temperatures for long periods and the processes required for the sterilization of canned foods are based on the time and temperature calculated to destroy the most heat-resistant spores. Sporing bacteria, when allowed to multiply in foodstuffs, may be responsible for spoilage decomposing the food with gas production. A few species, for example *Clostridium perfringens*, cause food poisoning when the bacilli are ingested in large numbers particularly in cooked meat dishes where the spores have survived and germinated. Most rapid growth of cells occurs at temperatures between 20 and 50°C (maximum 43–47°C). *Clostridium botulinum*, another sporing organism and the cause of botulism, produces a highly poisonous toxin while growing in food.

Although most food-poisoning bacteria cause symptoms only when eaten in large numbers after multiplication in food, they do not usually alter the appearance, taste or smell of the food. The types of bacteria which break down protein so that there is spoilage or putrefaction detectable by smell do not usually cause food poisoning. The early onset of obvious spoilage is a safeguard against the consumption of heavily contaminated foods.

It is generally impossible by visual inspection to know whether or not food is dangerously contaminated with food-poisoning organisms. Animals infected during life may reach the market or slaughterhouse excreting small numbers of food-poisoning bacteria yet without symptoms. Even in the early stages of illness there may be no obvious signs of changes in the carcass or offal to warn the meat inspector or veterinary surgeon of the potential danger.

Growth and multiplication

Bacteria multiply by simple division into two, and under suitable conditions of environment and temperature this occurs every 15–30 minutes. Thus one cell could become over 2 million in 7 hours and 7000 million cells after 12 hours continuous growth. When each cell has grown to its maximum size, a constriction appears at both sides of the centre axis (Fig. 2.2), the outside membrane or envelope of the cell grows inwards and forms a division which finally splits, releasing two new twin cells.

When the available nutrient in food or laboratory medium has been exhausted or the waste products of growth make the environment unsuitable, for example, by the production of acid, growth ceases and the cell dies. The length of life of a bacterial cell varies according to the food or medium on or within which it is growing or resting, and also according to the type of organism. The spores produced by certain bacteria can survive in a dormant condition for long periods of time under adverse conditions, but when suitable conditions of food, moisture and temperature return they are able to germinate into actively growing bacterial cells.

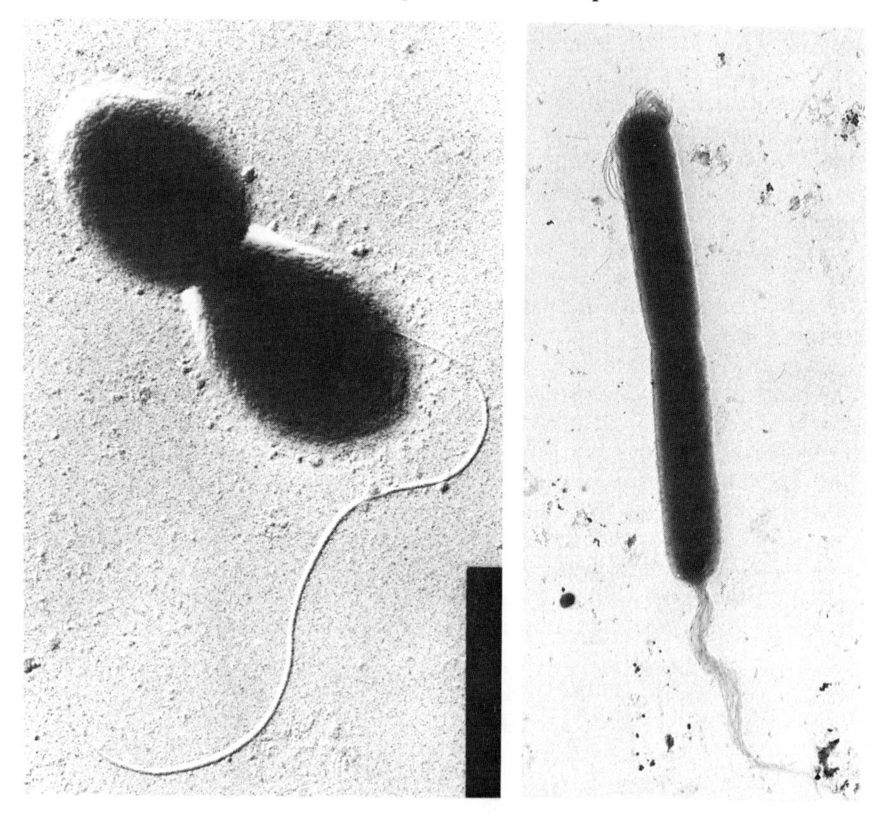

Fig. 2.2 Division of bacteria. (Left) *Rhodopseudomonas sphaeroides* (× 22 320). (Right) Freshwater isolate (× 12 090). (By courtesy of Dr. Crawford Dow, University of Warwick)

Conditions for growth and multiplication of bacteria

In the laboratory a variety of media are made to suit the growth requirements of different types of bacteria. Most of them have a meat-broth base set into a firm gel by means of agar, a substance which is extracted from seaweed. Agar has special properties; it melts at a high temperature and sets at a low temperature and it is thus more suitable for bacterial media than gelatin which melts at temperatures below those required for the growth of many bacteria. Blood, serum, milk, or other protein matter may be added for enrichment.

When bacteria are spread on the surface of agar media in a petri dish and left overnight at a suitable temperature, such as 37°C (98.6°F) – blood heat – they start to grow. By division of each cell into two every 15–30 minutes a small heap of bacteria is formed, consisting of millions of cells, which is called a colony. Every kind of bacterium has a typical colony form; the size, shape, colour, and consistency of these colonies on particular culture media help in identification (Fig. 2.3). Another method by which we can identify the different types of bacteria, as well as by their appearance under the microscope and on the surface of agar media, is to observe their biochemical activity in liquid media containing different sugars. Sugars are fermented with the production of acid and gas, or acid only, while some bacteria produce neither acid nor gas. There are chemical tests, also, which are often useful for distinguishing between different bacteria.

The individual members of bacterial groups may be divisible also by serological methods. The agglutination of organisms brought about by the altered blood serum of human beings or animals infected with the same organism may be used for this purpose (antigen-antibody reactions).

An even finer differentiation into types is sometimes possible by means of bacteriophages, which are viruses parasitic on bacteria. The bacteria are identified according to the types of invading phage, which are specific to the bacterial host cell. Cultivated phages are inoculated on to bacterial cultures which they destroy. *Salmonella typhi*, some other salmonellae, and staphylococci are typed in this way.

Many years of laboratory work have led to the development of special kinds of culture media which will enhance the growth of certain types of bacteria and depress other types. To isolate food-poisoning bacteria from foodstuffs there are media which will suppress the growth of harmless

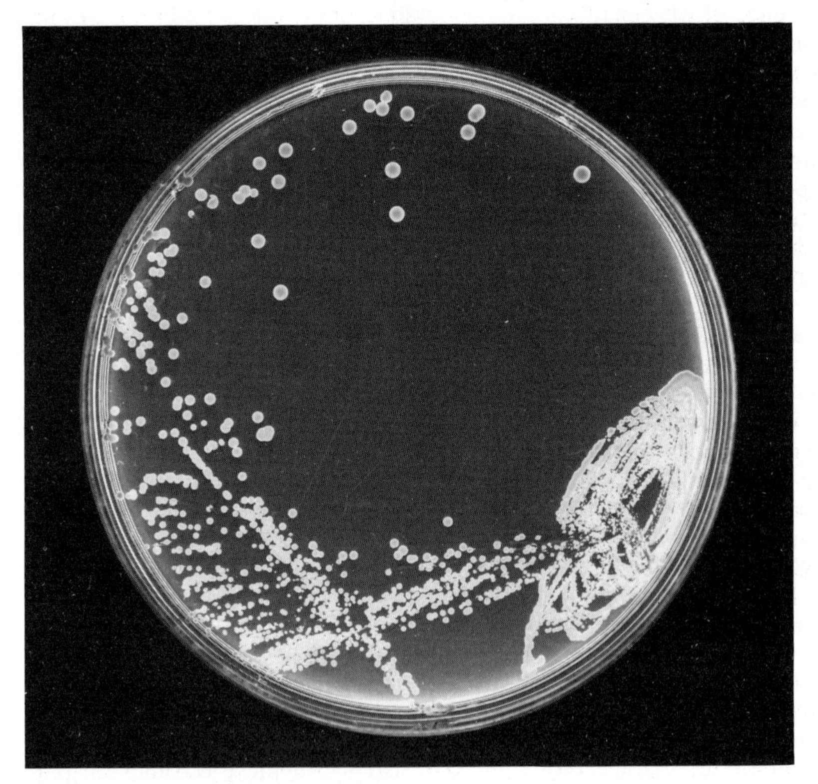

Fig. 2.3 Colonies of staphylococci

or non-pathogenic bacteria, and encourage the harmful or pathogenic organisms to grow.

'Rapid methods' have been developed for the identification of bacteria and viruses. They are often complicated to perform, but yield results within a few hours. They may be serological such as the enzyme linked immunosorbent assay (ELISA), biochemical, for example gas liquid chromatography when metabolites are recognized from organisms growing in defined media, and the genetic probe technique as in DNA/RNA hybridization.

Bacteria will live and multiply in many foodstuffs; sometimes the type of food and the atmospheric temperature and humidity of the kitchen provide conditions similar to those used for cultivation in the laboratory. Meats and poultry are good examples, whether raw or cooked they are excellent media for bacterial growth; pies, stews, and gravies resemble laboratory media. Milk and egg products, including custards and trifles, soft cheese and cream, are all good media for growth. Bacteria will multiply in these foods when they are stored in the shop or kitchen without refrigeration; in the same way, they multiply in specially prepared media inoculated in the laboratory and incubated for warmth. Thus food poisoning occurs more frequently in the warmth of summer than in the winter (Fig. 2.4).

Most bacteria require air to live actively, but some can multiply only in the absence of oxygen; they are called anaerobes. Included in this group are the sporing organisms *C. perfringens* causing food poisoning, and *C. botulinum* causing botulism. *C. perfringens* flourishes under the conditions found in the centre of rolled joints of meat, in poultry carcasses, and in boiled masses of meat such as beef, cut up or minced and in stews, because the oxygen is driven off by cooking. Care must be taken to destroy the spores of bacilli when meat, vegetables, and other non-acid foods are canned. Adequate heat treatment may not be given in the home, thus amateur canning or bottling of meats is discouraged.

Pathogenic or harmful organisms grow best at the temperature of the body, which is 37°C (98.6°F), although the majority will multiply between 15°C and 45°C (59–113°F). Except for *C. perfringens*, which grows well at temperatures up to 47°C (117°F), and slowly up to 50°C (122°F), the ability of most bacteria to multiply falls off rapidly above 45°C (113°F) and only a few groups can grow at temperatures above 50°C (122°F). Non-sporing cells of food-poisoning bacteria are killed at temperatures above 60°C (140°F), the length of time (10–30 minutes or more) required depending on the types of organisms (Fig. 2.5). For example, to make milk safe, i.e. free from harmful bacteria, it is pasteurized at 62.8°C (145°F) – 65.6°C (150°F) for 30 minutes at least, or at not less than 71.7°C (161°F) for at least 15 seconds or at higher temperatures for an even shorter time, such as 132°C (270°F) for not less than 1 second as laid down in the Milk (Special Designation) Regulations, 1989 for 'ultra heat-treated' milk. This milk must be put into sterile containers using aseptic precautions and is said to have a shelf life of several months.

'Sterilized' milk must be filtered or clarified, homogenized and heated to a temperature of not less than 100°C (212°F) either in bottles or by a continuous flow process followed by aseptic filling into bottles which must be sealed and airtight. All or nearly all bacteria are destroyed and the milk will remain close to sterility for an indefinite period while sealed in the container.

Methods of heat treatment, e.g. pasteurization, can be applied to other foodstuffs such as fresh cream, imitation cream, ice-cream mix and liquid whole egg and albumen (see Chapter 4).

Boiling kills living cells, but it cannot be depended upon to kill all bacterial spores and thus to sterilize. To destroy spores, temperatures above boiling can be obtained by the use of steam under pressure.

Fig. 2.6 illustrates the effect of temperature on the vegetative cells of bacteria. The toxic substance produced by staphylococci in foods needs boiling for 30 minutes or longer before it is destroyed, but the toxin of *C. botulinum* is destroyed more readily by heat.

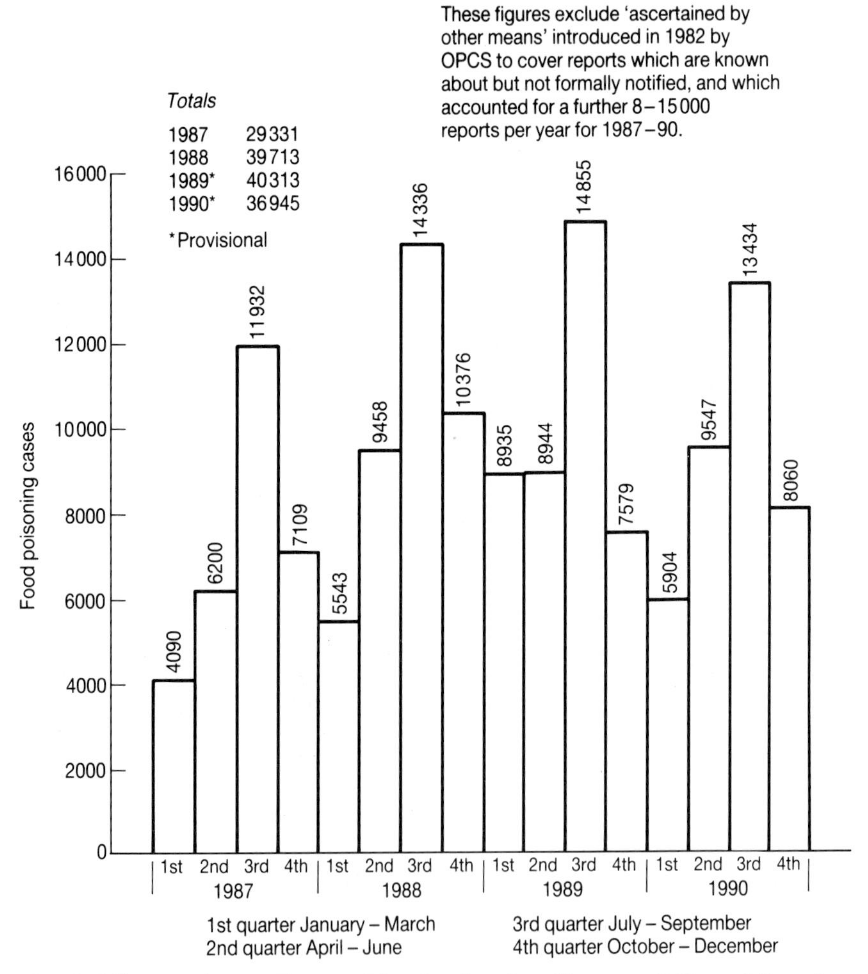

Fig. 2.4 Seasonal prevalence of food poisoning (Source: Office of Population Censuses and Surveys for England and Wales)

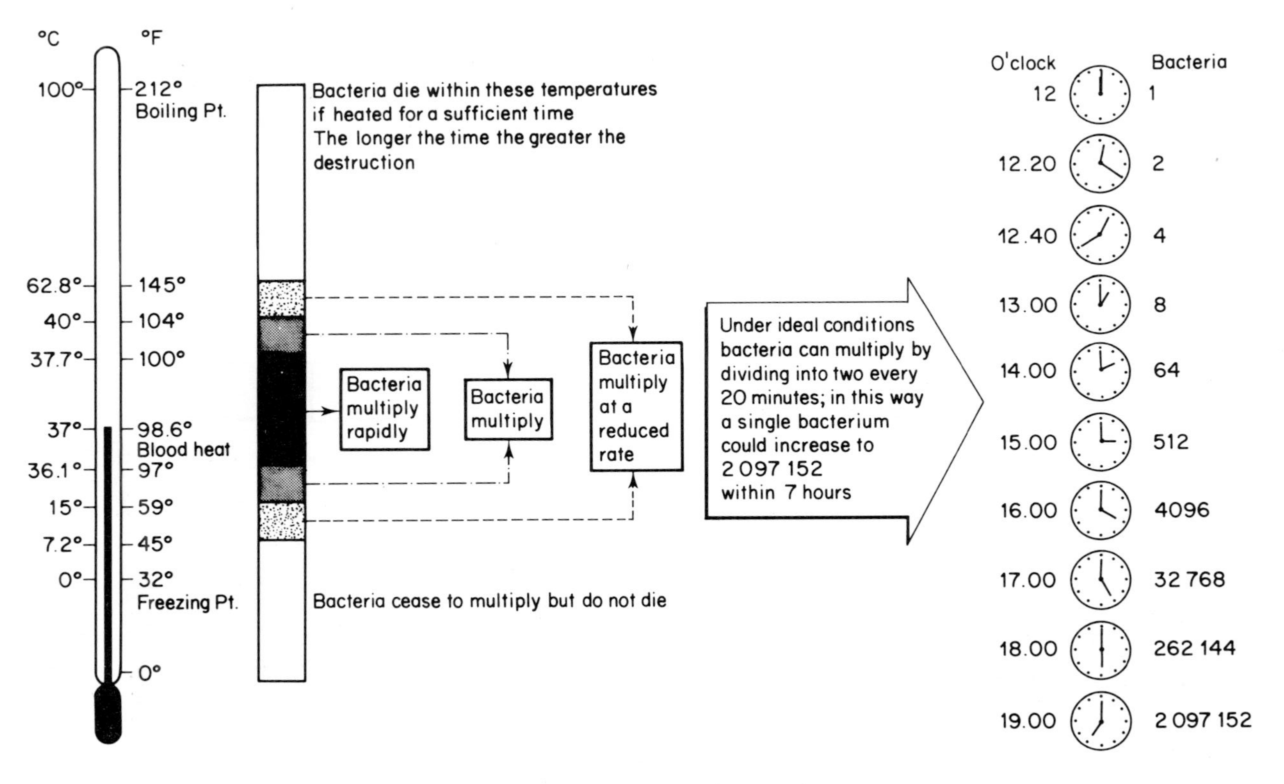

Fig. 2.5 Effect of temperature and time on the growth of bacteria. Safe and dangerous temperatures for foodstuffs

Extreme cold will not kill all bacteria, but it will prevent them from multiplying and for this reason foods which support bacterial growth should be stored at low temperatures. Domestic refrigerators not only delay the spoilage of foods but also prevent the growth of harmful bacteria. Freezing kills a proportion of cells. Safety measures for ice-cream include both heat treatment and cold storage with rapid cooling in-between.

Bacteria cannot multiply without water. Cold cooked meats contain sufficient moisture to support growth, whereas in dehydrated products such as desiccated soup or milk powder bacteria will survive but remain dormant until there is sufficient water added to revive them.

The addition of various salts, acids and sugar to laboratory media is carefully controlled. The presence of these substances in varying amounts in different foodstuffs exerts an effect on the ability of organisms not only to grow but to survive. Their use as preservatives and the effect of antibiotics, irradiation, and fumigation are discussed in Chapter 8.

Expressed simply, bacteria are single-celled living organisms which are present almost everywhere, often in large numbers. Mostly they are harmless, but some cause spoilage in foods and others give rise to disease in the human and animal body. Both humans and animals may harbour disease-producing pathogenic bacteria which may be passed to foodstuffs and multiply when conditions of temperature, time and moisture are suitable. These living bacterial cells are rarely able to multiply in the cold

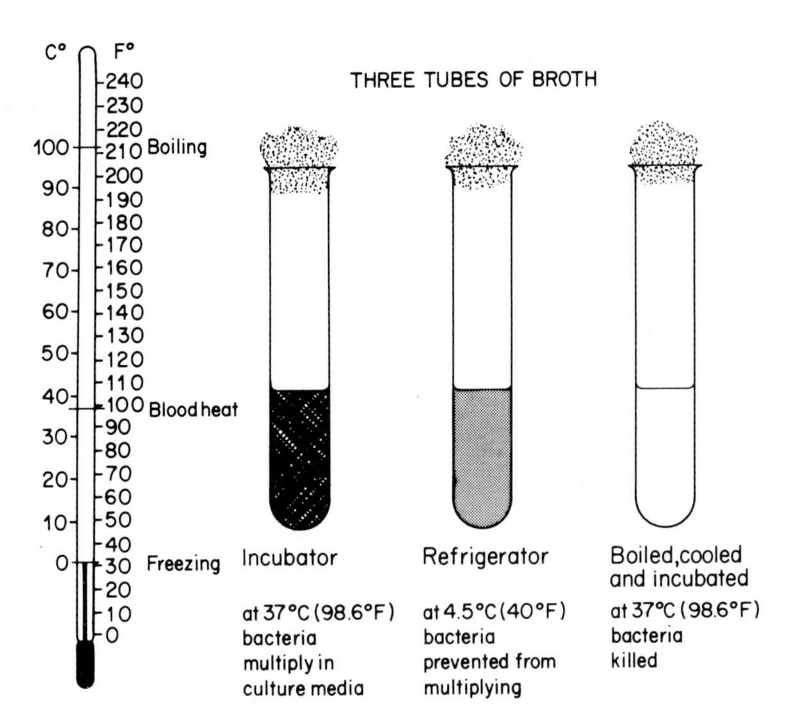

Fig. 2.6 The effect of temperature on bacterial cells

and die if they are too hot, although, as already mentioned, the structures known as spores, produced in some cells by certain bacteria, require steam under pressure for destruction.

The next chapter describes the bacteria which cause food poisoning, particularly where they live, their reservoirs, and their behaviour.

3

Bacterial and other microbial agents of food poisoning and food-borne infection

Bacterial food poisoning, acute gastroenteritis, is a disturbance of the gastrointestinal tract with abdominal pain and diarrhoea with or without vomiting and with or without fever. The time of onset of symptoms may range from less than one to more than 48 hours after eating contaminated food. Usually large numbers of organisms actively growing in food are required to initiate symptoms of infection, (invasion of and multiplication in or on body tissues) or intoxication, (poisoning by toxin produced in the food or in the body). There are many other food-borne diseases which are distinguishable from acute bacterial food poisoning. Water, milk and foods are vehicles of transmission of the (microbial) agents responsible, usually in relatively small numbers: multiplication in food as a medium for growth as in gastroenteritis of the food-poisoning type is usually unnecessary.

The food-poisoning organisms will be considered first. There are at least 12 groups which are well recognized and a miscellaneous group, members of which are described from time to time as agents of food poisoning:

Salmonella species
Staphylococcus aureus
Clostridium perfringens
Clostridium botulinum
Bacillus cereus and other aerobic sporing bacilli
Escherichia coli
Campylobacter jejuni
Yersinia enterocolitica
Vibrio parahaemolyticus
Aeromonas hydrophila
Streptococcus species
Listeria monocytogenes
Miscellaneous organisms

Salmonella, *Staphylococcus aureus*, *Clostridium perfringens* and *Campylobacter jejuni* are known to be common in many countries. Botulism from *Clostridium botulinum* is a hazard in northern countries and Japan, but is more rarely reported from southern Mediterranean areas. There are no recorded incidents from India. In Japan, food poisoning from *Vibrio parahaemolyticus* occurs frequently. *Bacillus cereus* is recognized as the causative agent of a toxin type of illness similar to that caused by staphylococcal enterotoxin; it can also produce symptoms similar to those of *C. perfringens* food poisoning. Certain types of *Escherichia coli* are known to cause diarrhoea in adults, which is often referred to as traveller's diarrhoea; this organism is also a common agent of infantile enteritis. Streptococci are occasionally incriminated in food poisoning. Other organisms, particularly the common Gram-negative bacilli, are reported from time to time, but are not always proven agents.

The duration of illness is mostly short, from 1 – 3 days, but patients with salmonella infection may not feel well for a week or more.

Clinical details, incubation periods and information about the foods eaten, the ingredients and how they were cooked are helpful to those examining suspected material in the laboratory. When an investigation is delayed or specimens are not available, relevant and accurate information may be the only means by which the cause of an outbreak can be established.

The general characteristics of the various types of food poisoning are described in the following pages and summarized in Table 3.1.

Salmonella

Organisms of the salmonella group are divisible into those that cause enteric fever, *Salmonella typhi* and *Salmonella paratyphi* A, B and C, and those that are agents of food poisoning. The salmonellae that cause food poisoning are classified into more than 2000 serotypes able to invade and infect the body of both man and animals.

They reach food directly or indirectly from animal excreta at time of slaughter, from human excreta, or water polluted by sewage; also, in the kitchen they may be transferred from raw to cooked foods by hands, surfaces, utensils and other equipment. Illness is more likely to occur when large numbers of organisms are ingested, after multiplication in food allowed to stand at atmospheric temperature for some hours. The immune system of the body may be able to control small numbers of ingested organisms.

The onset of illness occurs within 6–36 hours or even longer after eating the contaminated food. The symptoms are characterized by fever, headache and general aching of the limbs, as well as by diarrhoea predominantly, and vomiting. The duration of illness is from 1 – 7 days, or longer. The different types of salmonellae can be classified by serological methods – also, several types can be more finely divided by means of bacteriophage (see p. 20). The subdivisions are helpful for the investigation of outbreaks. All types can produce disease in man or animals. The commonest serotypes are *Salm. typhimurium* and *Salm. enteritidis*. Since 1987, the incidence of *Salm. enteritidis* phage type 4 has risen with cases and outbreaks throughout the UK, in parts of the USA and also in Europe; poultry and eggs are the common vehicles. Breeding stocks and feedstuffs are thought to harbour the organism. Suggestions for control included irradiation of poultry products and stringent measures, including irradiation, for feedstuffs. The wholesale destruction of flocks caused much hardship. Organisms of the Arizona group, which are classified with the *Salmonella*, are also described as agents of disease in man and animals.

Table 3.1 Bacterial food poisoning: incubation period and duration of illness

Cause of food poisoning	Incubation period (hours)	Duration
Salmonella (infection)	6–36 usually 12–24	1–7 days
Staphylococcus aureus (toxin in food)	2–6	6–24 hours
Clostridium perfringens (toxin in intestine)	8–22	24–48 hours
Clostridium botulinum (toxin in food)	12–96 usually 18–36	Death in 24 hours to 8 days, or slow convalescence over 6–8 months
Bacillus cereus (toxin in food)	1–16	12–24 hours
Escherichia coli (infection and toxin)	12–72	1–7 days
Campylobacter jejuni (infection)	3–5 *days*	Days or weeks
Vibrio parahaemolyticus (infection)	2–48 usually 12–18	2–5 days
Streptococcus (toxin in food? and infection)	3–22	24–48 hours
Yersinia enterocolitica (infection)	24–36 (3–5 *days*)	3–5 days
Aeromonas hydrophila (infection)	?2–48	2–7 days
Listeria monocytogenes	48 hours – 7 weeks	?

Staphylococcus aureus

Staphylococcal food poisoning follows the consumption of food contaminated with large numbers of certain types of *Staph. aureus* which produce a poisonous or toxic substance in the food. The skin of hands and the nose frequently harbour staphylococci, some of which produce toxins in cooked foods. Meat and poultry intended to be eaten cold, and prepared foods such as custards, trifles and cream products, are subject to such contamination. Since the toxin is formed by the organism growing in food before it is eaten and not after it has entered the body, the incubation period may be as short as 2 hours, but in general it is 4–6 hours. There is a rapid onset of symptoms characterized predominantly by severe vomiting, with diarrhoea, abdominal pain and cramps, sometimes followed by collapse. Recovery usually occurs within 6–24 hours.

The typing of staphylococci both by bacteriophage and serological methods has enabled the pathogenic species to be divided into groups, some of which are more common than others in food-poisoning incidents. There are several types of enterotoxin which can be extracted from culture filtrates and foods and identified serologically. Human and animal feeding experiments have been used also. Whereas the staphylococcus itself is fairly readily destroyed by the heat of pasteurization and normal cooking procedures, the toxin is more resistant to heat: it is destroyed gradually during boiling for at least 30 minutes. It may remain active after light cooking.

Clostridium perfringens

C. perfringens is a common organism frequently found in excreta from humans and animals and in raw meats, poultry and other foods, including dehydrated products. It can survive heat and dehydration by means of spores which remain dormant in food, soil and dust. Illness occurs after eating food contaminated with large numbers of *C. perfringens* grown from spores which have survived cooking; the spores are activated to germinate into bacilli by heat shock. Multiplication of the cells takes place during long slow cooling and warm storage of food in the kitchen or canteen; cooked meat, poultry, fish, stews, pies and gravies are excellent media for growth at temperatures up to 50°C (122°F). There is little growth below 15°C (59°F). The spores of some strains of *C. perfringens* can withstand hours of boiling; others survive for a few minutes only. The organism may be isolated in large numbers from boiled, stewed, steamed, braised, or even roasted foods, particularly those cooked in bulk, when stored unrefrigerated for a few hours.

Symptoms occur from 8–22 hours after consuming the contaminated food; they include abdominal pain, profuse diarrhoea and nausea, but rarely vomiting; they may continue for 12–48 hours. The symptoms result from the activities of a large dose of organisms swallowed in food, which produce an enterotoxin in the intestine; an effective amount of toxin is not found in the food before it is eaten.

There are five types of *C. perfringens* (A to E) classified according to the various toxins they produce. Two produce the enterotoxin able to cause gastroenteritis in man. Of these, type A is the more common agent of food poisoning, while type C causes a more serious but rare condition called enteritis necroticans. Type C was first isolated from home-canned rabbit responsible for illness in Germany, and it has been found in spit-roasted pork causing similar outbreaks in New Guinea. So far it has rarely been associated with human illness in the UK. The type A strains can be further divided into many serotypes by the use of specific antisera, a useful epidemiological tool.

Clostridium botulinum

The toxin of *C. botulinum*, another anaerobic spore-bearing bacillus, is a highly poisonous substance formed by the organism as it grows in food. It affects the nervous system and frequently results in fatal illness. Outbreaks and cases of botulism rarely occur in the UK but they are reported from some other countries and depend on the feeding habits of the population (see p. 74). The species is divided into seven types (A to G) according to the toxin produced; five (A, B, E, F and possibly G) are known to affect man. The spores are highly resistant to heat and survive boiling and higher temperatures. The toxin is sensitive to heat and in pure form is destroyed by boiling. Nevertheless, it may be protected when mixed with protein and other material in food. The toxin is lethal in minute doses and gives rise to symptoms different from those of the organisms just described. The incubation period varies from 24 hours or less to 96 hours but is usually 18–36 hours. The first signs of illness are lassitude, fatigue, headache and dizziness. Diarrhoea may be present at first, later the patient is obstinately constipated. The central nervous system becomes affected and there is a disturbance of vision, speech becomes difficult and there is paralysis of the throat muscles. The intoxication reaches its maximum within 24 hours to 8 days and death often occurs by paralysis of the respiratory centres. If, after 8 days, the patient survives, convalescence is slow. The prognosis is improved if antitoxin is given as soon as possible, within hours of eating food containing the toxin. Improperly preserved foods are usually responsible for outbreaks and cases. The toxin can be identified by mouse protection tests.

Bacillus cereus

Certain members of the *Bacillus* group of organisms, and especially *B. cereus*, are increasingly reported in food-poisoning incidents. They are common aerobic sporing organisms. The spores are often found in cereals and other foods; some spores will survive cooking and subsequently germinate into bacilli which, under warm storage conditions in cooked food, grow and produce toxin. A wide variety of foods has been associated with outbreaks, particularly cornflour sauce in Norway and boiled and fried rice in Britain and other countries. As with *C. perfringens*, long moist storage of warm cooked food in which spores are still alive encourages the growth of the organisms to large numbers and the consequent formation of toxin to amounts able to cause illness. Two toxins are known to be responsible.

There is a serological typing scheme for *B. cereus*: certain serotypes occur more commonly in outbreaks than others and there are two different patterns of symptoms. The outbreaks arising from rice, particularly, have a short incubation period of 1–3 hours with sudden onset of symptoms, acute vomiting and some diarrhoea; the incubation and symptoms resemble those of staphylococcal enterotoxin food poisoning. The other pattern resembles that of *C. perfringens* food poisoning with an incubation period of 9–18 hours and diarrhoea as the chief symptom. The toxin responsible for the vomiting syndrome, which is probably preformed when the organism grows in food, is extremely heat resistant: it is not destroyed after 1.5 hours at 121°C. The diarrhoeal toxin is quite heat labile. Other aerobic sporing organisms such as the *B. subtilis-licheniformis* group have also been reported as agents of food poisoning with similar characteristics to those of *B. cereus* (see p. 118).

Escherichia coli

The majority of *E. coli* in the gastrointestinal tract are harmless unless they are displaced to other parts of the body such as the urinary tract or meninges where they may cause disease.

The pathogenic or enterovirulent *E. coli* are divided into five groups according to their action in the body; virulence, interactions with intestinal mucosa, clinical symptoms, epidemiology and serotyping are all considered:

(1) Enteropathogenic (EPEC) are responsible for severe infantile diarrhoea, with an incidence of 8–24% in India and 7–30% in developed countries. Virulence is unrelated to either heat labile (LT) or heat stable (ST) enterotoxins or to shigella-like invasiveness and they belong to specific serotypes. The mechanism of pathogenicity is not clear.

(2) Enterotoxigenic (ETEC) produce an LT or ST toxin; both may be produced by the same organism. The colonizing factor antigen (CFA) is also produced. They are agents of the following conditions:
   (i) paediatric diarrhoea;
   (ii) severe cholera-like illness in adults in cholera areas;
   (iii) traveller's diarrhoea which can be food- or water-borne.

(3) Enteroinvasive (EIEC) with invasive properties for the mucosa causing ulceration and inflammation of the large bowel similar to that shown by the shigellae. Pus cells and occasional red cells are found in the stool. There is cross agglutination with polyvalent flexner antiserum. They are restricted to a limited number of serotypes with little known occurrence.

(4) Enterohaemorrhagic (EHEC) also known as verotoxigenic (VTEC) are responsible for bloody diarrhoea and colitis somewhat distinct from bacillary dysentery since fever is not prominent and the bloody discharges are copious rather than scanty. One serotype 0157:H7 has been predominant.

(5) Enteroadherent (EAEC) according to the pattern of adherence to cells.

The pathogenic *E. coli* can be differentiated from the less harmful varieties by immunological and other methods including the serotyping essential for epidemiological studies.

Investigations carried out in the Punjab (Shobha Ram, PhD thesis, Punjab University, Chandigarh, India) on 2661 diarrhoeal samples showed that the most prevalent group was ETEC, 609 (22.8%), followed by EPEC, 347 (13%) and EIEC, 57 (2.1%). Thirteen isolates of serotype 0157 were obtained but they were not 0157:H7, neither did the strains produce verotoxin. It was suggested that strains of *E. coli* 0157:H7 with the VT positive clone were rare in the north of India. For ETEC, EPEC and EIEC strains the peak incidence occurred in the hot dry season (37.9°C) and at low humidity. There was an inverse relationship between the age of the patient and diarrhoeal incidence. Infants and children up to 10 years of age of both sexes showed the highest susceptibility to all types of diarrhoeagenic *E. coli*. Males were more prone to infection than females.

The incubation period for *E. coli* gastroenteritis is 12 hours to 3 days. The symptoms consist mainly of diarrhoea sometimes with blood and mucus in the stool. Infants become infected by direct spread in maternity units and by contaminated feeds. Enterotoxigenic *E. coli* are prevalent in infants in tropical countries. Food and water, in addition to faeces, are sources of these organisms.

It appears that symptoms in adults are initiated by large doses of enterotoxigenic *E. coli* producing the heat stable toxin in food. *E. coli* may enter kitchens in many raw foods and are readily passed to cooked foods by the usual means: hands, surfaces, containers and other equipment; they are also water borne. During epidemics, human excreta may play a part in direct spread, although, like most Gram-negative organisms, *E. coli* will not survive long on the skin and the organisms are readily washed away.

Since *E. coli* are always present in faeces, they are used as markers for the faecal pollution of water, milk and food. Intestinal pathogens such as *Shigella* and *Salmonella* may be found intermittently in water and sewage; they will be present in small numbers only and are thus difficult to find. The laboratory enumeration of *E. coli* helps to assess the potential danger of water and food supplies.

Campylobacter jejuni

Reports from 1977 onwards confirm that *Campylobacter jejuni* is responsible for many incidents of gastroenteritis affecting both children and adults. There are large numbers of sporadic cases and comparatively few outbreaks. The organisms is said to be isolated from patients with diarrhoea at rates greater than for *Salmonella*. Surveys on campus personnel of colleges in the USA showed that isolation rates were 10 and 46 times more frequent than those for *Salmonella* and *Shigella* in young adults. *C. jejuni* has been associated along with other organisms in cases of 'traveller's diarrhoea'.

Raw or imperfectly pasteurized milk has been the most commonly implicated vehicle of infection. Milk-borne outbreaks can be extensive because of the homogeneous distribution of the organism in the fluid medium. Water-borne outbreaks are reported occasionally. *C. jejuni* is readily isolated from poultry, and cross contamination in the kitchen from raw to cooked food is the most likely means by which the organism reaches man; there may be direct infection from raw poultry to the food handler. The unusual conditions required to grow *Campylobacter* have sometimes hindered isolation from foods, milk and water and delayed epidemiological investigation. Person-to-person spread has been demonstrated, thus the dose level needed to initiate symptoms may be small and therefore perhaps this organism belongs to the category of food-borne infections, as growth in food has not been demonstrated.

The onset of symptoms may be sudden with abdominal cramps followed by the passage of foul-smelling and often bile-stained stools; blood and mucus may be present also. The diarrhoea, which may persist for 1–3 days, is sometimes preceded by fever, headache and dizziness for a few hours to several days. The central abdominal pain may persist after the diarrhoea has eased; symptoms may be more severe in adults than in children. There may be nausea but rarely vomiting. The incubation period is probably 3–5 days but is fairly wide ranging, both shorter and longer. Cats and dogs may be also affected with spread between animals and owners or acquired from the same source. There are many serotypes.

Yersinia enterocolitica

The commonest manifestation of *Y. enterocolitica* infection is mild diarrhoea, often unrecognized; symptoms may also include abdominal pain, fever, nausea and vomiting. The duration of excretion in Canadian school children extended from 14–97 days (mean 42 days). Symptoms may persist for 5–14 days and occasionally for several months. The incubation period ranges from 1–11 days. Ileitis with or without mesenteric adenitis may be diagnosed as appendicitis. Temperature may be slightly raised or normal. Vomiting, debility or acute dehydration are rare. Small scale epidemics have been reported, the infection originating from the same source. The importance of this organism compared with other agents of diarrhoea varies from one country to another. In the Punjab, *Y. enterocolitica* was isolated from 3% of 235 patients with diarrhoea; for children 1–5 years of age the rate was 7.4%; isolations were highest in December. Reports from European countries give 2.5% rates of isolation. It may rank higher than *Campylobacter* and *Shigella* but usually lower than *Salmonella* or approximately the same as *Shigella, Campylobacter* and *E. coli*. In the UK, the incidence of *Y. enterocolitica* is low compared with that of *Campylobacter* infection, less than 1000 cases compared with more than 30 000 for *Campylobacter*. A report from the UK gave 3.5 isolations from 4585 samples predominantly in age groups 1–14 years, the milk drinking years. Serotypes vary between countries, predominantly O8 in North America and O3 and O9 in Europe.

The organism may persist in mucosal and lymphoid tissue, simulate an immune response, cause reactive sequelae and give rise to episodes symptomatic of cryptic bacteraemia. Other invasive pathogens such as *Salm. typhi* may be found in blood culture without symptoms, transient bacteraemia can occur from mild *Campylobacter* diarrhoea. Donors with *Y. enterocolitica* in the blood may initiate transfusion infection, although the recipient may appear to be unaffected. Growth can occur in blood stored in the cold when sufficient haemin is released to allow profuse multiplication of the organism, which is iron-dependent. *Y. enterocolitica* was isolated from a fatal case of septic shock and also from the donated blood. The donor had a history of mild gastroenteritis and antibodies were found in blood samples. The species *Y. frederiksenii* is possibly associated with post-antibiotic diarrhoea. Animals are important reservoirs and *Y. enterocolitica* has been isolated from all vertebrate species examined.

The organism has been isolated from dairy products such as untreated and also pasteurized milk, chocolate milk, dairy cream and ice-cream, as well as from many vegetables and types of meat including 50% of retail pork tongue samples, hare, beef, lamb and black pudding.

Vibrio parahaemolyticus

In Japan, *Vibrio parahaemolyticus* is reported to cause 50% or more of food-poisoning incidents. In the warmer weather it can be isolated from fish, shellfish and other seafoods and also from coastal waters. Reports of outbreaks and isolations of the organism from sea creatures and inshore waters come from many countries. Both raw and cooked sea foods, for example, crab, lobster, shrimp and prawn are vehicles of infection. Cooked foods may be contaminated by the raw products in kitchens and even at picnics when raw crabs have been transported in baskets on top of cooked crabs. Imports of frozen cooked seafoods from eastern countries are frequently found to be contaminated with vibrios including *V. parahaemolyticus*. Food poisoning following dinner parties and banquets with prawn on the menu is not uncommon. The illness is like a mild form of cholera, with an average incubation of about of 15 hours. There is rapid onset of symptoms with profuse diarrhoea often leading to dehydration, some vomiting and fever. There is acute abdominal pain. The duration of illness is usually 2–5 days although the ill-effects may linger. There are many serotypes.

Aeromonas

Aeromonas species are widely distributed in nature; they are found frequently in many foods including seafoods, meats, dairy products and poultry. They are largely aquatic and there is clinical potential for cold-blooded creatures such as fish and frogs; high counts are found in fish. The wide temperature range of growth (4–30°C) varying with species, enables the organism to grow at refrigerator temperatures and also to invade and thrive in the mammalian body, principally the gastrointestinal tract. Although *Aeromonas* is not normally found in the stools of healthy persons, asymptomatic human carriers could be sources of the organism. The incidence of *Aeromonas* in human disease may be underestimated because there are no clear markers and colonies may be easily confused with *Pseudomonas* or coliform organisms and discarded as such. There are three types of human illness associated with *Aeromonas* including extra-intestinal and wound infections and food-associated gastrointestinal infections that may be present as (a) toxigenic rice water small intestinal diarrhoea, (b) classical dysentery involving the large intestine or (c) a combination of both. Two family outbreaks and nine sporadic cases were reported from the UK (1986–1988). It is suggested that the organism may be involved in 'traveller's diarrhoea' and a duration of 10 days diarrhoea has been noted. Reports from India, Sri Lanka and Ethiopia (Swedish workers) stated that *Aeromonas* was isolated from the stools of adults and children with diarrhoea and that the cultures produced enterotoxin. It was thought that the mechanism of reaction was similar to that of *Vibrio parahaemolyticus*.

Indirect evidence of the role of *A.hydrophila* and *A.sobria* in gastroenteritis has come from long-term studies of diarrhoea in children and adults compared with control groups in a number of countries, as examples, the ratios varied from 11–4% in the US Navy, 9–4.3% in Peru, 11–0.7% in Australia and 2% to less than 1% in the north of India. It appeared, therefore, that both *A. hydrophila* and *A. sobria* played significant roles. Water, milk and seafood were considered to be basic sources and refrigerated meat and poultry emphasized as potential vehicles. In some instances enteropathogenicity of isolated strains was confirmed in animal models. Heat stable cyotoxicity and enterotoxin activity were demonstrated in *A. caviae* strains from stools and the environment.

Aeromonas bacteraemia after bowel infection occurs in immuncompromised hosts with underlying malignancies such as leukaemia, the fatality rate has been reported as 61%; skin conditions, for example, ecthyma gangrenosum may be associated with septicaemia (fatal cases have been reported). Aggressive infection does not respond to treatment with antibiotics. Incomplete abortions unresponsive to treatment may require extreme surgical debridement and maximal doses of antimicrobials. *A.hydrophila* is also the agent of red leg disease in frogs. *A. punctata* appears to be pathogenic for frogs also. *A. salmonicida* causes disease in salmon and other fish: these diseases are not transmissable to man.

Listeria monocytogenes

Listeria monocytogenes was first described in 1927 when it was isolated from rabbits with large mononuclear monocytosis. This organism occurs widely in the environment and is found in many animal species both wild and domesticated. The same serotypes occur in man and other animals. The important clinical forms of listeriosis, meningitis, abortion and meningoencephalitis have been reported from many countries. *L. monocytogenes* type 4b was isolated from the cerebrospinal fluid of a meningitis patient 12 hours after pork meat containing the same serotype had been eaten. Chloramphenicol and penicillin were effective. *L. monocytogenes* has been isolated from liver abscesses in diabetics (rare). Although *L. monocytogenes* occurs commonly in the environment, listeriosis is comparatively rare, but the case fatality rate is high, that is approximately a third of cases have been fatal or resulted in still births. There is increasing concern, particularly when illness is related to food-borne infection. Persons most at risk are pregnant women, the unborn child, alcoholics, drug abusers, those receiving treatment that alters the immune system, AIDS patients and the elderly. The organism may be transmitted to man by means of contaminated foodstuffs at any point in the food chain from the source to the kitchen. The organism is found in milk and other dairy products and in approximately 10% of a wide range of food samples including meat, poultry, vegetables, salads and seafoods: for example survival was shown on lettuce during packaging and distribution and 3–5% of prepacked ready-to-eat sliced meat and poultry products were found to be contaminated with *Listeria*. Multiplication is slow in the refrigerator at 4–6°C and the generation time at 3°C is 1–2 days. Storage of chilled foods for any length of time is hazardous. Total elimination of *Listeria* from foods is impracticable except by pasteurization, irradiation, cooking or pickling, but recontamination can occur. *L. monocytogenes* is known to be moderately resistant to heat, and it was suspected that large numbers in milk may not be totally destroyed by pasteurization. Leucocytes remaining in milk after filtration may act as a shield from heat for intracellular organisms. Recontamination is now thought to be the most likely reason for the presence of *L. monocytogenes* in pasteurized milk.

The first well-described outbreak of listeriosis occurred in Boston (Mass) in 1979. It was assumed that raw celery, lettuce and tomatoes were the vehicles of infection for 20 people. In 1981, an outbreak in Canada was traced to contaminated coleslaw. In 1983, in the USA, listeriosis affected 49 persons, 7 in foetuses or infants and 42 in adults with suppressed immune systems, 14 (29%) persons died. Forty isolates of *L. monocytogenes* were of the same serotype. The vehicle was suspected to be a particular brand of pasteurized whole or 2% milk. Listeriosis was diagnosed in dairy cattle on farms supplying the milk. Multiple serotypes were isolated from milk samples collected from the relevant farms after the outbreak. In 1985, also in the USA, listeriosis was associated with Mexican-style cheese. There were 86 cases, 58 in mother and infant pairs with 29 deaths, 8 were neonates and 13 still births. Additional cases were reported throughout the country. 43 (63%) were mother and newborn infant pairs. Most (70%) of the women had a prior febrile illness or were febrile on admission to hospital. The onset of illness in 42 of the neonatal patients was within 24 hours of birth; all isolates of *L. monocytogenes* were of the same serotype as that isolated from cheese. A relatively small dose of the organism was thought to initiate symptoms. Samples of Mexican-style cheese from three different manufacturers were examined and four packages containing two varieties of cheese grew *L. monocytogenes*. Expiry dates indicated continual contamination and the batches were recalled. Dairy herds, plants and cheese processes were investigated. It was thought that unpasteurized milk was probably incorporated into the cheese. In 1987 listeriosis was again traced to soft cheese and there was general concern about the frequent occurrence of *Listeria* in certain soft cheeses; in a survey, about 10% of samples of soft cheese contained *L. monocytogenes*.

The first cases of foetal listeriosis associated with the consumption of food were detected in the UK in 1988, linked to chicken and vegetable products. That year 291 cases of listeriosis were reported and 26 newborn or unborn babies died. In 1989, there were reports of 250 cases, and in 1989 also, a survey of 18 000 samples of a wide variety of foods indicated that 10% contained *Listeria* and 6% *L. monocytogenes*. Another PHLS survey in 1989, showed that 9% of soft mould-ripened cheeses, 28 of 52 samples of displayed pâtés and 9 of 21 unopened packs were positive for *L. monocytogenes*. The predominant serotype isolated from the pâté was 4b, the same as that responsible for 80% of human cases. The estimated counts were 100/g in 23 samples and 10 000/g in seven samples. Two sporadic cases of listeriosis in the UK, in previously healthy individuals, were associated with soft mould-ripened cheese. Soft unripened cheeses, like cottage and cream cheese are free from *Listeria*. Twenty per cent of precooked ready-to-eat poultry grew *L. monocytogenes* and 60% of raw poultry were contaminated. The HTST process for the pasteurization of milk effectively eliminates *L. monocytogenes* from naturally contaminated milk; also when food is cooked to 70°C for 2 minutes *L. monocytogenes* is reduced from 10^6/g to a non-detectable level. While the serotype 4b is predominantly related to human illness, type 1/2 is more frequently found in food, although amongst foods so far examined pâté has yielded 4b as the predominant type.

More knowledge is required on the significance of human and animal excreta, the environmental spread in sewage and surface water and the role of vermin. The precise source of *L. monocytogenes* and the points of entry into the food chain are unclear. Control measures are, as usual, concerned with minimizing survival and growth in foods.

Diagnosis of listeriosis depends on culture. Identification of the organism is simple if the possibility is considered and isolates not discarded as diphtheroid contaminants. In septicaemia and meningitis the organism may be isolated provided sufficient amounts (1ml of cerebrospinal fluid) are cultured. Foetal blood, meconium and gastric and tracheal aspirates are required if early neonatal listeriosis is suspected. Special selective and enrichment media are necessary for the isolation of *Listeria* from foods and faeces.

Streptococcus

Certain streptococci, including those of Lancefield Groups A and D and *Strep. viridans*, have been isolated in large numbers from incriminated foods suspected of causing illness. The Group A β-haemolytic streptococci have been found in foods associated with outbreaks of both upper respiratory tract infection and gastroenteritis. Raw milk, custard, boiled eggs in salads, ham and shrimp cocktail have all been incriminated, but in most incidents they were implicated on epidemiological evidence only (no bacteriological studies were made on the suspected foods). Usually similar strains of streptococci were isolated from patients and from food handlers. In many of the outbreaks of septic sore throat, in which foods other than raw milk were the vehicles of infection, abdominal pain, vomiting or diarrhoea or both occurred in 10–20% of the cases.

Outbreaks of streptococcal food poisoning have occurred in which large numbers of α- or non-haemolytic streptococci were isolated from the suspected foods. Most of the streptococci are identified as *Strep. faecalis*, but

Strep. viridans was the cause of a toxin-type school outbreak in which the implicated food was chocolate pudding. The association of enterococci and food poisoning is not always clear. It is thought that toxins are produced giving symptoms similar to but less acute than those of staphylococcal enterotoxin.

Miscellaneous organisms (*Proteus*, and scombrotoxic fish poisoning)

Various Gram-negative bacilli including *Proteus, Providencia, Citrobacter* and *Pseudomonas* have been described as agents of food poisoning from time to time. Usually these organisms are isolated in large numbers from foods, but not always from the faeces of patients so that part of the evidence for their causal role is missing. Serological tests may be used to support the diagnosis.

Proteus is involved in scombroid fish poisoning. The organisms synthesize histamine from histidine, present in certain fish such as tuna and mackerel; the toxin consists of two components, histamine and saurine. Three conditions must be met: (a) The fish must contain abundant free histidine; (b) microorganisms that can produce the enzyme histidine decarboxylase must be present, and; (c) time and temperature conditions must allow production and accumulation of histamine from the histidine in the fish. The symptoms are nausea, vomiting, flushing, puffiness around the eyes, swelling of the lips, tongue and gums with cyanosis, rash, severe itching, headache and respiratory distress developing within a few minutes to a few hours and persisting 8–12 hours (See p. 123–4). Other organisms may be implicated also.

Incidence

Table 3.2 gives the number of cases of food poisoning reported by laboratories in England and Wales for the years 1988–1991.

Table 3.3 gives the distribution of causal agents according to the place where the outbreak occurred, and also divides the episodes into general outbreaks, family outbreaks and sporadic cases. The table includes not only laboratory reports, but also those made by medical officers for environmental health and environmental health officers. An 'outbreak' involves two or more related cases in persons of different households; a family outbreak involves two or more related cases or persons in the same household; a sporadic case refers to a single case unrelated to any other. Some sporadic cases may be part of an unrecognized outbreak. Sporadic cases and outbreaks may be missed if the food is eaten in hotels and restaurants. Symptoms may appear only after the guests have dispersed.

Tables 3.2 and 3.3 show that organisms of the salmonella group and particularly *Salm. typhimurium* and *Salm. enteritidis* are responsible for the highest number of cases and outbreaks, although it should be noted that sporadic cases account for a large number (94%) of the incidents due to salmonellae. Prior to 1968, *Salm. typhimurium* was the predominant salmonella serotype responsible for more than 50% of cases of infection; gradually all the other serotypes (combined) began to take precedence over *Salm. typhimurium*. However, since 1977 there has been a continuing increase in isolations of this organism; in 1985 *Salm. typhimurium* accounted for 41% of all salmonellae reported by laboratories.

During the mid-1980s it became apparent that another serotype, *Salm. enteritidis* was becoming increasingly involved in cases of food poisoning. In 1986, 27% of all reported cases of *Salmonella* infection were caused by *Salm. enteritidis*, there was an increase to 33% in 1987 and 55% in 1988, a slight decrease to 53% in 1989 but a continuous increase to 62% and 63% in 1990 and 1991.

C. perfringens is the next most common agent with *Staph. aureus* ranking third.

The number of cases due to all bacterial causes reported between the years 1969 to 1974 was 8000 to 9000, except for 1972 when the numbers dropped to 6020. In 1975 more than 11 000 cases were recorded and the number remained around this level until 1982. In that year, 12 684 cases were reported and there was a substantial increase to 15 168 cases in 1983. The increase was largely due to the rising incidence of salmonella food poisoning. However, in 1984 there was a slight increase only, to 15 312 cases and in 1985 there was a fall to 12 837 cases. This decrease was not maintained and until 1988 there was a marked increase in the number of cases reported from 15 214 in 1986, 18 573 in 1987 to 25 662 in 1988. There was a similar level of cases in 1989 of 25 849 reports which was not maintained in 1990 (26 960 reports) although the figures for 1991 (23 548 reports) showed a decrease again.

Table 3.2 Causal agent of bacterial food poisoning (laboratory reports, England and Wales, 1988–1991)*

Causal agent	1988 Number	%	1989 Number	%	1990 Number	%	1991 Number	%
Salmonella								
Salm. typhimurium	5566	22	6080	24	4655	17	4459	19
Salm. enteritidis	13051	51	13013	50	15636	58	14368	61
Other salmonellae	5204	20	5560	22	5010	20	3832	16
Clostridium perfringens	1312	5	901	4	1442	5	733	3
Clostridium botulinum			27	<1				
Staphylococcus aureus	111	<1	104	<1	55	<1	61	<1
Bacillus cereus and *Bacillus* spp.	418	2	164	<1	162	<1	95	<1
TOTAL	25662	100	25849	100	26960	100	23548	100

*Public Health Laboratory Service figures (Communicable Disease Surveillance Centre, to be published).

In addition the following numbers of reports of Campylobacter infection were made.

	1988	1989	1990	1991
Campylobacter enteritis	28761	32359	34556	32160

Table 3.3 Causal agents of bacterial food poisoning (England and Wales 1989–91). Outbreaks by location, and sporadic cases*

	Salmonella	*Clostridium perfringens*	*Staphylococcus aureus*	*Bacillus cereus* other *Bacillus* spp.	*All* causes
General outbreaks					
Restaurants † and receptions	202	58	5	40	305
Hospitals	21	18		2	41
Institutions §	52	35	1	1	89
Schools	9	7		1	17
Shops	32	1	1	1	35
Canteens	9	9	1	3	22
Farms	3				3
Infected abroad	12				12
Other	44	6	5	3	58
Unspecified	8	12	3	6	29
All general outbreaks	392	146	16	57	611
Outbreaks in private houses	2374	6	9	16	2405
Sporadic cases	51845	12	8	37	51902

* reported by laboratories, MOsEH and EHO's.
† includes 'take away' restaurants.
§ includes old people's homes, hostels, halls of residence, nurseries and children's homes.

Public Health Laboratory Service figures (Communicable Disease Surveillance Centre, to be published).

Over the years 1973 to 1989, cases of staphylococcal food poisoning fluctuated from approximately 100 to 500 annually; they represented only a small percentage of the whole. On the other hand, cases of *C. perfringens* food poisoning recorded from 1973 to 1984 varied from 918 to 2924 annually, between approximately 9 and 32% of the total number of cases recorded for all bacterial agents. The proportion of cases due to *C. perfringens* dropped sharply to 10% in 1978 and remained between 9 and 14% up to 1984. Between 1985 and 1991, although the annual number of cases of *C. perfringens* remained at about the same level, the proportion of the total cases of food poisoning fell steadily from 11% in 1985 to 4% in 1991. This is a reflection of the dramatic increase in cases of salmonella food poisoning reported over those years.

In addition to those caused by the common bacterial agents and the general and family outbreaks mentioned earlier in this section, incidents of food poisoning caused by a range of other agents were reported (in 1985–1991) including scombroid fish (scombrotoxin), contaminated shellfish (small round viruses and hepatitis), rotavirus, red kidney and haricot beans (haemagglutinin), *L. monocytogenes*, *Plesiomonas shigelloides*, *V. cholerae* non-01, *V. parahaemolyticus* and copper. There were also a number of outbreaks of *Campylobacter* gastroenteritis associated with the consumption of unpasteurized milk and poultry, and a number of incidents each year for which a microbial cause was not determined.

Food-borne disease other than food poisoning

Many of the more serious food-borne outbreaks of disease are associated with polluted water, milk and products from milk and eggs. Shellfish are dependent on water for life, and filter out and retain the organisms from the water around them. The contents of cans may be contaminated by polluted cooling water sucked in through imperfect seams and closure faults after heat treatment while cooling. Organisms from hands may enter cans in the same way when handled wet and warm while they are under vacuum and hence infect the contents.

Stream water, polluted by sewage from residential houses on or near pastures, has been used for washing dairy equipment and also the udders of cows before milking. Wells can be polluted with surface water and sewage when the surrounding walls are broken or too low to stop flood water. Such infection in rural areas can be widespread. Compared with foodstuffs in which organisms may be irregularly placed in the mass and where conditions for growth may vary, milk and water provide surroundings for even distribution of organisms and thus may give rise to outbreaks throughout a population.

These are some of the ways in which the typhoid and paratyphoid bacilli, the cholera and related vibrios and sometimes *Shigella* spread and cause disease amongst small groups or large numbers of persons. Shigella dysentery can be water-borne or food-borne, but in the UK it is more often spread from person to person either directly or through the environment.

Salm. typhi and *Salm. paratyphi* A, B and C are agents of the enteric diseases typhoid and paratyphoid fevers. *Salm. paratyphi* B is divisible into serotypes, which give rise to: (a) an enteric illness and; (b) gastroenteritis or food poisoning. The incubation period for typhoid fever is 7–21 days; it may be as short as 3 days and as long as 18 days, according to the dose of organisms consumed and the state of the immune system of the individual. The incubation period for the paratyphoid fevers is 7–10 days and the symptoms are less severe. *Salm. paratyphi* A and C are comparatively rare in western countries, but both occur under tropical conditions.

The enteric fevers may begin insidiously. Intestinal symptoms may or may not appear early but they are not prominent until the second or third week of fever. The organisms may be isolated from the blood during the first week and antibodies may be demonstrated in the blood by the Widal test after 10 days with a rising titre. The Widal test should be interpreted with caution when vaccine against the enteric fevers has been administered during the last year. The common means of spread, water, milk, ice-cream and some foods are considered in more detail under the separate headings.

Listeria monocytogenes (see p. 35–37) could also be included under this heading.

Water

The water-borne diseases include the enteric fevers, other members of the *Salmonella* group, cholera, shigellosis (rarely), and viral agents including poliomyelitis and hepatitis B. The role of water is important in the spread of diarrhoeal disease in the tropics and polluted water may be given to infants to drink. The use of polluted water for food manufacturing, and in kitchens for food preparation and the washing of vegetables, equipment and premises, is dangerous. Infection with food or water will increase in hot climates because of rapid growth of bacteria in food and the necessity to drink large volumes of water. More ice-cream and salad vegetables will be eaten, and the conditions in markets and bazaars worsen when local impure water is used to freshen vegetables and fruit.

Salm. typhi and *V. cholerae* outbreaks have frequently been traced to polluted water. In 1854 the well of the Broad Street pump in London was polluted with sewage; as a result there was a large-scale outbreak of cholera and many deaths. The epidemiologist investigating the outbreak was the famous John Snow. He denounced the well water and aroused public opinion, drawing attention to the necessity for hygienic water supplies, efficient sewage systems and general sanitation.

After some years, early in the twentieth century, Alexander Houston introduced the means to make water supplies safe by chlorination. It was not a popular suggestion at the time, but has come to be adopted as a safety measure in most parts of the world. Except for minor incidents of typhoid fever from well water and unchlorinated rural supplies, the next large outbreak occurred in 1937 near London. Three hundred and forty one people were ill after drinking water polluted by *Salm. typhi* from the urine of a symptomless excreter working on repairs to a deep well. In 1963 many local people as well as holiday makers at Zermatt in Switzerland developed typhoid fever because sewage seeped into the mains water supply through an undetected leak in the pipe.

There have been instances in many countries when adults and children have swallowed polluted river and stream waters intentionally or accidentally and developed typhoid fever. One of two small boys contracted typhoid fever after playing in a roadside stream. *Salm. typhi* was traced up the stream and led to the examination of stools from the occupants of a mental hospital; eleven symptomless excreters were found. The sewage effluent from the hospital discharged into the stream approximately half a mile away from the place where the boys drank the water. The patients excreting *Salm. typhi* were segregated into one ward and the sewage outflow from this ward was treated with chloride of lime. The rate of excretion of *Salm. typhi* varied from patient to patient and from time to time, but continued intermittently for many years.

Oysters from the estuary of the river at West Mersea were responsible for cases of typhoid fever in 1958. The river was polluted by sewage from an excreter living upstream. There is a rapid circulation of water through the body of oysters and mussels. Bacteria and viruses may be filtered off into the gut of the shellfish or even cling to the flesh in other parts of the body including the gills.

Another indirect source of *Salm. typhi* from sewage-polluted water can occur when the water is used for cooling heat-sterilized cans in food manufacture. Minute faults in the can structure allow water and organisms to be sucked in while the can is under vacuum. *Salm. typhi* has been isolated from canned cream and epidemiologically known to be present in sealed cans of corned beef and tongue. The outbreaks are described later. (See Chapter 6).

Milk

Milk is an ideal medium for bacterial growth and means of contamination are numerous. It may be already dangerously contaminated when taken from the cow. Tuberculous cattle excreting tubercle bacilli in milk used to be one of the main sources of tuberculosis in children and adults. Formerly, 4000 or more new cases of tuberculosis from bovine tubercle bacilli were recorded yearly in Britain. The eradication scheme, including the tuberculin testing of cattle and the pasteurization of milk, has reduced the incidence almost to vanishing point. In countries where tuberculosis is common, the spread of the organism is assured in the ill-nourished occupants of overcrowded and unhygienic rural and urban dwellings.

Brucella abortus, the bacillus responsible for abortion in cows, may be excreted in milk and give rise to an infection known as undulant fever, in those who consume contaminated raw dairy products such as milk, cream and 'cream' cheese. Farmers and veterinarians working in close proximity to infected animals may also acquire the disease. Goats and their products may be similarly infected with *B. melitensis*, particularly in the Mediterranean countries. There is a long incubation period and the illness is protracted and debilitating.

Undulant fever in man is not often fatal but may cause ill health over a long period of time. The disease has been eradicated from cattle in Scandinavia and the USA and similar methods for eradication are now statutory in the UK. Little is known of brucella infection in rural areas of countries such as India.

In the intimate association between farm workers and their animals, diseases are spread from one to the other and the organisms causing infections in farmers and their helpers may spread to milk. Diphtheria bacilli from human cases and carriers may be implanted on ulcers on the teats of the cows and cause contamination of the milk, although this is rare.

In the past there were widespread outbreaks of scarlet fever and tonsillitis from streptococci passing into milk from infected lesions in the udder and on the teats of cows. Occasionally, milkers with streptococcal infections have been involved. In one large outbreak, pupils in two schools were infected by drinking raw milk. Haemolytic streptococci of the same type were found in the throats of 87 children with scarlet fever, the throats of the milker and his two children, and in the milk. The danger from the spread of infection by those handling cows and milk is greatly lessened by the automatic machine method of milking. Outbreaks of dysentery, typhoid, and paratyphoid fever, and of salmonella food poisoning have been spread by milk contaminated by carriers amongst farm and dairy workers, but the cow herself is also a source of pathogenic organisms. Widespread outbreaks of paratyphoid fever have occurred; more than 1000 people were infected in one outbreak by drinking raw milk from cows excreting *Salm. paratyphi B*.

Pails, milking machines, and milk bottles may be washed with water polluted with human or animal sewage. Recommended chlorine compounds are available for use on farms, and milk bottles should be rinsed, washed in detergent and rinsed again in very hot water followed by cool mains water. Chlorine is sometimes used in rinse waters also.

The faeces of cows infected with campylobacters or salmonellae contaminate milk directly. There has been an increasing number of reports of *Campylobacter* infection in which the vehicle is raw or imperfectly pasteurized milk. The milk outbreaks are probably due to recontamination of the milk after heat treatment, possibly from the bottling plants. Cows may shed *L. monocytogenes* in their milk, but the organisms should be eliminated by HTST pasteurization. The same is probably true for *Y. enterocolitica*, which is not known to be particularly heat resistant. There is also a water-borne cycle of *Campylobacter* infection. The ways of spread and behaviour of this organism in food are ill-defined probably because its growth requirements are unusual.

Infections in the cow's udder, known as mastitis, may lead to excretion of food-poisoning organisms such as staphylococci, streptococci, campylobacters and even salmonellae in the milk.

In spite of all the hygienic precautions which may be taken at farms, bottling depots, and dairies, the only method to safeguard milk for delivery to the consumer is heat treatment before or after it has been bottled and capped. By this means all pathogenic organisms introduced from the cow, from subsequent contacts, or other sources will be destroyed and there will be no further danger of contamination until the bottled milk reaches the consumer's kitchen. Here the milk is subject to the hazards of household use, but with reasonable care and cold storage it can be kept safe.

There are three commercial methods for heat treatment: by pasteurization, which is usually carried out in a plant; by disinfection at a temperature of 100°C (212°F) or over; and by direct or indirect applications of steam which necessitates aseptic filling but provides milk with a long shelf life. (See Chapter 8).

The following description of a typhoid outbreak illustrates some of the points mentioned.

The extensive episode of milk-borne disease occurred one autumn in the adjacent towns of Christchurch, Poole, and Bournemouth in Dorset. There were 518 cases among the local inhabitants and about 200 persons were infected while on holiday in the district, although the disease did not develop until they returned home; there were about 70 deaths. Men, women and children of all ages and occupation were affected. This suggested infection by a common food of wide distribution such as milk. All the primary cases had consumed raw milk from one particular dairy. Immediately, steps were taken to pasteurize the supply. A search for the carrier among the employees of the firm, including the twelve roundsmen, was unsuccessful. Investigations spread to the 37 farms that produced the milk to be mixed and distributed by the firm under suspicion. At one farm the housewife was ill with enteric fever; she died and her son developed the disease. This farm contributed 91 litres (20 gallons) of milk each day to the mixed supply of the retailer. Yet there was another puzzling feature in that a number of people had been infected before the farmer's wife became ill.

Further enquiries revealed the fact that, some years ago, a fatal case of typhoid fever had occurred in a house adjoining the farm. The water supply was common to both houses, eight others in the vicinity and also to the dairies. It came from a deep well situated about 91 metres (100 yards) from a small stream. This stream was liable to pollution from storm water and the sewage effluent from a large house. Typhoid bacilli were found in the effluent and their origin traced to an excreter in the house.

It was next demonstrated, by chemical means, that there was a connection between the stream water and the well. Furthermore, the farmer's cows and those of another producer, who also contributed to the same dealer's milk supply, used pasture alongside the stream. The cows drank from and stood in the stream with their udders in the contaminated water. Spread of infection was considered to be water-borne to the cows followed by contamination of the raw milk supply consumed by the population. The milk was thought to have been contaminated with typhoid bacilli for 31 days before pasteurization was adopted. Outbreaks of enteric fever and food poisoning due to *Salm. paratyphi B* in raw milk are not uncommon and the sequence of events is similar. The efficient pasteurization of milk eliminates the hazards of milk-borne disease.

Ice-cream

The ice-cream heat treatment regulations were made statutory in 1947, 1959 and 1963 (see Chapter 17), earlier ice-cream was a common vehicle for incidents and outbreaks of food poisoning and food-borne disease. The mix was usually contaminated during preparation and when stored for many hours and even overnight at ambient temperatures contaminants were encouraged to grow.

Salm. typhi from a symptomless urinary excreter contaminated ice-cream which was responsible for approximately 210 cases and four deaths in a holiday town in Wales. The vendor of the ice-cream had been ill with typhoid fever many years earlier and declared free of *Salm. typhi* on recovery, but he continued to excrete the organism in his urine. Instead of heating the ice-cream mix immediately before cooling and freezing, the vendor used a cold mix powder requiring no heat treatment of the rehydrated mix. Though not recorded, the time between mixing and freezing must have been several hours at the summer temperature. The warm weather encouraged increased consumption of ice-cream by holiday makers.

Salm. paratyphi B fever was reported from year to year in visitors at a north Devon holiday resort with a long sandy beach. Most people used one end of the beach for bathing and it was assumed that the paratyphoid bacilli discharged into the sea from the main sewer nearby. The bacteriologist in charge of investigations disbelieved this explanation. He suspended cottonwool swabs attached to lengths of string at various points in the main sewer, its tributaries, and sewage pipes from individual houses. Results showed that one house only was discharging paratyphoid bacilli; the house was occupied by a local ice-cream manufacturer and his wife. Examination of stool samples revealed that the wife was excreting *Salm. paratyphi B*. Their ice-cream barrow was frequently on the bathing beach. It was considered that contaminated ice-cream was the vehicle of infection rather than the sewage in the sea.

Shigella flexneri 103Z was excreted by a child ill with dysentery. It transpired that 2 days previously the child had visited the Pets' Corner of a large general store. There were several cages of monkeys and while the child was standing near one of the cages, a monkey suddenly tried to seize the ice-cream cornet she was holding. The child continued to lick the cornet before the mother threw it away. The same type of *Shigella*, commonly found in monkeys, was isolated from one of the monkeys in the shop. The dose of organisms must have been directly from the monkey's paw, as there was no time for multiplication on the ice-cream.

Food

Typhoid fever after the consumption of canned ox tongue led to the supposition that the typhoid bacilli were present within the can when it was opened. The Argentinian establishment responsible for canning used untreated water from the polluted River de la Plata to cool the cans after processing; typhoid bacilli of the same phage type were isolated from the river. The cooling water was chlorinated for a few years until the chlorination plant broke down. In 1964 there was another outbreak of typhoid fever. A can of corned beef from the same establishment was responsible for a trail of contamination amongst cold cooked meats in a small supermarket in Scotland. Four hundred case records indicated that for 3 weeks cold meats were purchased by persons who subsequently developed typhoid fever. The spread of the organisms was encouraged by a number of factors. The can opener, slicing machine, pedestals for window shows, price tickets and trays for overnight storage in the refrigerator were common to all canned meats sliced during the day. The slicing machine was washed each day but plastic buckets containing non-disinfected water were used for cloths and scrubbing brushes.

With closure of the cold meat counter and finally of the whole shop cases ceased to occur. Earlier there were three outbreaks of typhoid fever all associated with canned meat from another establishment in the Argentine, also using non-chlorinated cooling water. In one outbreak the sale of corned beef was stopped immediately. In the other two instances the corned beef was sold in supermarkets with a high standard of hygiene and plentiful use of refrigerators, so that the number of cases was limited.

Following these five outbreaks records were searched for information about earlier episodes of typhoid fever. Many were associated with cold meats bought at provision stores or retail butcher's shops. Symptomless excreters were not found so that contamination of the meat after removal from the can was unlikely. Leakage of cooling water through defective seams was probably the explanation. Typhoid bacilli neither produce gas nor spoil the meat even though growing abundantly in corned beef under both aerobic and anaerobic conditions. Thus, the presence of these organisms is not manifested in the blowing of contaminated cans, unless other contaminants are present. The chlorination of cooling water at canning factories is now strictly imposed.

Vibrio cholerae is a small curved highly motile Gram-negative bacillus, divisible into classical and eltor biotypes. There is sudden onset with profuse watery diarrhoea (rice water stools) and vomiting; asymptomatic infections are common, particularly with the eltor biotype and in children. Dehydration must be counteracted immediately with electrolyte solutions to correct the sodium/potassium imbalance due to loss of fluid. Incubation is a few hours to 5 days, usually 2–3 days. The vibrio is primarily water-borne and in certain areas epidemics are common and affect large numbers of people because of water pollution. The main reservoir is man, but little evidence exists of person-to-person spread; one member only of a family may be infected, perhaps because more water was consumed and therefore a larger dose of organisms. The disease spread from India to most parts of the world. For the first time cholera has appeared in South America and currently there is a pandemic of more than 200 000 cases that have spread from Peru to many other south and central American countries. Active immunization is of limited value because of its short duration.

The *Shigella* are Gram-negative, non-motile, non-sporing bacilli of several biotypes. They give rise to diarrhoea, cramps, vomiting and fever of varying degree. The most severe symptoms are caused by *Sh. dysenteriae* (Shiga's bacillus) when there is blood, mucus and pus in the profuse stools;

cases of septicaemia may occur. The mildest form and the commonest in the West is *Sh. sonnei*. *Sh. flexneri* is of intermediate severity and found more frequently in the East. Water-borne outbreaks are rare and the organisms can be spread from person to person or via fomites such as pencils, rubbers, toilet seats and pull chains, particularly amongst children in day-care nurseries and paediatric wards. The organisms can be food-borne also. Fluid and electrolyte replacement are important. Incubation is 1–7 days, usually less than 4 days. The duration may be 2–3 weeks and the period of excretion 3 weeks or longer. The reservoir is man and higher primates. It is thought that the number of organisms necessary to initiate symptoms is small.

Shigella sonnei infected at least 248 of 600 pupils in a school. The symptoms appeared about 12 hours after a meal which included a pie made with mutton. After preparation the pie was stored in a defective refrigerator and warmed up before serving. *Sh. sonnei* was isolated from the remains of the pie, from 139 pupils and from four of 20 food handlers in the school kitchen.

The same organism was responsible for the illness of approximately 100 children 24–36 hours after a meal that included blancmange prepared the previous day and stored in the kitchen until served. The children were in two schools and only those who ate the blancmange were ill. *Sh. sonnei* was isolated from the stools of 29 children and all of six kitchen staff. There were no remains of the blancmange for examination but packets of the dried powder were available; they were negative for *Shigella*. The pudding was thought to be contaminated during preparation.

Shigella dysentery was reported in a number of people who ate pease pudding prepared in a small provisions store. The dysentery bacilli were isolated from the remains of the pudding. The day the pudding was made, a nursemaid and a small girl 3 or 4 years of age visited the shop. While the nurse was talking to her friend in the kitchen the girl crammed a handful of pease pudding into her mouth. The hole in the pudding and the evidence on the child's face were remembered vividly by both the nurse and her friend. The child had recently recovered from an attack of diarrhoea caused by the same shigella dysentery organism as that responsible for the outbreak. She was still excreting the organism in stools and it was assumed to be present on her hands also.

Virus

There are at least two groups of virus associated with non-bacterial gastroenteritis (NBGE): (a) a reovirus-like particle approximately 70 μm in diameter termed *rotavirus* because of its wheel-like shape or *duovirus* which describes the double-shelled capsid structure; (b) the so-called *Norwalk Agent*, a parvovirus-like particle, 27 μm in diameter. Other agents, including *Hawaii* and *Montgomery County*, have also been reported; the Hawaii agent is similar to, but immunologically distinct from the Norwalk agent. Rotavirus is an important agent in childhood diarrhoea. The presence of large numbers of rotavirus in the stools of children with NBGE diarrhoea is reported from many countries, and the virus is thought to be responsible for more than 50% of the world's cases of infantile enteritis. In adults the infection is usually mild or asymptomatic. The virus multiplies in the cytoplasm of the cells lining the villi of the small intestine from the duodenum to the jejunum. After 2–4 days incubation enormous numbers of virus particles (10^{10}/g faeces) are excreted. The profuse watery stools, sometimes accompanied by fever and vomiting, continue for 5–8 days. Death may result from untreated dehydration. Lactose intolerance results from the attachment of the rotavirus to the lactase enzyme in the intestine. Serological evidence of infection can be obtained and kits are available. The Norwalk and related agents usually give rise to self-limiting disease of 24–48 hours duration characterized by nausea, vomiting, diarrhoea, abdominal pain, malaise, low-grade fever or any combination of these symptoms. These agents may be responsible for the so-called 'winter vomiting disease' or 'epidemic diarrhoea'. The viruses may be water- and food-borne. They have been demonstrated in oysters and other shellfish from polluted water. They cannot multiply in food but must be passaged in small doses by other means. The approach to the study of viruses differs widely from that of food- and water-borne bacteria. Their presence is demonstrated by electron microscopy and immunological techniques. Some may be demonstrated in tissue culture but they cannot be grown on media as used for bacteria.

The slow viruses are responsible for progressive and fatal infections without immune response; scrapie in sheep is the best known example. Three other such diseases are 'kuru', Creutzfeldt–Jacob and bovine spongiform encephalitis. The scrapie virus causes a degenerative disease of the central nervous system in adult sheep, 2–4 years old; it may also attack goats. The infection starts early in life and is thought to be encouraged by intimate contact between ewe and lamb and spread via the nasopharynx and intestinal tract. Clinically, there is apprehension, cutaneous irritation, tremor, incoordination of gait, aimless wandering, visual impairment, weight loss, terminal prostration and death in 6 weeks to 6 months. The virus occurs in the lymph nodes and spreads lymphatically. 'Kuru' and Creutzfeldt–Jacob diseases are similar and infection occurs through cannibalism. The first case of a similar disease in cattle was reported in 1986 and called spongiform encephalitis (BSE). This viral disease occurs sporadically among dairy cattle and was thought to follow the feeding of cattle with sheep offal and other infected feed ingredients. There are fears about the possible pathogenicity for humans eating beef or beef products. Cases have been reported in cats and in one pig (experimental). The consumption of beef, not only from infected herds, was reduced in the UK and also in Europe. Controls were imposed on the disposal of infected cattle.

Mycotoxins

Some fungi produce toxic substances which are poisonous for man and animals. *Aspergillus flavus* growing in ground nuts and other cereals forms aflatoxin which affects turkey poults, ducklings, cattle, pigs, sheep and trout. The induced disease in laboratory animals gives rise to hepatic changes associated with liver carcinoma. Aflatoxins have been detected in many foods including peanuts, peanut butter, unrefined oil, corn, wheat, cotton seed, sweet potato, cassava, rice, sorghum, soybeans, milk and milk products and liver. Evidence that mycotoxins can cause chronic disease in man is based on epidemiological studies; there has been an association between liver cancer and the consumption of high levels of aflatoxin. The toxin in animal feed may lead to detectable levels in the liver and musculature of swine. Feed for cattle containing more than 60 mg aflatoxin/kg can cause contamination of milk. Toxic substances from *Fusarium graminarum* in corn can cause illness in pigs. Many of the substances produced by mould growth are found to have adverse effects on animals. It has been strongly recommended that more attention should be given to the dangers of eating mouldy foods, whether contaminated from bad practices or intentionally for texture and flavour purposes. All fungi used for commercial practices should be carefully screened for toxicity.

Protozoa

The protozoans *Giardia* and *Cryptosporidium* are zoonotic causes of enterocolitis and diarrhoea in mammalian species. The cryptosporidia infect the entire bowel, but predominantly the lower small intestine, with extensive mucosal changes. There is partial atrophy, fusion and distortion of villi resulting in maldigestion in the brush borders of intestinal epithelial cells and malabsorption. The symptoms include:

(1) In persons with intact immune systems, profuse watery diarrhoea with mild epigastric cramping pain, nausea, anorexia and dehydration for 10–15 days.
(2) In immunocompromised patients, for example, those with AIDS or receiving immunosuppressive therapy, more prolonged and severe symptoms persist for several weeks, months or even years; there is no therapy.

An Australian survey (1983) found 4.1% of 884 patients with gastroenteritis due to cryptosporidiosis and the highest incidence from February to May. In another survey of 220 stools from children less than 3 years of age, isolation rates were for *E. coli*, 44.5%; rotavirus, 14.1%; shigellae, 3.6%; *Giardia*, 2.7%; *Salm. typhimurium*, 1.3%; *Entamoeba histolytica*, 1.8%; and *Cryptosporidium*, 1.4%. In India, the incidence of cryptosporidia in children less than 3 years of age was 1.3–13%. The organism could thus be regarded as an important aetiological agent of diarrhoea in children. Cases and outbreaks occur in day-care centres, among international travellers and in various communities due to polluted water. The organism is associated with diarrhoea in calves, lambs and deer, and diarrhoea can be experimentally induced in lambs, pigs and calves by means of the oocysts. The organism multiplies by both asexual (schizogony) and sexual (sporogeny) cycles. The sporulated oocyst is the infective form and can be ingested from drinking water, food and vegetables. The source of human infection may be infected cattle (calves), pigs, dogs, cats, sheep and other humans. The diarrhoea can threaten the life of immunocompromised persons regardless of age.

4

Reservoirs and vehicles of infection and ways of spread

Bacteria occur widely in the environment. They can be isolated from water, food, soil, air and on and within animals and man. Mostly they are harmless, but a small proportion can infect man, animals and plant life. Under certain conditions they are able to grow and multiply in the tissues of the body, some in one tissue and others in another. During illness, germs can be transferred from one person to another, from animal to animal, and from animal to man or man to animal, either directly or by means of a medium such as food. The fresh host may become ill, or may resist the invasion and show no symptoms, although the organisms may be harboured in the body for a variable period of time. In this way bacteria and viruses which depend for life on the conditions provided by the human or animal body maintain their existence. They may survive in the nose, throat or bowel of healthy persons or other living creatures.

Some food poisoning is caused by the toxic products of bacteria growing actively in food and not by invasion of the body, for example, staphylococcal and *Bacillus cereus* food poisoning and botulism.

It was early in the twentieth century that bacteria were first recognized as agents of food poisoning and of other food-borne disease. Animals were reported as the main reservoirs of the salmonella group of intestinal pathogens responsible for the majority of food-poisoning incidents. Early investigations led workers to diseased animals as the source of contaminated meat eaten by those affected; subsequent workers showed that the human and the animal body could harbour organisms of the salmonella group without showing signs of disease. Also, they found that human nose, throat and skin lesions were the main sources of staphylococci.

Foodstuffs of animal origin may be regarded as primary sources of many food-poisoning bacteria. Organisms carried by the live animal may be found in the raw meat after slaughter and may be passed to other foods. The contaminated raw food as received is an even greater hazard in food preparation areas than food handlers who are carriers and temporary excreters, providing they take simple hygiene precautions.

Table 4.1 Human reservoirs of food-poisoning organisms

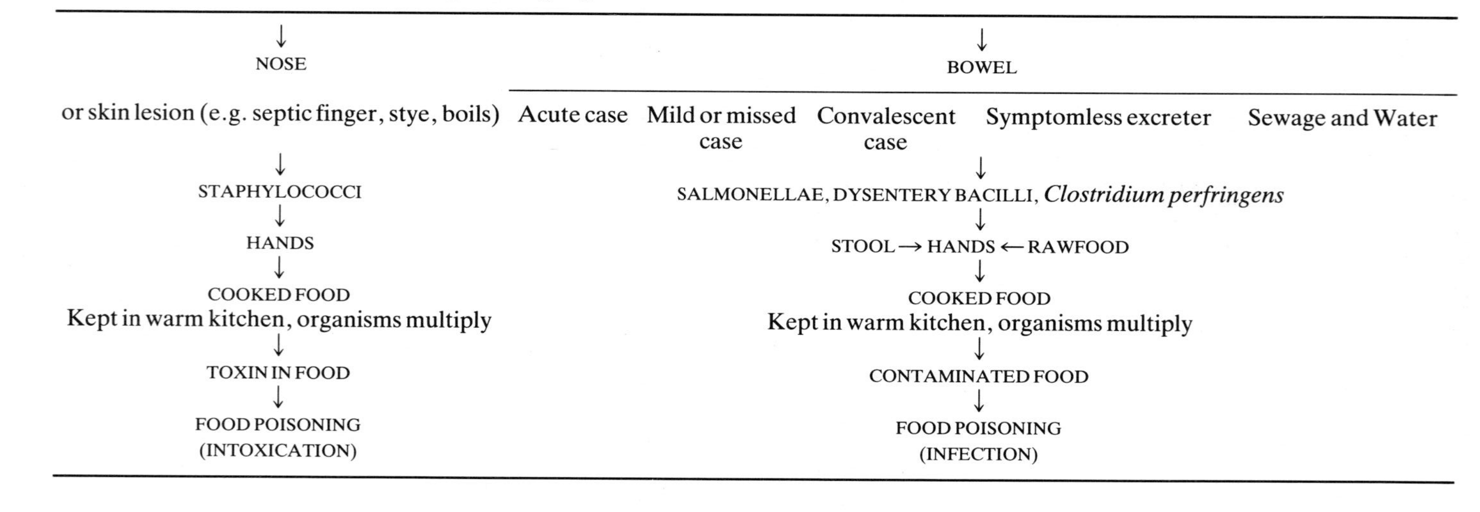

↓
NOSE
or skin lesion (e.g. septic finger, stye, boils)
↓
STAPHYLOCOCCI
↓
HANDS
↓
COOKED FOOD
Kept in warm kitchen, organisms multiply
↓
TOXIN IN FOOD
↓
FOOD POISONING
(INTOXICATION)

↓
BOWEL
Acute case | Mild or missed case | Convalescent case | Symptomless excreter | Sewage and Water
↓
SALMONELLAE, DYSENTERY BACILLI, *Clostridium perfringens*
↓
STOOL → HANDS ← RAWFOOD
↓
COOKED FOOD
Kept in warm kitchen, organisms multiply
↓
CONTAMINATED FOOD
↓
FOOD POISONING
(INFECTION)

The animal and human reservoir

Tables 4.1, 4.2 and 4.3 show the reservoirs and likely paths of spread from animals and man to food for the three main groups of food-poisoning organisms which frequently reside in living creatures.

The animal reservoir

Table 4.3 names the animals known to be occasional or persistent reservoirs of organisms of the salmonella group. Animals may be infected clinically or become transient symptomless faecal excreters of salmonellae.

Farm animals become infected through eating contaminated feeding meals, grazing on contaminated pasture land, or by contact with animal, human or bird excreta on the farm, during transport, or in the lairage of markets and abattoirs. It has been observed, for example, that the proportion of animals excreting salmonellae amongst pigs and sheep rises steeply under conditions of stress, excitement, fear or privation of food and drink.

The mass production of animals and birds on farms using intensive rearing methods encourages not only the spread of organisms between animals and birds, but also a build-up of potential infection in the environment. The number of calves and adult cattle infected with *Salmonella typhimurium* has been shown to increase during the close contacts of mass production.

A large number of animals actively excreting salmonellae may overcome the normal hygienic precautions carried out in the abattoir with the result that many contaminated carcasses will be distributed for retail sale and also for manufacturing purposes. In such instances a rise in the number of outbreaks and incidents of salmonella food poisoning in the local population may be expected.

Table 4.2 Human and animal reservoirs of *Clostridium perfringens*

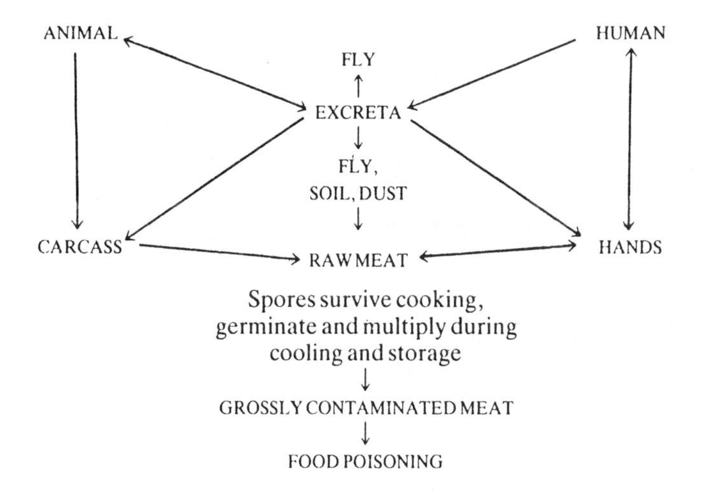

Table 4.3 Animal reservoirs of salmonella organisms

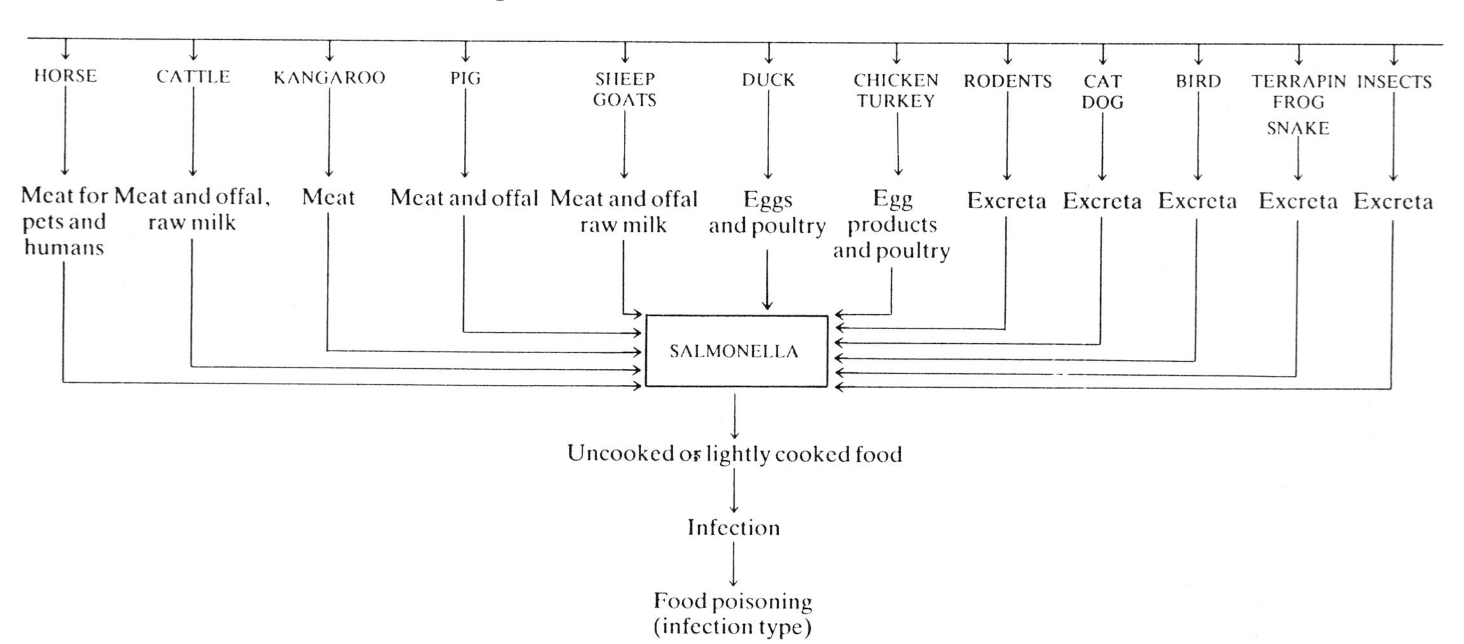

Table 4.4 *Salmonella* from frozen and chilled chickens*

Year		Number examined	Number positive	%	Number of different serotypes
1971–76†	frozen	519	175	34	16
1979–80	frozen	100	79	79	18
1987	frozen	101	65	64	28
	chilled	103	56	54	
1990	frozen	146	79	54	25
	chilled	146	62	42	

* Data from the Food Hygiene Laboratory and Roberts, D. (1991) *Lancet* **337**, 984–5.
† Chickens from 13 producers.

The rapid growth of the broiler industry for chickens, turkeys and ducks is apparently responsible for the increased infection of birds and carcasses with salmonellae. A higher proportion of poultry carcasses will be contaminated during bulk handling than with small-scale operations. The extent of contamination may be increased by retail sale, cooking, storage and packaging.

Salmonellae have frequently been isolated from both frozen and chilled retail chickens (Table 4.4). Although the proportion of carcasses with salmonellae shows some decrease in the later surveys, the number of different serotypes isolated has increased. In the 1987 and 1990 surveys there was one predominant type, *Salm. enteritidis* PT4, which was found in 20% and 23% of frozen and chilled chickens respectively. This particular type was not isolated in the earlier surveys. Detailed monthly reports from the Salmonella Reference Laboratory in South Australia (see p. 89) indicate that the poultry reservoir is a predominant source of salmonellae.

Organisms from infected carcasses of animals and birds may be transferred to other meats in the abattoir, in transport and in the retail shop or factory. Widespread outbreaks of salmonella food poisoning have occurred through the distribution of contaminated raw meats and by the contamination of cooked meats from the raw carcass meat. More than 1000 cases of *Salm. typhimurium* food poisoning with 12 deaths were reported from Britain. The cases spread south by means of contaminated carcasses and from minced meat originating from the flesh of infected pigs. In another outbreak, legs of pork were cooked in a factory working with large quantities of contaminated pork from infected pigs. The meat was distributed over a wide area, and was also eaten in the factory canteen.

The continual danger of cross-contamination from one raw meat to another was demonstrated in an outbreak of salmonellosis in which thousands of cases and a few deaths were reported. The spread between

Table 4.5 *Salmonella* from packets of pork sausages and sausagemeat from large and small manufacturers

Source of samples	Year	Number of of packets examined		Number positive		%		Number of different serotypes
Large manufacturer A	1968–74*	312		7		2.2		
Large manufacturer B		854		413		48.4		
Medium manufacturer C		101	1467	4	435	4.0	29.7	38
Medium manufacturer D		100		11		11.0		
Small manufacturers		100		0		0		
Various	1981†	208		25		12.0		14
Various	1988–89†	203		21		10.3		9

* Roberts, D., Boag, K., Hall, M.L.M. and Shipp, C. (1975). The isolation of salmonellas from British pork sausages and sausagemeat. *Journal of Hygiene* **75**, 173.

† Results from the Food Hygiene Laboratory, Colindale.

carcasses was encouraged by the use of wiping cloths in slaughterhouses, now forbidden (Slaughterhouse Hygiene Regulations 1977; Chapter 17).

The raw meat was handled at the same time as canned meat and other cold cooked meats using the same balances and knives. Home-killed infected veal used for manufactured products and also for retail sale was responsible for many cases of salmonellosis.

In general, when various serotypes and phage types of salmonellae are isolated from abattoir drains the same types are found in human cases of salmonellosis in areas served by the abattoirs. In 31 separate incidents of food poisoning there was convincing evidence that the meat and meat products from pigs, cattle and calves were vehicles of infection. Salmonellae were found in 21% of 4496 swabs from abattoir drains. Isolations were most frequent where a large number of cattle were slaughtered.

Table 4.5 gives the rates of isolation of salmonellae from packets of pork sausages and sausage meat from large, medium and small manufacturers.

The results indicated the likelihood of higher rates of isolation from manufacturers gathering in large numbers of animals from a wide area than with those contracting out locally with small numbers of animals. From the various sources the average isolation rate was 30% of 1467 samples. In 1981, salmonellae were found in 14% of 208 samples of sausages. In 1988/89 a further survey on the incidence of salmonellae in sausages showed that 10% of 203 samples contained the organism. This survey in north London and Hertfordshire during October 1988 to May 1989 indicated that branded sausages more often contained salmonellae than those prepared on the premises of retail shops. Nine serotypes were isolated including *typhimurium*, *enteritidis* phage type 4, *saint paul*, *agona*, *infantis*, *derby*, *montevideo* and *livingstone*. The 1981 isolates included 14 serotypes, for example, *derby*, *typhimurium* and *saint paul*.

In 1989 an outbreak of salmonellosis occurred in North Wales with

Table 4.6 *Salmonella* in raw meat, 1961–78

Year	Raw meat and offal	Number of samples examined	Number positive	%
1961–65	Wholesale (imported)			
	Frozen boneless			
	veal	1328	192	14.4
	beef	3420	410	11.9
	beef offal	102	9	8.8
	mutton	1107	123	11.1
	mutton offal	155	4	2.5
	horse	12 406	4516	36.4
	horse offal	1031	421	40.8
	kangaroo	929	410	44.1
	Retail (English and imported)			
	butcher's shop meat	262	4	1.5
	butcher's shop offal	52	0	
	pet shop meat	433	120	27.7
	pet shop offal	148	25	16.9
	knacker's yard meat	64	7	10.9
1977–78	Retail			
	minced beef and pork	279	21	7.5

600 cases; sausages were the vehicles of infection. Twenty of 203 (9.8%) of samples were positive for salmonellae; 38% were prepared in retail premises.

Imported boneless meats have shown high rates of contamination with salmonellae from time to time (Table 4.6). Imports of meat, particularly from new sources, should be monitored for salmonellae. Hitherto unrecognized serotypes have become common as agents of food poisoning in the importing country.

The prophylactic use of antibiotics in feeds for calves to control intestinal infections has resulted in the development of drug-resistant strains of salmonellae. The treatment of the human population with antibiotics is jeopardized when antibiotic-resistant strains of salmonellae are passed from animals to foods intended for human consumption. Furthermore, resistance may be transferred to other organisms causing human infections. In 1969, the Joint Committee on the Use of Antibiotics in Animal Husbandry and Veterinary Medicine made the following recommendation: antibiotics used for treatment of humans should not be used in feed for animals nor for prophylaxis.

As well as animal-to-animal spread of faecal matter, there may be soiling of food, such as from cow to milk and meat, pig to meat, and poultry to the shells of eggs and broken-out egg melange. Vermin and domestic pets may soil various kinds of food. The rate of contamination of carcass meat and poultry will depend to a large extent on the number of animals or birds excreting salmonellae as well as on the hygienic practices in meat and poultry establishments and also in shops.

The general picture of salmonellosis in tropical countries indicates that it is more common in urban than in rural areas. The slaughter of individual animals in villages results in far less spread of infection than the general abattoir practice in the towns where cross-contamination of carcasses will be inevitable.

Feeding meals

Farm animals, cattle and pigs as well as poultry are exposed to infection from feeding meals. Fish, bone and meat meals are known to contain salmonellae in a proportion of samples. Consumption may lead to transient excretion and sometimes to actual disease. Imported meals come from various countries including South America and India. Home-produced material from waste products of the meat trade recycles salmonellae back to animals. Serotypes imported in feeds soon become established in the farm animal population and are known to spread to the human population by means of meat and poultry products. The rise in incidence of *Salm. brandenburg* in the population in relation to pigs, feeds, abattoirs and sausage meat is described on page 94.

Countries that require by statutory law that animal feeds, whether home-produced or imported, are free from salmonellae, and which monitor the products, appear to have a lower incidence of salmonellosis in the population.

The Zoonoses and Protein Processing Orders were designed especially for this purpose, yet the isolation rate for salmonellae in poultry carcasses and the incidence of human salmonellosis continue to rise. Opportunities for spread and increase in numbers of salmonellae occur in the lairage and in wet feeding troughs.

The salmonellae isolated from 2–14% of 5637 pigs in the UK were similar in serotype to those found in feeding meals and responsible for salmonellosis in the human population. A comparative study with another country, which legislates for treatment to free feeding meals from salmonellae indicated lower figures for the incidence of salmonellae in pigs and of salmonellosis in the population. Furthermore, the majority of serotypes found in pigs in the UK were not isolated from the human population of the second country. Thus the widespread distribution to farm animals of feeding meals contaminated with salmonellae may lead to transitory infection of a proportion of animals over a wide area. If animal husbandry is good, spread may be reduced but when general care is poor, infection may spread through herds. The feeding of raw or inadequately heated offal from sheep, poultry, cattle and other animals will inevitably lead to infection from *Salmonella, Campylobacter* and viruses such as the bovine spongiform encephalitis virus, thought to have come from the raw or imperfectly cooked offal of sheep suffering from scrapie. (See p. 49). The feeding of such raw material or such material inadequately heated contravenes the EC Directive on the Disposal and Processing of Animal Waste and the Prevention of Pathogens in Feedstuffs, also the Diseases of Animals Protein Processing Order. (See Chapter 17).

Pet food

Horses and kangaroos may excrete salmonellae; frozen boneless meat from these animals is imported for pet foods, but designated fit for human consumption. The organisms may spread in domestic kitchens unless the meat is carefully handled and cooked and care taken to disinfect, by heating, the utensils used. The pets may also become excreters and in turn infect children playing with them; furthermore, the excreta of infected pets may infect flies which invade the house. In many countries horse meat is eaten by the human population and although sold mainly for pets, it may be displayed raw alongside other meats and also fish. Condemned meat from knackers' yards is sold raw for pets.

Outbreaks of salmonella food poisoning have been caused by cross contamination from pet food to domestic food in the kitchens of families owning pets. The Meat (Sterilization and Staining) Regulations, 1982, amended 1984, ensure that condemned meat intended for animals, whether imported or home-produced, is cooked before sale and sold for animal feed only. Regulation of the animal feed industry is under review (See Chapter 17). Canned pet foods, sterilized by heat treatment, should be regarded as safe.

Vermin and birds

Rats and mice can suffer from infection and become symptomless excreters of types of salmonellae known to infect man. The organisms may be picked up on farms, in sewers and from garbage. Poisons used to kill rats and mice should not contain virulent organisms of the salmonella or any other group of organisms. The danger is obvious, from accidental contamination of human food and the promotion of excreters amongst the animals that survive. The sale of such material is forbidden in the UK.

Wild birds may excrete salmonellae if they eat food contaminated with these organisms. Scavengers such as seagulls and even pigeons feeding on urban waste from tips and sewage outflows are frequently infected, whereas salmonellae are rarely found in country birds in sparsely populated areas.

Pets

In Britain approximately 1–2% of cats and dogs have been known to excrete salmonellae without symptoms. In a survey carried out in the USA 15% of dogs fed meat and bonemeal (rarely free from salmonellae) were found to be excreters. Nearly 10% of dogs in one area of Australia were excreting salmonellae. There were 12 serotypes corresponding with those isolated from cases of human salmonellosis in the same area. These results suggest that food hygiene regulations justifiably ban cats and dogs from food shops.

The sources of infection on farms are likely to be four fold:

(i) animal excreters as foci of infection;

(ii) feeding stuffs;
(iii) environmental factors including man, various animals, vermin, birds and water;
(iv) movement of animals such as calves for breeding purposes and pigs for mating.

It may be assumed that if the feeds were safe, the importance of (i), (iii) and (iv) would be reduced also.

The human reservoir

The right-hand side of Table 4.1 shows the chain of spread of salmonellae from the human intestine. In any outbreak of infectious disease there will be the possibility of one of four types of reaction: acute illness; ambulant cases with mild symptoms which may be ignored or attributed to other indispositions; convalescent carriers who will continue to excrete the organisms in their faeces for some time after recovery from illness; and temporary carriers or symptomless excreters who may harbour the infecting organism for a short time without exhibiting symptoms. The excretion of *Salm. typhi* in stools may persist for many years, while other salmonella serotypes and dysentery bacilli are excreted for a few weeks only, rarely months and only exceptionally for a year or more. Treatment of persistent excreters with antibiotics tends to prolong the period of excretion and encourage resistance of the organisms.

Water droplets that spray from flushing lavatories, soiled seats, pull chains, door handles and taps may pass infection from person to person. The more fluid the stool the more danger from spread. Contaminated hands may pass infection to food. Most intestinal organisms are readily washed from the skin by soap and water and they are not harboured in the skin like staphylococci.

The human nose, hand and skin are the primary habitat and natural home of *Staphylococcus aureus*; they reside in the mucous membranes of the nose and the skin of man and animals. The left-hand side of Table 4.1 indicates the manner of spread of staphylococci from the human reservoir to food. 30–50% of the general public carry staphylococci in the nose, in patients and staff in hospitals the nasal carriage may be as high as 60–80%.

Staphylococci can be isolated from the hands of 14–44% of persons. There are many types of staphylococci but they do not all produce the toxins responsible for food poisoning. Nasal secretions contain large numbers of bacteria; in some persons a large proportion will be staphylococci which will contaminate the hands. These organisms can penetrate into the deeper layers of the skin, where they live and multiply in the pores and hair follicles.

Hands infected in this way may be washed and scrubbed without removing the organisms. In hot steamy atmospheres the staphylococci rise to the surface of the skin. Antiseptic lotions, hand creams and soaps may help to reduce the skin carriage of staphylococci. There are also preparations available for the treatment of nasal carriers. Food handlers known to be habitual carriers of staphylococci in the skin of the hands should avoid working with foods such as cooked meats, poultry, egg and milk dishes which encourage bacterial growth.

The pus from staphylococcal skin lesions, for example, boils, carbuncles, whitlows and sycosis barbae (barber's rash), as well as septic cuts and burns, will contain innumerable organisms and a small speck of pus could inoculate food with millions of staphylococci. The organisms like to live in the moist serous fluid of cut surfaces. Therefore, the smallest cut or abrasion, however clean and healthy it appears to be, may act as a focus of infection. There are many examples of outbreaks of staphylococcal food poisoning initiated through the contamination of foods by staphylococci from boils, ulcers, abrasions on the hands as well as from healthy hands (see p. 103–111).

As well as being a direct source of contamination, hands will also transmit bacteria from food to food. Fig. 4.1 shows fingerprint cultures of colonies of bacteria from unwashed and washed hands and from hands after holding a moist scouring sponge and after touching raw chicken. It can be seen that there are moderate numbers of bacteria on the dry hand before washing, but that many and various bacteria are picked up by the hands from the kitchen sponge and raw chicken flesh; immediately after washing the hands were by no means sterile.

The sources of *Clostridium perfringens* are animal, human and environmental. The cycle movement is suggested in Table 4.2.

Environmental ways of spread

In some countries flies are important vectors in the spread of disease, particularly where horses and cattle work alongside the human population, so that streets and lanes are soiled with excreta. In countries with efficient sanitation systems and where transport is mechanized, flies are of little concern, but there may still be direct contamination from animal sewage to food. When sanitation is poor and intestinal disease such as dysentery and typhoid are endemic, flies in abundance are a menace. There will be ready access to infected excreta from cases and carriers and bacteria can be transferred to foodstuffs. Rivers and other water courses polluted by sewage from man and animals will flood fields and vegetable crops and thus increase fly infestation.

In western countries with good sanitation there are still untidy dustbins with ill-fitting lids, and boxes of refuse in the back premises of private houses, blocks of flats, restaurants, hotels, canteens and other food establishments; they may be placed outside kitchen windows. The house fly and the stable fly can feed on unwrapped food waste and regurgitate meals on to foods. Particles of infected material will be carried on the feet. *C. perfringens* has been isolated from many batches of green and blue bottle flies.

Other insects, such as cockroaches and ants, can carry food-poisoning bacteria from place to place. An outbreak of salmonellosis occurred amongst children in an hospital where mice, cockroaches and flies were all found to be carrying the agent of infection.

In general, cockroaches are less likely to harbour infection than flies because they breed in wall cracks and paper rather than refuse and manure heaps. Similarly, wasps, bees and spiders are unlikely to harbour pathogenic bacteria because of their breeding and feeding habits. Ants in large numbers are more significant vectors.

Inanimate objects such as towels, pencils, door handles, crockery and cutlery may serve as intermediate objects of transfer of infection, particularly when inadequately cleaned. Staphylococci from boils may spread by means of contaminated towels. Paper towels, hot air dryers and paper

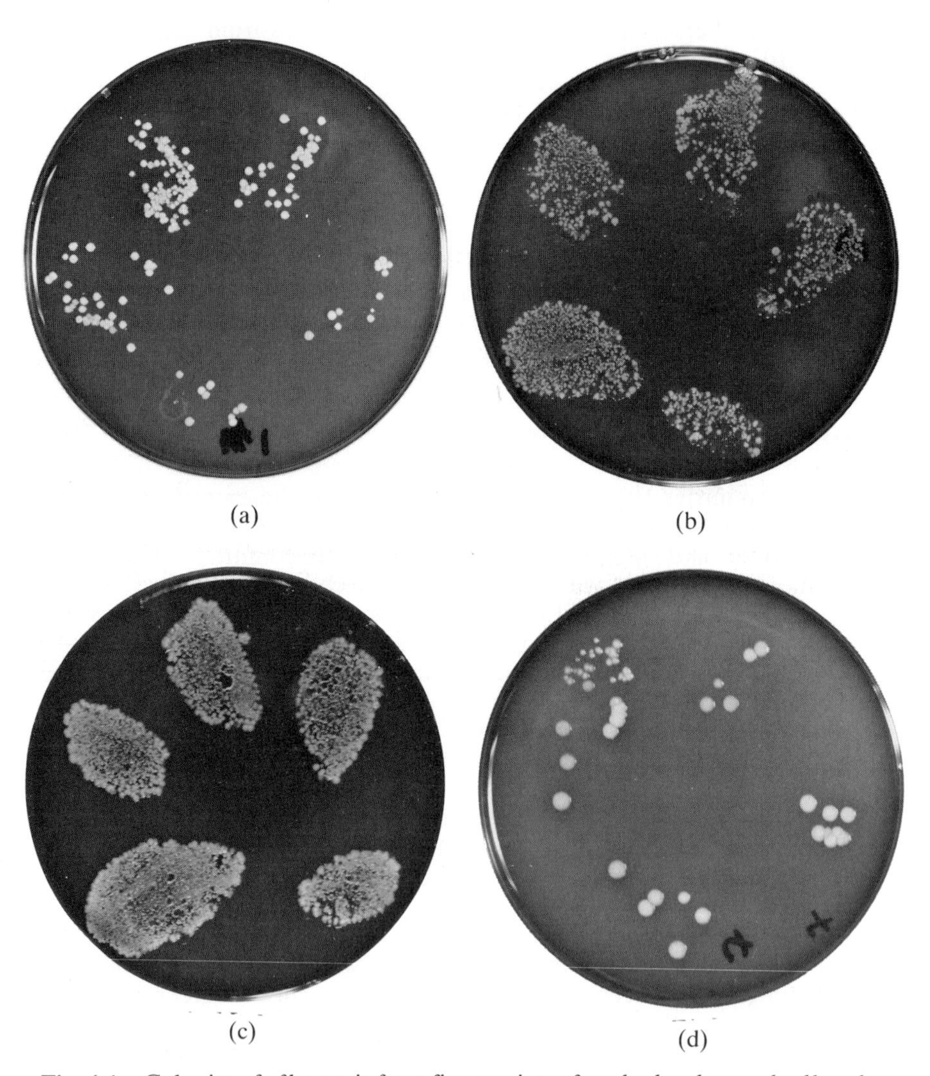

Fig. 4.1 Colonies of of bacteria from fingerprints of washed and unwashed hands: (a) Dry hand before washing (b) Hand after touching moist scouring sponge (c) Hand after touching raw chicken (d) Hand immediately after washing

handkerchiefs will help to reduce the spread of infection. WC equipment should be constructed of material which will not encourage survival of bacteria. Where possible, metal should be used in preference to wood, rubber or plastic. Pedal-operated water supplies both for hand washing and for toilet flushes should be installed where practicable.

Coins of copper, silver or nickel discourage bacterial survival, but bank notes, newspaper and other used paper may harbour living bacteria. Licking fingers to pick up paper should be discouraged.

Foodstuffs

Meat and poultry

Records indicate that in more than 74% of incidents of food poisoning for which a food vehicle is established a meat or poultry dish is incriminated. Table 4.7 summarizes the food vehicle responsible for 405 general outbreaks of bacterial food poisoning in England and Wales between 1989 and 1991. If freshly cooked, roasted, boiled or fried meats were always eaten hot the incidence of food poisoning would be considerably reduced. The spread of organisms from raw to cooked foods and the periods of storage between preparation and eating contribute largely to contamination. The source of the majority of salmonella serotypes that reach cooked food is the raw product. Poultry and comminuted meats may be contaminated with salmonellae before they enter the kitchen and the spread of infection between raw and cooked materials takes place by means of hands, surfaces of equipment such as chopping blocks, cutting boards, slicing machines, various utensils and cloths. Food handlers may be victims of the foods they touch and thus become sources of infection also.

Cooked foods eaten cold or warmed are predominant vehicles of food poisoning. The spores of C. *perfringens* survive cooking and germinate into cells which multiply actively during long slow cooling and storage. Pies and pasties may be made with pre-cooked meat. Cured and uncured meats eaten cold are subject to contamination from the hands. Staphylococcal food poisoning occurs mainly from cold cooked meats including manufactured cured products; *Staph. aureus* grows readily in meats with a relatively high salt content from curing solutions. The organisms usually originate from food handlers and grow while foods stand in the kitchen, shop or at table pending banquets.

The possibility of survival of spores, such as those of *C. perfringens*, through cooking procedures is not generally recognized. Temperatures higher than 100°C (212°F) are required to ensure that spores are killed. Heat penetrates slowly into meat so that there is a greater hazard in large cuts, more than 2.7 kg (6lb) for example. When meat is rolled the contaminated outside will be folded inside where the temperature reached may be inadequate to destroy spores or even vegetative cells. Inside rolls and crevices of meat when oxygen is driven off the atmosphere will be sufficiently anaerobic to encourage the growth of *C. perfringens*. Tables 4.8, 4.9 and 4.10 list outbreaks caused by the growth of *C. perfringens*, *Staph. aureus* and *Salmonella* in meats and other foods.

Table 4.7 Vehicles of infection in general outbreaks of bacterial food poisoning (England and Wales, 1989–91)*

Vehicle of Infection	Presumed causal agent: *Salmonella*	*Clostridium perfringens*	*Staphylococcus aureus*	*Bacillus cereus* and *Bacillus* spp.	All bacterial agents Number	%
Poultry	44	18	4	4	70	17
Meat	43	67	6	2	118	29
Rice			1	19	20	5
Milk/dairy products	5			1	6	1
Egg	50		1	1	52	13
Sweets/puddings	21		1	4	26	6
Other/mixed foods	61	23	1	27	113	28
Total	224	108	15	31	405	
Not known	156	37	1		194	
Total outbreaks	380	145	16	58	599	

* Reports by laboratories, MOsEH and EHOs, including outbreaks where a food vehicle was suspected.

† Does not include 12 outbreaks from abroad or one outbreak of botulism from yoghurt.

The food responsible was established in most (75–100%) of the general outbreaks due to *Clostridium perfringens, Staphylococcus aureus, Bacillus cereus* and other *Bacillus* spp., but in a smaller percentage (59%) of those due to *Salmonella*.

Public Health Laboratory Service figures (Communicable Disease Surveillance Centre, to be published).

Table 4.8 Major outbreaks of *Clostridium perfringens* food poisoning in England and Wales (1987–1990)

Year	Type of food	Location	Number of persons affected	at risk
1987	Minced ham/beef burgers	Psychiatric institute	60	400
	Beef	Club	57	93
	Mince	Geriatric home	26	49
	Brisket of beef	Geriatric home	21	63
	Pork	Geriatric home	21	21
	Braised steak	Army camp	22	100
	Chicken	Club	75	113
	Beef ghoulash	Hotel	118	120
	Turkey	Wine bar	17	21
1988	Turkey	Works canteen	49	200
	Pork	Restaurant	70	96
	Beef	Hotel	40	260
	Chicken	School	65	140
	Chili con carne	Party	19	21
	Shepherd's pie	Party	11	40
1989	Pork	School	64	146
	Steak and kidney pie	Hospital	32	750
	Rolled lamb	Function	49	150
	Beef	Restaurant	48	116
	Gravy	Canteen	100	412
	Shepherd's pie	Meals-on-wheels	30	100
1990	Chicken	Hotel	51	114
	Beef	Function	42	54
	Pork	Wedding	116	245
	Gravy	School	31	37
	Mushy peas (bacon joint juice)	School	31	37
	Minced beef	Geriatric home	200	200
	Turkey	Restaurant	22	48

The irradiation of poultry, boneless and other meats known to be subject to contamination with salmonellae would eliminate the hazard of cross-contamination in the kitchen.

Processed and canned meat and other foods

In general, the standard of commercial canning is high. Faults in processing or structural defects leading to leakage during or after processing are rare. Foods for canning should be fresh and of good quality. They are packed, processed and sealed in airtight containers. Cans are mostly lacquered to prevent interaction between the foodstuff and tinplate. Products such as corned beef, steak, soups and vegetables are heat treated at temperatures well above boiling and calculated to destroy heat-resistant bacterial spores,

Table 4.9 Major outbreaks of *Staphylococcus aureus* food poisoning in England and Wales (1987–91)

Year	Type of food	Location	Number of persons affected	at risk
1987	Chicken	Home for the blind	15	44
	Ham	Reception	22	99
	Quiche	Christening party	18	47
	Egg mayonnaise sandwiches	Sandwich bar	10	
1988	Japanese meal	Mobile caterers	72	
	Egg mayonnaise sandwiches	Factory canteen	18	20
	Ham	Home	4 families	
1989	Canned ox tongue	Home	40	
	Ham	Shop	4	4
	Chicken in sauce	Boarding school	16	30
	Duck	Home	2	2
1990	Kedgeree	Pleasure boat	7	107
	Chicken in sauce	Picnic	24	66
	Corned beef	Home	3	
	Ham	Café	8	
1991	Chicken in sauce	Party	12	39
	Chocolate pudding	Caterer	26	
	Turkey	Home	3	3

such as those of *C. botulinum* and spoilage clostridia and bacilli. Some canned meats including chopped pork, veal and tongue are processed at lower temperatures in order to preserve the maximum bulk and aesthetic qualities of appearance and flavour. Other products such as certain soups, mushrooms and fish are processed at borderline temperatures for the same reasons. The contents of these cans receiving lighter heat treatment are rarely sterile and their shelf life is limited.

Another category of canned meat includes the shoulders and hams of pork given pasteurization temperatures, again to avoid shrinkage and loss of flavour, in combination with curing salts, sodium chloride, nitrate and nitrite. Good manufacturers will ensure that the lowest possible bacterial population survives by care in production, the use of meat from freshly slaughtered animals and the correct concentration of salts in curing solutions. Although some heat-resistant spores may survive they are unlikely to develop in well-cured products. All non-sterile canned meats should be labelled 'perishable, keep in a cold place'. International recommendations for heat treatment, salt, nitrite and nitrate content are available.

There is concern over the role of nitrosamines in carcinoma and a desire to reduce the concentration of nitrate and nitrite in cures. The suppression of *C. botulinum* growth in cured meats depends on the interplay of various

Table 4.10 Major outbreaks of *Salmonella* food poisoning in England and Wales (1987–90)

Year	Type of food	Location	Number of persons affected	at risk
1987	Chicken liver pâté	Hotel	169	553
	Turkey	Restaurant	35	102
	Trifle	Wedding reception	60	102
	Beef and turkey	Hotel	60	400
	Chicken	Restaurant	49	59
	Variety of foods	In flight	53	88
	Pork	Wedding reception	50	90
	Turkey	Public house	30	79
	Beef	Wedding reception	24	180
	Chicken	Restaurant	50	80
1988	Turkey curry	Canteen	65	400
	Chicken liver pâté	Hotel	169	553
	Baked Alaska	Function	18	33
	Chicken mayonnaise	Club	30	70
	Home-made ice-cream	Village hall	18	75
	Turkey	Restaurant	34	92
	Chicken pie and ham	Club	101	191
	Almond parfait	Conference	96	472+
	Turkey	Hotel	35	53
	Coconut mousse	Training centre	51	123
1989	Pie	School	60	130
	Sausage meat	Nursing home	100	107
	Prawn cocktail mayonnaise	Restaurant	54	110
	Chicken and quiche	Wedding reception	66	102
	Mayonnaise	College	72	100
	Turkey	Wedding reception	67	159
	Beef	Wedding reception	106	222
1990	Meringue	Camp	42	48
	Mayonnaise	Hotel	91	110
	Chicken	Function	86	93
	Coleslaw and turkey	College	42	86
	Cold buffet	Wedding reception	80	100
	Chicken	Camp	38	70
	Turkey	Hotel	50	210
	Chicken mayonnaise	Firm	90	363

chemicals including nitrite. To lower the concentration may upset the balance in favour of *C. botulinum* outgrowth; the use of sorbic acid together with nitrate has been suggested for cures. Cans should be stored in a cool dry place as high temperatures may alter flavour and colour; dampness will encourage rust and may lead to perforations. Once opened, a can of meat or other food should be kept covered and cold for a short time only. The belief that foodstuffs must not remain in opened cans arose from the fear of interaction between foodstuff, container and air. Tinplate has much improved and the lacquer sprayed on the inside is an added protection. Cans showing bulges which are hard should be discarded.

Infrequently, small leaks develop in strained seams during processing. Sometimes they close again, but even with pinhole leaks impure water used for cooling after heat treatment may be sucked into the can while it is under vacuum. Staphylococci from hands have been sucked through leaks during transport of wet cans by hand. Typhoid bacilli have been found in canned cream and the growth of this organism in canned corned beef cooled with impure river water was responsible for hundreds of cases of typhoid fever (see Chapter 3). *Salm. typhi* was found to survive in the cans for at least 8 years. Spoilage by gas-producing organisms may be exhibited by hard bulges at one or both ends of a can, but organisms such as staphylococci and *Salm. typhi* do not produce gas nor cause changes in appearance or flavour of food. Thus illness occurs without prior warning of growth within a can. Over-packed cans may not receive the correct heat in the centre and *C. botulinum* was found in canned mushrooms packed too tightly. Cans of veal and carrots in gravy retrieved from one of Sir Edmund Parry's expeditions in search of the Northwest Passage were wholesome after 115 years. In general, sterilized packs of meat may be kept for 5 years, also canned fish in oil for 5 years, but the limit for fish in tomato sauce is one year. There may be slow chemical changes in canned foods, milk should be stored for no longer than 1 year and vegetables for 2 years. A regular yearly turnover for all canned food is advisable; cans should be coded and the date of purchase written on the can.

Dairy products (milk, fresh cream, cheese) and eggs

Milk

A large portion of England and Wales and all of Scotland requires the heat treatment of milk. In England and Wales there are areas in which milk may still be sold 'untreated formerly designated "tuberculin tested"' but only direct to the consumer (see Chapter 17). Yearly there are outbreaks of food poisoning from salmonellae and *Campylobacter jejuni* in milk not receiving heat treatment or imperfectly pasteurized. In some episodes the same phage types of *Salm. typhimurium* or even *Salm. paratyphi B* have been found in calves, cows and milk as well as in sick persons and excreters without symptoms; salmonellae have been found in farm manure and water and examples of outbreaks are given in Chapters 3 and 6. The source and paths of spread of the *Campylobacter* are difficult to follow.

Staphylococci may be isolated from the udders of cows, goats and sheep. The animals may suffer from mastitis due to *Staph. aureus*. The organism can be isolated from most samples of raw milk, and may be found in unheated or lightly heated dairy products. Although the phage types of *Staph. aureus* isolated are amongst those known to produce enterotoxin,

outbreaks of staphylococcal food poisoning from milk and cream are uncommon. Dehydrated milk requires care in preparation. Spray-dried milk powder has been the vehicle in staphylococcal enterotoxin food poisoning. In one extensive outbreak amongst school children *Staph. aureus* grew in the evaporated milk during an unusually long period of storage at warm atmospheric temperature; excess of milk was said to be the reason for the lapse. Sludge in the balance tank immediately before homogenization was thought to encourage the growth of residual organisms probably from the milk itself. The spray drying process failed to reduce the millions of *Staph. aureus* so far as could be ascertained. Many children were ill shortly after school meals in which uncooked dried milk had been added to fortify the calorie content. Salmonellae in dried milk caused a large interstate outbreak in the USA; 17 families were affected. The source of contamination was thought to be one of 800 farms supplying raw milk to the dehydration plant. Similarly, an outbreak of *Salm. ealing* infection amongst infants in the UK was traced to a spray-dried infant milk formula.

Tests for detecting traces of antibiotics in milk are statutory in some countries. In England and Wales action may be taken under the Food and Drugs Act; a concentration of penicillin greater than 0.05 International Units per ml is not permitted.

Fresh cream

Outbreaks due to fresh cream appear to be more common in the USA than in the UK. Cream may be prepared from raw milk, pasteurized milk or pasteurized after separation but before bottling. A minority of creameries carry out in-bottle pasteurization. Cream may be filled into cartons and bottles at the creamery, or it may be transported in cans or churns for filling at distribution centres. A survey of samples from retail shops indicated post-pasteurization contamination, except in cream receiving a second heat treatment after pasteurization.

The introduction of a grading scheme and the notification of pasteurization on the carton have improved supplies and drawn attention to unsuitable methods of processing, storage in bulk, transportation and retail sale. Legislation relating to the heat treatment and microbiological testing of cream was introduced in the UK in 1983 (see Chapter 17).

Cheese

Commercially prepared natural cheese is made by the acidification of raw or pasteurized milk using bacterial cultures combined with clotting by rennet. The use of antibiotics for the treatment of disease in animals, including staphylococcal mastitis in cows, results in trace antibiotics in milk and the development of resistant strains of *Staph. aureus*. Thus starter cultures may fail to grow and give place to the staphylococci from raw or inadequately heated milk, the organisms may survive and multiply freely. Staphylococcal food poisoning has occurred in factories and hospitals from cheese classified as second grade because failure of the starter culture resulted in abnormal flavour. Such cheese is imported for blending and processing and is not intended for direct consumption. The pasteurization of all milk from cows, buffaloes, goats and sheep for drinking or for the production of cream and cheese reduces, if not eliminates, the hazard of food poisoning and other food-borne disease. In 1963, New Zealand introduced legislation to compel the pasteurization of milk for making cheese. The soft mould-formed cheeses are not infrequently contaminated with *L. monocytogenes* and constitute a hazard (see p. 37)

Cream cheese constitutes a risk when prepared from raw milk, but most proprietary brands are made from heat-treated milk.

Eggs

Poultry are known to be major reservoirs of salmonellae and the shells of hens' eggs may be contaminated through contact with faeces in the cloaca or in the nest or battery. The organisms may penetrate the shell under certain conditions of humidity and temperature. In flocks infected with *Salm. enteritidis* phage type 4, it has been shown that the infection can be transmitted via the ovary. Thus eggs are laid with salmonellae already present in the yolk and/or albumen. Though this may be uncommon in hens' eggs it is more common in ducks' eggs. The oviduct of ducks, and hens also, may be infected and thus the egg yolk (see p. 98–99). The duck is particularly vulnerable to infection through swimming in ponds occupied by other birds and possibly rats. The drake may also spread infection from duck to duck. Furthermore, ducks may lay their eggs in wet and muddy places, and the egg shells are more porous and thus more susceptible to bacterial penetration than those of hens' eggs. Recommendations for the use of eggs in uncooked and lightly cooked foods should be observed. The use of hen's eggs in lightly cooked or uncooked foods such as mousse and custards has given rise to cases and outbreaks of *Salm. enteritidis* phage type 4 reported since 1987; warnings given to the public do not seem to have reduced the incidence. (See p. 39).

Available records in the literature indicate that isolations of *Salm. typhimurium* and *Salm. enteritidis* and *Salm. thompson* from ducks' eggs varied from no isolations from 2000 eggs to 19% (3 of 16 eggs). For hens' eggs, *Salm. thompson, Salm. paratyphi* B, *Salm. typhimurium, Salm. give, Salm. senftenberg, Salm. infantis, Salm. anatum* and *Salm. senegal* were isolated from various batches of eggs in different countries. Results varied from negative findings from 3648 eggs to about 15% from 428 eggs. Even one infected egg may contaminate batches of liquid egg broken out in bulk for freezing or drying. Fragments of shell and the hands of those breaking out the eggs may contaminate the equipment and the mix. With regard to egg melange, before compulsory pasteurization was introduced, many batches of liquid and frozen whole egg mix both from hens and ducks were found to be contaminated with salmonellae. Investigations of frozen whole egg mix in 1961 and 1962, before pasteurization, showed that salmonellae could be isolated from 16% of 1369 samples and from frozen whites approximately 9% of 862 samples. Outbreaks of paratyphoid fever and salmonella food poisoning were eventually traced to Chinese frozen whole

Table 4.11 *Salmonella* from imported egg products (1961–1987)*

Type	1961–1970 Number examined	Number positive	%	1971–1987 Number examined	Number positive	%
Whole dried	4289	226	5.3	745	2	0.3
Albumen granules, flake, powder and frozen	8588	253	2.9	2910	35	1.2
Total	12 877	479	3.7	3655	37	1.0

* Updated from Gilbert, R.J. (1982). The microbiology of some foods imported into England through the Port of London and Heathrow (London), In: *Control of the Microbial Contamination of Foods and Feeds in International Trade: Microbial Standards and Specifications*. Kurata H. and Hesseltine C.W. (eds), pp. 105–119. Tokyo: Saikon Publishing Co. Ltd.

egg used in bakeries for making cakes including imitation cream cakes. The cream was the immediate vehicle of infection contaminated inadvertently in the bakery by the egg mix used at the same time on the same surfaces and with the same equipment and hands (see Chapter 6).

The salmonella contamination of spray-dried whole egg was noted in the war years 1939 to 1945; outbreaks of salmonellosis were traced to this source. The same types of salmonellae were found in the faeces, mesenteric glands and flesh of pigs fed on dried whole egg. Of 1009 samples of the egg powder approximately 13% were positive for salmonellae. Of dried white approximately 15% of 317 samples contained salmonellae; a similar percentage was found for 772 samples of flaked white, salmonellae were found in nearly 19% of 91 samples of dried yolk.

This large-scale investigation of imported and home-produced egg products indicated that the use of such contaminated ingredients in bakeries could lead to widely distributed outbreaks of paratyphoid fever and salmonella food poisoning. The examination of 829 samples of pasteurized whole egg from four countries showed that the isolation rate of salmonellae was markedly reduced – only five (0.6%) gave salmonellae – whereas of 800 raw unpasteurized samples 136 (17%) were positive for salmonellae.

The salmonella contamination rate of imported whole dried egg and egg white as granules, flake, powder and frozen for 1961–1970 and 1971–1980 are given in Table 4.11. It should be noted that pasteurization was effective from 1964, so that prepasteurization and postpasteurization results are mixed in the 1961–1970 figures and the improvement apparent after pasteurization as shown in the 1971–1980 figures is not so marked as might be expected. Also results for the albumen products are mixed together.

In late 1963 the Liquid Egg (Pasteurization) Regulations came into force. In England and Wales there is compulsory pasteurization of liquid whole egg at 64.4°C (148°F) for 2.5 minutes. Destruction of the enzyme α-amylase, present in the yolk of eggs, is used as a test for the effectiveness of heat treatment; a similar test is used for pasteurized milk when the enzyme phosphatase is destroyed by heat. Imported liquid whole egg must also conform to the α-amylase test. Other nations pasteurize bulked liquid egg also, but the processing times and temperatures vary. All egg products should be pasteurized prior to dehydration.

Liquid egg white coagulates at a lower temperature so that pasteurization is difficult, but not impossible. A maximum temperature of 57.2–57.8°C (135–136°F) can be used or higher if the egg white is stabilized in some way. Pan-dried flaked albumen may be heated in the dry state at 54.4°C (130°F) for 9–10 days to kill salmonellae, but the treatment of spray-dried albumen by heat is unsatisfactory because of the low moisture content. As there is no α-amylase in egg white the test as used for whole egg is invalid. Although there is improvement in the isolation rates of salmonellae from bulked egg products, salmonellae can still be found.

Irradiation of frozen whole egg mix of the white and yolk with small doses of gamma rays will kill salmonellae. Irradiation methods for the treatment of foods are approved, but not fully sanctioned.

Cold sweets

It is possible that custard, trifle and other lightly cooked milk and egg dishes may be touched by hand; they are suitable media for the growth of staphylococci, streptococci and other organisms. Outbreaks of food poisoning from these foods are mostly due to staphylococcal enterotoxin and initiated by contamination from boils, ulcers, whitlows, abrasions on the hands and also from healthy-looking hands. A trifle responsible for acute toxic poisoning amongst hospital patients was prepared by a chef with a varicose ulcer on the leg. The same phage type of *Staph. aureus* was isolated from the ulcer and also found to have grown in the trifle.

Many Indian sweets are made from milk casein and sugar, for example, barfee, gulab jamun; outbreaks of staphylococcal food poisoning not infrequently occur from these popular sweets. The concentrated milk or casein, after the extraction of whey in the production of cheese, is contaminated with staphylococci from the hands of the workers. Long storage at atmospheric temperature encourages growth and toxin production. The sweets are sold and distributed over wide areas, particularly at times of festivals when gifts of boxed sweets are given to friends and relatives. Dahi is a fermented milk similar but not identical to yoghurt. It is commonly eaten with curry or mixed with sugar as a sweet dish. It is prepared in most homes and the inoculum is passaged daily to prepare the fresh supply. Sometimes chopped vegetables, onion, tomato and cucumber are added when fermentation is complete. If the pH rises above 5.0 growth of pathogens can occur such as staphylococci or even salmonellae and shigellae introduced by hands or ingredients. Cold sweets such as home made ice-cream, mousse and meringue prepared from raw or lightly cooked egg have been implicated in outbreaks of *Salmonella* infection.

Ice-cream

The Ice-cream (Heat Treatment, etc) Regulations, 1959, amended 1963, originally introduced in 1947, require ice-cream mix to be pasteurized by holding the mix at 65.6°C (150°F) for at least 30 minutes, at 71.1°C (160°F) for at least 10 minutes, or at 79.4°C (175°F) for at least 15 seconds. After heating, the mix must be cooled to 7.2°C (45°F) within 1.5 hours and not allowed to rise above this temperature until frozen; it must remain frozen until sold. A mix which has been held at 149°C (300°F) for 2 seconds and transferred aseptically into sterile airtight containers need not be reduced in temperature until required for freezing. Such a mix is distributed for use in the preparation of 'soft' ice-cream, which is frozen at the counter immediately prior to sale. 'Soft' ice-cream is also prepared from the pasteurized mix stored at 7.2°C (45°F) until required for freezing at the counter. The responsibility for the safety of the unfrozen mixtures passes from the specialist manufacturer to those working at the counter. They have to be conscious of the need for vigorous cleaning followed by disinfection of the freezer and other apparatus in contact with the ice-cream. Ice-cream prepared with a cold-mix powder evaporated and dried from a liquid mix after approved heat treatment need not be heated again, but must be frozen within an hour of reconstitution. Before the 1947 regulations ice-cream mix was frozen at the convenience of the manufacturer and bacterial multiplication could take place during long hours of storage at ambient temperature between mixing and freezing. There were many outbreaks of infection both from staphylococcal enterotoxin and from salmonellae (see Chapters 3 and 6). After 1947 the hygienic quality of ice-cream was much improved in the UK and also in other countries with similar regulations.

The introduction of a grading scheme based on tests for bacteriological content further helped to improve the hygienic conditions of manufacture and sale. Thorough cleaning and disinfection of equipment is essential. The bacteriological examination of samples taken from different points throughout the processing plant indicates sources of contamination occurring from failures in cleaning procedures, particularly in the cooler, in pipes and buckets which convey the cream from the cooler to the ageing vat, in the vat, freezer and conservator. Imperfect control of 'holder' type pasteurization may allow survival and multiplication of organisms from the ingredients. Various detergents are available for preliminary washing; steam, hypochlorite or other suitable disinfectant should follow.

Imitation cream

The ingredients of the different types of confectionery cream are variable. Usually they contain dried milk products emulsified with fat and sugar. Manufactured imitation cream is pasteurized before distribution in cans and the bacterial counts are low. In contact with choux pastry and other confectionery, for example, sponge cakes and custard products, nutriment is absorbed which will encourage bacterial growth at the interface.

Egg products are rarely incorporated into creams, but splashes from liquid egg and airborne dehydrated egg may reach cream in bakery practice. This was an important source of contamination in the days before the compulsory pasteurization of whole egg mix. Many outbreaks of paratyphoid fever and salmonella food poisoning were traced to imitation cream cakes and other confectionery prepared in bakeries using contaminated egg products. Such outbreaks illustrate the ease with which food poisoning bacteria can spread from contaminated ingredients by means of mixing bowls, utensils and other articles to cream mixes and finished confectionery. Failure to disinfect piping and Savoy bags each time they are used is another factor in the spread of contamination from hands and ingredients.

Fresh cream

Fresh cream has largely replaced the artificial material. It has an inbuilt safety factor in souring which not only makes it unpalatable but also reduces the pH to levels unsuitable for the growth of pathogens. (See *Milk*.)

Fish and other seafoods

Salmonellae have been isolated from fish taken from polluted river water and also from fish taken from the holds of ships washed out with polluted dock water. Contamination may take place when the fish are gutted and filleted at the quayside or even in shops. Otherwise, fish recently caught at sea and freshly cooked are improbable vehicles of food poisoning. Made-up items of food such as fish cutlets and fish pies may be contaminated with staphylococci from the hands or from equipment used to mash potatoes in the mass production of fish cakes.

C. perfringens food poisoning has occurred from imperfectly thawed frozen salmon steamed or boiled, left overnight in the liquor and eaten cold. *C. botulinum* type E may be found in some fish. Many outbreaks of botulism in Canada, the USA and Japan were traced to the consumption of uncooked, under-cooked, stale or fermented fish and also to seal and whale meat. Canned salmon was responsible for two deaths among four cases in the UK (see Chapter 6).

Food poisoning from *Vibrio parahaemolyticus* is common in Japan where fish and other seafoods may be infected in coastal waters. Uncooked seafoods and also cooked fish and crab meat recontaminated from the raw materials have been reported as vehicles of *V. parahaemolyticus* infection from other countries also (see Chapter 6).

Frozen cooked seafoods

Frozen cooked prawns, shrimps and lobster tails are processed by a number of countries for international trade. They may be handled extensively after cooking and before freezing. There have been few isolations of salmonellae from samples examined from the port; even raw frozen samples are rarely found to contain salmonellae. Staphylococci are frequently isolated from frozen cooked seafoods, but they are infrequently reported as agents of food poisoning from these imports. *V. parahaemolyticus* and other pathogenic vibrios are sometimes isolated from frozen cooked prawns from the East. If the prawns are allowed to thaw for some hours in warm water before they are required for the table, and served as prawn cocktail or seafood salad there is the possibility of bacterial growth.

Microbiological guidelines for the worldwide industry and also for port health authorities may be of value in stabilizing the hygiene of the product. Viable counts of less than 100 000/g, less than 10/g of *Escherichia coli* and not more than 1000/g of *Staph. aureus* should be the aim and readily attainable with care in manufacture. The responsibility of maintaining the product in good condition rests with the user. The time and temperature of storage between thawing and eating is important, not only for seafoods, but for all frozen food.

In certain countries, for example India and Africa, seafoods such as fish, shrimps and prawns are sun-dried, spread or hung over beaches. Thus they are targets for contamination by bird droppings and by faecal matter from other creatures. The dried material may be consumed in the dehydrated state or used as ingredients for various animal feeds. Sundried materials are known to have a high rate of contamination with salmonellae; they have been responsible for the contamination of mixed feeds and the infection of flocks and herds of birds and animals.

Shellfish

Oysters and mussels may be bred and fattened in sewage-polluted waters of tidal estuaries; oysters are usually eaten uncooked. From time to time cases and outbreaks of enteric fever and viral infection are attributed to shellfish including cockles gathered from polluted waters. Plants for cleansing oysters and mussels consist of a series of interconnected tanks with recirculated water passing under ultraviolet light or the water may be continuously flowing from rivers or sea. The oysters are laid out on trays and remain in the tanks for 36–48 hours. Sometimes oysters are lifted from polluted areas and relaid in sea or river waters known to be free from pollution. Freshly chlorinated sea water may be used in large tanks. Although the intestinal contents are flushed out by these methods, if heavily polluted water is used in small tanks and the ultraviolet light source is low, organisms such as vibrios and viral particles may adhere to the flesh of oysters and accumulate in the tanks and pipelines.

There have been outbreaks of gastroenteritis of recognizable aetiology from the consumption of oysters cleansed in tanks of water treated with UV light. The examination of samples gave low and satisfactory *E. coli* counts, but high counts of other organisms in the liquor and body of the shellfish. Vibrio-type organisms were found, the *Campylobacter* were not recognized at that time and viral aetiology was unproven. Subsequent outbreaks of gastroenteritis from cockles both raw and cooked have been strongly suspected to be viral in origin. *Vibrio fluvialis* has been isolated from patients with acute watery diarrhoea in middle and far eastern countries, and is thought to be associated with fish and shellfish.

Cockles are collected into huts, cooked and sometimes salted before distribution. Incidents attributed to cockles have been thought to be due to under-cooking and cross-contamination. There are few recorded incidents of bacterial food poisoning from mussels, since they are usually cooked by the consumer.

The regulations dealing with shellfish in England and Wales came under the Public Health (Shellfish) Regulations, 1934. These regulations allowed local authorities to impose orders. The Conway Mussel Fishery (Amendment) Order, 1948 states that any person taking mussels from the Conway River and failing to deposit them forthwith at the Purification Tanks shall be guilty of an offence. There are more than 40 Prohibition Orders made under the Public Health Acts and kept in force by the Food and Drugs Act, 1955. They prohibit the sale of shellfish for food either absolutely or unless they have been treated in some way. They cover a large proportion of British coasts and the areas suitable for growing shellfish. Prohibition Orders refer to the gathering and distribution of shellfish for sale; they do not prevent people gathering them for their own consumption. The law is broken if the shellfish are sold or otherwise distributed to other persons. It is not illegal for itinerant pickers to gather mussels from shores, but they may not legally sell them if they come from a Prohibited Area. Directive 91/492/EEC gives health conditions for the production and marketing of live bivalve molluscs and deals with the condition of harvesting, relaying, purification, processing and distribution of molluscan shellfish such as oysters, clams, mussels and cockles. (See Chapter 17.)

Dehydrated foods (including cereals and bakery products)

Dried food substances are not sterile, some microorganisms and particularly bacterial spores will survive the dehydration process. There will be no growth until the addition of water and the temperatures are suitable.

Proprietary cake mixes and meringue powders which incorporate unpasteurized powdered egg products either whole or white may contain organisms of the salmonella group. Outbreaks of salmonellosis following the use of contaminated cake mixes have been reported in Canada and the USA. Pasta with egg has been responsible for incidents of both salmonella and staphylococcal food poisoning.

The milled products of wheat and other corns and also rice may contain sporing organisms such as *B. cereus* and *C. perfringens*. Cornflour sauce and cooked rice, particularly, are described as vehicles of toxic food poisoning due to *B. cereus*. This organism and other sporing organisms are frequent contaminants of dairy plants and may cause spoilage. Spices, dehydrated soups, baby foods and other convenience dried foods will all contain spores, usually in small, but can be larger, numbers. Nevertheless the addition to meat of spices, for example, will introduce spores and add to the possibility of *B. cereus* and *C. perfringens* food poisoning.

Spores may survive in bread baked with cereal products particularly if dough is undercooked. When bread is wrapped too warmly after baking, the warmth and moisture will encourage bacterial and mould spores to germinate; mould on sliced bread is not uncommon and growth of bacilli can lead to 'ropy' bread. Mycotoxins may develop from mould growth on stale bread. The crust of well-baked bread is a barrier against the invasion of bacteria and moulds, but intestinal pathogens could be passaged from hands to sliced bread. Home-baked bread prepared by a baker unable to wash his hands properly because of adhesive dressings was the vehicle for hepatitis A virus. Unusual incidents of staphylococcal enterotoxin and *C. perfringens* food poisoning occurred from a fruit cake stored moist in its wrapper in the refrigerator and handled from time to time while slicing. It was presumed that the staphylococci came from the hands, and the *C. perfringens* from heat-resistant spores in the flour or other ingredients. Bran infested with round worms was imported from West Africa.

Gelatin

Gelatin is derived from the skin, connective tissue and bones of animals. It is processed in sheet and powdered form.

Salmonellae have been isolated from sheet gelatin and also from molten gelatin. *Salm. senftenberg* was found in the nozzle of a machine used for inserting gelatin into pies after cooking. It was assumed that the gelatin was responsible for the food poisoning outbreak due to *Salm. senftenberg* which followed consumption of the pies.

Sheet gelatin in liquid form was found to be contaminated with *Staph. aureus* responsible for a large outbreak of enterotoxin food poisoning. The gelatin was used as a glaze coating on meat rolls distributed over a wide area. The molten gelatin was kept warm for several hours while the operation of coating was in progress. Both the strength and the temperature of the gelatin were increased to prevent a recurrence.

An aspic glaze used to coat meats served on aircraft was the vehicle of infection in a large outbreak of *Salm. enteritidis* food poisoning affecting many travellers on several flights of one airline.

Salmonellae have been isolated from marshmallow sweets; gelatin is used as a coagulant for marshmallow. Pathogens may survive through mild cooking processes or gain entry by recontamination when cooked materials are whipped or mixed in containers used for raw products.

Spores are known to survive in gelatin and may be introduced with the molten product into pies and other foods. If they remain as spores they are harmless, but if they germinate into vegetative cells when conditions encourage proliferation during storage, food poisoning may result both from aerobic and anaerobic sporing bacilli.

The need for care in production and use of gelatin in foods is shown on p. 105. Microbiological standards have been suggested for powdered gelatin.

Fruit including coconut

Coconuts are probably the most vulnerable of the fruits. They are allowed to remain on the ground after picking or falling from the tree and before harvesting; the fall may crack some nuts. The excreta of grazing animals, vermin and lizards will be in contact with the shell and may penetrate through damaged areas. Desiccated coconut is used to garnish and flavour cakes, trifles and other custard dishes. Many tons, graded according to the particle size, are imported annually into the UK for industrial use in cakes and biscuits, and for making sweets. Cases of typhoid fever occurred in Australia due to *Salm. typhi* in coconut. *Salm. typhi* and *Salm. paratyphi* B were isolated from 8–10% of import samples before legislation was introduced to improve the hygiene of production. By this measure the proportion of contaminated samples was reduced to negligible levels and supplies for export are now monitored. Earlier cases of paratyphoid fever in children were traced to coconut confectionery and salmonellae including *Salm. paratyphi* B were isolated from sweetmeats and other confectionery coated with coconut.

In acid fruit, for example cooked or canned apples and plums alone or in tarts, pies and cordials, the growth of pathogenic organisms will be inhibited by the low pH. Organisms may survive and moulds may grow. Soft fruit such as melons, guava, mango and papaya are not acid and they are subject to spoilage. The skins are relatively soft and may be penetrated by pathogenic as well as spoilage bacteria. Cases of *Salmonella* infection have been reported in the USA associated with the consumption of cantaloupe melons. The contamination rate for such fruits was 1% and a range of serotypes were isolated; many of the cases were due to *Salm. poona*. Dessert fruits which are eaten uncooked and possibly unpeeled should be washed preferably in water containing a small amount of hypochlorite in a concentration of 60–80 ppm calculated on the available chlorine in the substance used. Many kinds of fruit are imported from countries with a high incidence of diarrhoeal disease and many flies. Washing will remove traces of pesticide also.

Vegetables

Vegetables eaten freshly cooked are safe. They may be contaminated after cooking or be subject to spore germination and outgrowth if cooled slowly and stored warm. Salad vegetables, lettuce, tomatoes, radishes, cucumber and watercress should be washed in water with added sodium hypochlorite (60–80 ppm available chlorine) for not less than 30 seconds, as recommended for dessert fruit. This is important in countries where crops are sometimes flooded with water polluted with human and animal sewage. *Salm. typhi, Salm. paratyphi* A, B and C, other salmonellae, shigellae, protozoa such as *Entamoeba histolytica* and even *V. cholerae* and other vibrios may be found in flood water. The cysts of *E. histolytica* require heat for destruction. Bean shoots (mung beans) eaten raw or lightly cooked were responsible for cases of *Salmonella* infection in both the UK and Sweden. The original beans were shown to harbour the organisms, which survived and proliferated during germination of the seeds in a warm, moist environment. A code of practice for the production of bean shoots has been prepared to improve the safety of the product.

Raw vegetables may also be a source of *L. monocytogenes*. The organism is isolated more frequently from chopped ready-to-eat mixed salads than from entire single salad vegetables.

Other foods

Fats such as lard, margarine and butter may be disregarded as media for the growth of food poisoning organisms. Small numbers of typhoid and paratyphoid bacilli may survive for many days and possibly cause infection. An unusual outbreak occurred in the USA from staphylococcal enterotoxin in an emulsion of butter in milk.

Jam preserves, *honey* and *syrup* usually contain 60% or more of sugar which will not permit the survival of most bacteria. There are reports of the isolation of *C. botulinum* from honey fed to sick babies described as 'floppy'; the toxin of *C. botulinum* has been demonstrated in the intestine of such babies.

Moulds may grow on the surface of jam especially when the sugar concentration is lower than required for preservation. Mouldy jam has not been regarded as a vehicle for food poisoning, but when certain fungi grow on any food substances, mycotoxins should be considered.

Sauces and *mayonnaise* are dressings which include emulsified oil and vinegar; they are usually too acidic to support the growth of organisms other than those which may cause spoilage. Those made without vinegar, such as Hollandaise sauce, may contain eggs, as well as fat and flavouring material. They are suitable for bacterial growth and they should be eaten within a short time of preparation, 1–2 hours, unless stored in the refrigerator. Some populations like a bland mayonnaise with a pH of 5–7 instead of below 4.5, others prefer an acid flavour and add vinegar or lemon juice until the pH is less than 4.5. When mayonnaise has a neutral pH, salmonellae from egg products or staphylococci from food handlers can grow; many mayonnaise outbreaks have been reported. Ingredients of uncooked, non-acid foods should not include ducks' eggs or unpasteurized hen egg products.

Pickles made from vegetables and fruits may be disregarded as vehicles of food poisoning, because they are markedly acidic.

5

Epidemiology

Epidemiology, as the name implies, is the study of disease upon the people, its frequency, distribution, causes and control, and likewise its relation to disease in animals.

As in many branches of science, progress depends upon the formulation of hypotheses from an orderly sequence of recorded observations leading to conclusions. One of the early (1880–1938) epidemiologists, W.H. Frost, described his science in the following statement: epidemiology 'at any given time is more than the total of the established facts. It includes their orderly arrangement into chains of inference which extend more or less beyond the bounds of direct observation. Such of those chains as are well and truly laid guide investigation to the facts of the future; those that are ill-made fetter progress'. The investigation of food poisoning and other food-borne disease is concerned with the follow-up of such chains. It is important for several reasons: ongoing outbreaks may be terminated by withdrawal of contaminated food. Bad catering practices or environmental contamination may be remedied to prevent future outbreaks. It is of national importance to highlight the need for government action, for example, the careful documentation of milk-borne outbreaks in Scotland led to the prevention of the sale of raw milk by Scottish legislation.

In the UK, epidemiological investigations and control are the responsibility of the National Health Service (NHS) and local authorities. The investigation of food poisoning and other food-borne diseases in England and Wales is facilitated by compulsory notification under the NHS and Public Health (Control of Diseases) Act, 1984. Suspected and actual cases must be notified to the local authority by the doctor concerned. Outbreaks and other important incidents are required to be reported on behalf of the Department of Health (DH) and Welsh Office to the Public Health Laboratory Service (PHLS) Communicable Disease Surveillance Centre (CDSC) and in Scotland to the Communicable Disease (Scotland) (CDS) Unit. The Food Poisoning Memo (188/Med. HMSO, 1982) on investigation and control of food poisoning in England and Wales issued by the then DHSS and Welsh Office gives regulations and notification procedures.

Outbreaks in any locality are notified immediately to the authorities.

The Medical Officer for Environmental Health (MOEH) or Consultant in Communicable Disease Control (CCDC) reports to the PHLS, CDSC or CDS Unit; when applicable it is the responsibility of food managers to notify the MOEH or appropriate proper officer. Speed of notification is particularly important when suspected food vehicles are distributed nationally or internationally.

Members of many disciplines take part in full investigations including environmental health officers, epidemiologists, community and port health specialists and food scientists together with clinicians, veterinarians, microbiologists and persons intimately involved with the food, such as cooks and other food handlers, food and feed factories and farm personnel. Careful observations are wanted both at the site of infection and also of secondary cases, if any. Detailed studies of case records from hospital patients are valuable; conversations with the patients themselves with regard to shopping, visits and eating habits may reveal facts of importance.

In case-control studies, data on possible sources of the agent and risk factors are obtained from both cases and controls retrospectively. Cases should be representative of those affected and controls of the normal population from which they came. Controls commonly used are taken from GP registers, hotel and reception guest lists, family members and neighbours of cases, and persons investigated by the laboratory, but whose samples gave negative results. It may be found necessary to obtain samples from selected controls to exclude asymptomatic infection. In studies of acute incidents it is usual to match controls for sex and age and often for neighbourhood of residence. An example of a case-control study is provided by the outbreaks of salmonellosis due to *Salm. typhimurium* DT124 in salami sticks (see p. 97) and also *Salm. ealing* in infant dried milk (see p. 103). Whereas the case-control study begins by identifying people with and without disease and then retrospectively identifies factors associated with the disease; cohort studies identify groups of people other than by the presence of disease and then seeks information on disease occurrence; the studies may be prospective or retrospective. *Salm. enteritidis* phage type 4 in home-made ice-cream provides an example (see p. 101).

General data

Certain standard data are necessary to study any communicable disease; they include the 'incidence' or frequency of occurrence, that is, the number of new events, episodes or outbreaks in a specified period, for example, ten *Clostridium perfringens* outbreaks in one month; and 'prevalence', the number of events of a given disease in a given population occurring at any one time compared with the number a year ago at the same time.

In describing an outbreak of food poisoning, the following facts must be included: the situation; the number of persons affected; the attack rate by age, sex and race; the number of persons who remained well (the non-attack rate); the epidemic curve for affected persons; the period between exposure to infection or intoxication and onset of symptoms (incubation); the clinical nature of the illness; the reservoir and source of the agent; the food vehicle and mode of transmission to food and victim.

Vulnerable situations need special attention, for example in hospitals, where food poisoning in geriatric and paediatric wards may lead to fatalities; also in aeroplanes, trains and ships where cases and outbreaks are indicative of food contamination and poor storage in flight kitchens and galleys.

Statistics include the number of cases and outbreaks, the distribution and frequency grouped according to the infective agents, the place, the incidence of general and family outbreaks and sporadic cases, and food vehicles, and are given either on an annual basis or ranging over a number of years. Tables in Chapters 3–5 illustrate the compilation of data.

Food history

Outbreaks and incidents of food poisoning and food-borne infection require careful histories of the food vehicles, with environmental studies of the areas of food production and preparation as far back as possible. Sites of infection and areas of spread may include the farm of origin, dealers, markets, processing factories, wholesale and retail outlets to catering establishments, restaurants, canteens and domestic kitchens. Transport conditions for live animals and for foodstuffs may enhance spread also. Changing national and international trends for retailed foods could increase the distribution of organisms with food-poisoning potential.

Cases scattered over a wide area sometimes prove to be part of a general outbreak or epidemic arising from a particular food product. Examples of foodstuffs which have caused widespread salmonella outbreaks and incidents are given on p. 70 (frozen whole egg mix), p. 77 (coconut), p. 57 (boneless bobby veal) and p. 55 (animal feedstuffs). It may be necessary to organize surveillance programmes for foods and animal feedstuffs, both home-produced and imported, to investigate the proportion and extent to which material is contaminated. This would require the examination of a large number of samples over a long period of time with consideration of seasonal changes. Estimation of the approximate numbers of salmonellae or other pathogens in at least some samples will help to assess the hazard in terms of dose for the consumer and the extent of spread throughout the environment. Unknown to food handlers, intestinal pathogens from raw foods can spread over food preparation surfaces and pass to cooked foods in the same area. Also, if salmonellae become part of the intestinal flora of food handlers they may temporarily excrete the organisms and add to the general spread. To monitor foods on a regular basis is arduous and expensive, but for foods known to cause trouble, it can serve to alert authorities that disease in the human or animal population is imminent unless precautions are taken to reduce the hazard. The presence of a small number of salmonellae in a low proportion of raw poultry and meat may be undesirable, but unavoidable; the organisms are not acceptable in any cooked food.

Spoilage organisms including *C. perfringens, Bacillus cereus* and *Escherichia coli* are frequently found in small numbers in both raw and cooked foods. When numbers rise to a million or more per gram in cooked and raw foods there is danger of food poisoning and faults in preparation and storage need investigation. In many instances the remains of suspected food vehicles are discarded leaving no samples available for laboratory examination. Even so, descriptions of the methods of preparation and storage of the foods eaten, together with the attack rate, will help to identify the bacterial agent and confirm association with a particular food. The history of the food preparation will also help to clarify confusion which might arise from cross-contamination after the function. Thirty cases of food poisoning were reported in a school population of 241 girls and staff. Food specific attack rates for the meal preceding the outbreak (see Table 5.1) showed that 30 of 161 who ate lasagne were ill, giving an attack rate of 18.6% compared with none of 51 who did not eat it. There was a choice of lasagne or chipsteak, so that eating chipsteak was significantly associated with being well.

In another staphylococcal outbreak, 30 children and three adults were ill after a school picnic. Attack rates were calculated from the food eaten by 83 of the 85 picnickers. Results showed that the illness was confined to those who ate the school packed lunch with 32 of 76 (42%) affected compared with none of the seven who did not eat the sandwiches. A more detailed analysis of the 76 who ate the packed lunch showed that luncheon meat sandwiches and egg sandwiches produced the highest differential attack rates (see Table 5.2). From an analysis distinguishing those who ate (a) both types of sandwich, (b) only one type or (c) neither, it was confirmed that illness was associated with both the egg and the meat sandwiches. The picnic lunch was prepared too far ahead and kept in the larder at a temperature of 15–27°C (59–81°F).

Table 5.1 Food specific attack rates for the meal preceding an outbreak of food poisoning in a girls' school

	Ate			*Did not eat*			
	Ill	Not ill	Attack rate %	Ill	Not ill	Attack rate %	χ^2 1 d.f.
Lasagne	30	131	18.6	0	51	0	$p<0.0005$
Chipsteak	0	27	0	30	154	16.3	$p<0.05$
Cream potato	8	49	14.0	21	125	14.4	NS
Pears	21	95	18.1	9	84	9.7	NS
Bread and jam	8	69	10.4	22	107	17.1	NS
Tea	13	60	17.8	17	121	12.3	NS

NS Not significant

Note: χ^2 1 d.f. = chi squared test with 1 degree of freedom

χ^2 in basic form is used to test whether the observed frequencies of individuals with given characteristics (e.g. they were ill/ate a particular food) are significantly different from the expected frequencies from some specific hypothesis.

Table 5.2 Attack rates of illness for all food items from a school picnic lunch

	Persons who ate specified food			Persons who did not eat specified food			
	Total	No. ill	Attack rate %	Total	No. ill	Attack rate %	Difference in attack rates p*
Meat Sandwiches	41	24	59	35	8	23	0.002
Egg Sandwiches	19	16	84	57	16	28	0.00002
Cheese Sandwiches	31	12	39	45	21	47	0.3
Biscuits	49	20	41	27	12	44	0.5
Bun	36	18	50	40	14	35	0.1
Apple	45	21	47	31	10	32	0.2
Orange Squash	65	32	49	11			

* Exact probability using Fisher's equations.

(Note – Fishers exact test is used to analyse 2 × 2 tables where any of the cells has an expected value of <4)

Laboratory studies

Proof of the cause of an outbreak depends on the isolation and identification of the causal agent from stools of as many patients as possible, from the food vehicle and sometimes from vomit samples; stools from food handlers should be examined also. The hands, nose and throat of food handlers may also be swabbed, if necessary, to see if *Staphylococcus aureus* might have contaminated cooked foods and initiated staphylococcal enterotoxin food poisoning. On the other hand, workers may be victims of the foods they touch and excrete organisms, salmonellae, for example, found in the raw foodstuffs such as poultry and meat.

The more quickly notifications of outbreaks reach the authorities, the sooner samples can be collected and examined and retrospective investigations begun. In practice, careful observation and the information given by food handlers and other personnel at the site of the episodes will help to identify the causal agent and the source. Later the laboratory skills used in the isolation, identification and typing of bacterial, viral and other agents will provide definitive proof of the agent of infection or intoxication. Epidemiological studies include tracking the pathways of the organisms using all the available methods of identification and typing. Biochemical, serological, bacteriophage and bacteriocin typing techniques are readily available, antibiotic sensitivities can also be used to 'fingerprint' the organism implicated. The identification of toxins and tests for pathogenicity need the resources of a research laboratory where more advanced immunological and genetic methods may be used also. Tests for pathogenicity are necessary to explore the role of organisms not hitherto recognized as agents of food poisoning. The laboratory can give guidance on antibiotic treatment; the development of multiple antibiotic resistant intestinal pathogens will render treatment ineffective for both man and animals.

Although not strictly belonging to epidemiological studies, the charac-

teristic behaviour of bacterial agents in food is important when preventive measures are devised. There are many factors to be considered such as the temperature limits for growth, the optimum generation time (rate of multiplication), the heat resistance in relation to cooking methods and initiation of growth, the survival and growth under different conditions of water activity (a_w), pH (acidity), salt with nitrate/nitrite and other methods of preservation, and the ease of sporulation and toxin production under different conditions. Results and general information from authentic incidents which have been thoroughly investigated using epidemiological and microbiological methods will provide reliable data for weekly, monthly and annual statistics.

Sources of data

Data on food poisoning in England and Wales are available from the Office of Population Censuses and Surveys (OPCS) in conjunction with the CDSC. These two organizations have recently amended the notification system, so that cases discovered but not officially notified are now reported by local authorities. In Europe, the World Health Organization (WHO) programmes, coordinated in West Berlin, cover the surveillance of food poisoning and food-borne infection.

In the USA the Centers for Disease Control (CDC) and in Canada the Canadian Department of Health and Welfare have surveillance systems that provide reports weekly and annually; they are summarized in the WHO *Newsletter*. Thus local investigations and events fit into the national and international coverage and control of most food-borne diseases in the western part of the world.

Reporting systems

Some countries have well-developed reporting systems for infection, producing regular weekly and/or annual publications. These include England and Wales, Scotland, USA, Canada, Australia and the Netherlands. Others are still at the developmental stage, for example South America, which provides a report at irregular intervals; it is compiled by a subcommittee of the International Commission on Microbiological Specifications for Foods (ICMSF) and gives useful accounts of incidents in Latin America that otherwise would not be known in other parts of the world.

A bi-monthly *Bulletin* from the State Public Health Laboratory in Pune, India, gives surveillance data on water-borne and food-borne outbreaks as well as the results from regular examination of samples of water and food in eight areas reporting to the central laboratory in the State of Maharastra. In India, morbidity and mortality returns for cholera, salmonella and shigella infections are compiled by the government from data received from government and other laboratories.

Data required for outbreak investigation may be categorized as follows:

(1) Microbiological: results from the examination of food samples, not only those remaining from the suspect meal, but also a selection

of ingredients, particularly raw materials such as poultry and other meats likely to be responsible for cross-contamination.

(2) Environmental: information on food preparation and storage, the kitchen surfaces, equipment and cleaning methods.

(3) Epidemiological: comparative data on the characteristics of affected persons (cases) and a similar group of healthy persons.

United Kingdom

Surveillance reports on food poisoning in England and Wales have been published by the PHLS since 1950, the latest location being the *British Medical Journal*. The CDSC provides a weekly *Communicable Disease Report* (CDR) of events related to all infectious disease and in 1991 introduced an additional CDR monthly review. The information is compiled from reports gathered in from the PHLS and hospital laboratories spread over England, Wales and Northern Ireland. Analysis of these data provide the basis of national surveillance. The sudden increase in numbers of reports of specific organisms provides an early warning so that epidemiological investigation can be set in action. Post-incident data are also reported. The sections of the CDR on gastrointestinal infections summarize surveillance reports from MOsEH and laboratories on food poisoning and salmonellosis (pp. 38–41). Isolations of salmonella serotypes are reported from man and food. Laboratories identifying salmonella infections submit strains to the PHLS Laboratory of Enteric Pathogens (LEP) for confirmation of serology and phage type. Short accounts of incidents are given in the CDR monthly review and outstanding outbreaks may be reported in detail. Cumulative totals of weekly reports are compared with those from the previous year. More detailed statistics and trends are given annually. The following examples illustrate how the early warning system functions: reports of the isolation of *Salm. brandenburg* from a gradually increasing number of sporadic cases and small outbreaks during a particular period were warnings that a distributed foodstuff might be causing food poisoning. Sausages and pork appeared to be the most likely vehicles of infection. Investigations led back to an abattoir where there was malpractice; pigs were stocked for up to a month before slaughter which magnified the minor infections in pigs from farms using salmonella-contaminated feedstuffs. The provision of contaminated carcasses to meat factories was stopped by the closure of the abattoir and by regulations prohibiting the holding of animals for longer than 72 hours before slaughter.

The isolation of large numbers of *Staph. aureus* and also salmonellae from samples of imported noodles and lasagne examined in a number of laboratories led to factory inspection. Improved hygiene of production undoubtedly prevented incidents of staphylococcal enterotoxin and salmonella food poisoning. Other examples of the rise in cases from unusual serotypes giving early warnings of impending outbreaks are provided by *Salm. typhimurium* DT 124 in salami sticks and *Salm. ealing* in infant feed, see p. 97 and 103.

Other gastrointestinal tract infections reported in the CDR include those due to *Shigella*, rotavirus and other viruses including hepatitis,

Cryptosporidium, Entamoeba histolytica, Giardia, C. difficile, Aeromonas, Plesiomonas, Vibrio cholerae, Listeria, Yersinia and *E. coli*. Comparable information on animals can be found in *Animal Health*, the annual report of the Chief Veterinary Officer of the Ministry of Agriculture, Fisheries and Food (MAFF). The MAFF Veterinary Centre at Weybridge in Surrey compiles figures for the isolation of salmonellae from animals from reports received from the Veterinary Investigation Service. The CDR also gives isolations of salmonellae from foods, and reports of animal-associated disease. The PHLS and the State Veterinary Service also produce a quarterly salmonella update, which reports on numbers of salmonella infections in both humans and animals and compares them with the previous year.

Scotland has a separate system from that of the rest of the UK, and outbreaks are reported to the CDS Unit in Glasgow. This Unit produces a weekly report on infectious disease, *Communicable Diseases Scotland* as well as an annual report on various aspects including food poisoning and other food-borne illness. The careful and thorough documentation of milk-borne outbreaks resulted in the Scottish law (1983) requiring the heat treatment of all milk before distribution. There is close communication and regular interchange of information between the CDS Unit and the CDSC; many of the reference laboratory facilities of the PHLS are used by hospital and other laboratories in Scotland. Quarterly updating from English and Scottish reports is published in *Community Medicine*, and the *Annual Report* of the Chief Medical Officer of the DH briefly reports on all important infections.

Europe

The WHO, founded in 1948, issues a weekly report from Geneva, the *Weekly Epidemiological Record*, in English and French, which covers all infectious disease. There is also a WHO surveillance programme for the control of food-borne infections and intoxications in Europe which has published a *Newsletter* three to five times a year, since 1982. It is intended to link programme management between West Berlin, Copenhagen, Geneva and cities in other countries participating in the programme. In 1990 32 countries collaborated including England and Wales and Scotland; Scotland was one of the earliest participants in the programme. The aims are described by WHO as follows:

(i) identification of the causes and epidemiology of food-borne diseases in the region;
(ii) distribution of the gathered and collated information;
(iii) cooperation with the national authorities in efforts to strengthen the prevention and control of food-borne disease in the regions.

This surveillance provides an early-warning or alert system for the participating countries allowing appropriate measures for control to be instituted without delay.

Examples of incidents that would merit an early-warning system, or alert to participating countries of the programme are: incidents of a severe nature or of an unusual type in the area; incidents due to internationally-distributed foods and associated with international carriers (air, surface, sea) or related caterers; incidents among tourists or introduced by tourists or immigrants; matters related to food-borne disease causing an emergency situation. Information is sent through designated contact points to collaborating centres. Summaries from published reports are given in the *Newsletter*. For example, in 1990 there were notes on diarrhoeal disease on cruise ships, an in-flight outbreak of salmonellosis, *C. botulinum* on fresh mushrooms, 'traveller's diarrhoea', Hepatititis A and gastroenteritis associated with raw shellfish, dinoflagellate poisoning from mussels in France, *Yersinia enterocolitica* outbreaks amongst families processing pig intestines, costs of food-borne disease in Canada and the Latin American network on epidemiological surveillance of food-borne disease. The *WHO Salmonella Surveillance Programme* is published by the WHO Collaborating Centre for Phage Typing and Resistance of Enterobacteriaceae, Laboratory of Enteric Pathogens, Central Public Health Laboratory, Colindale Avenue, London, UK. Reports contain information on salmonella isolations and serotype identifications during the reporting period for a large number of countries.

The FAO/WHO Collaborating Centre for Research and Training in Food Hygiene and Zoonoses organizes World Congresses on 'Food-borne Infections and Intoxications' under the title 'Safe Food for All', at which many reports of outbreaks and results of research and investigation are brought together, discussed and published.

Many countries within Europe collect data on food-borne illness and may or may not individually produce reports available to others. These reports may be in the language of the producing country and consequently difficult to interpret. The WHO collaborating system therefore fulfils an important task in drawing these reports together and distributing them in a single language. The Bureau of Hygiene and Tropical Diseases (London) issues monthly *Abstracts on Hygiene and Communicable Disease*. There is a section on Public Health News with items from journals, newspapers and other reports.

North America and Canada

The Centers for Disease Control (CDC) in Atlanta, USA collate information received from state public health systems and issue reports such as the *Morbidity and Mortality Weekly Report (MMWR)*. The *MMWR* covers both food-borne outbreaks and epidemiological investigations as part of a wider field also including alcoholism, smoking and accidents. Annual reports on both *The Surveillance of Food-borne Diseases* and on *Water-Borne Diseases* are also produced by CDC. The Health and Welfare Department in Canada has a similar system operating from the Laboratory Center for Disease Control, Bureau of Epidemiology, in Ottawa. The *Canada Diseases Weekly Report*, on infectious and other diseases, is published in both English and French, together with an annual summary on food-borne disease.

Other countries

Australia produces *Communicable Disease Intelligence*, a fortnightly report on food-borne illness compiled and issued by the Communicable Disease Branch of the Department of Health. The Australian Salmonella Reference Laboratory (Institute of Medical and Veterinary Science) issues reports on the salmonella case rate per month in South Australia. Tabulated details give human, animal, food and water sources (see p. 55). The Caribbean countries compile the monthly *CAREC Surveillance Report* from the Caribbean Epidemiology Centre (CAREC), Pan American Health Organization, Trinidad.

The preceding paragraphs do not give a comprehensive coverage of all the reporting systems in operation throughout the world, but the reports mentioned are fairly readily obtained in the UK.

The following chapter gives illustrative outlines of some typical outbreaks of food poisoning. The accounts are not presented with all the traditional epidemiological facts, but they serve to highlight faults leading to illness.

Short pro forma for food-poisoning outbreaks Date

Name of reporting officer ....

Name of local authority ....

Area/place of outbreak ....

Date and time suspected meal eaten ....

Number affected ....

Number at risk ....

Incubation period ....

Symptoms ....

Occupation/age group ....

Details of suspected meal ....

Foods eaten by affected persons ....

Number of meal sittings and times ....

Methods of cooking (particularly for meat/poultry) ....

Time and temperature of storage after cooking ....

Staff illnesses ....

General notes on facilities and equipment ....

6

Outbreaks of food poisoning and other food-borne disease

The occurrence of bacterial food poisoning depends on a peculiar set of circumstances and some or all of the following factors are present:

(i) the infecting organism (causal agent), in foodstuffs, in the food handler or in animals;
(ii) the hands of the food handler transmitting the organisms from raw to cooked food and to utensils, cloths and other kitchen tools or from the person of the food handler to cooked food;
(iii) surfaces contaminated by raw foods;
(iv) food suitable for bacterial growth;
(v) conditions favourable for warm storage over a period of 2 hours or more;
(vi) susceptible human subjects.

The investigation and prevention of food poisoning depends on the ability to examine the situation with these six factors in view.

The clinical symptoms and incubation period, as described in Chapter 3, will generally indicate whether the illness is of the infection or toxin type.

Immediately information is received the kitchen staff and factory manager should be warned not to discard any foodstuff. The cause of many incidents has been obscured because all relevant foods were thrown away before the EHO arrived on the scene. The isolation and identification of known pathogenic bacteria from the food and from patients provide clues for further investigation. The food vehicle is often prepared meat and poultry, warmed-up dishes, lightly cooked egg and milk dishes and other foods suitable for bacterial growth; they have usually been left for some hours without refrigeration. Any contaminated sources of raw materials which may be found will require future surveillance.

When the suspected food vehicle is not available, it may be helpful to examine ingredients, particularly raw meats and poultry known to be sources of salmonellae and other pathogens. If salmonellosis is suspected, other than enteric fever, enquiries should be made about sources of meat and poultry and other protein food and samples examined from kitchen, retail and wholesale stocks. Stool samples from patients and food handlers must be examined also. When the aetiology of an outbreak suggests that staphylococcal enterotoxin is in the food, vomit, stool and food samples should be examined. Swabs from nose, throat and hands of food handlers and also pustular spots and healed lesions may reveal the source of the staphylococcus.

To investigate *Clostridium perfringens* food poisoning, foods and faecal samples from patients should be examined. The organism is commonly present in many foods and it can be found in the stools of most people, usually in small numbers. After food poisoning, *C. perfringens* will be found in large numbers in food and faeces. There is little or no significance in the isolation of the organism from food handlers; preventive measures should be directed towards quick cooling and cold storage of foods not eaten immediately after cooking.

Whatever the source of food-poisoning bacteria, faults in preparation and storage will lead to multiplication in the cooked food vehicle immediately responsible for an outbreak. Efforts should be made to find the means by which the organisms spread from raw to cooked foods or from persons to cooked foods in relation to methods of food preparation; also at what stage after cooking the time lag occurred which allowed growth at temperatures convenient for the organism.

There are other organisms held responsible for diarrhoeal disease with less known about the epidemiology of origin and spread. *Campylobacter jejuni* must be sought in the stools of sick persons and animals, particularly cats and poultry. Milk, raw or inadequately pasteurized, and water are the usual vehicles, but isolation of the organism is difficult. *Listeria* in milk and cheese and *Yersinia* in pasteurized milk are associated with food-borne diarrhoea. *Aeromonas hydrophila* is occasionally reported as a causative agent in food poisoning, but so far without reference to the source, means of spread and behaviour in food. This organism may be overlooked in the laboratory amongst many other inhabitants of the intestine.

It cannot be assumed that food handlers found to be excreting the relevant organisms after an outbreak are necessarily the source of infection; they may have acquired the organisms from raw materials handled or even eaten. Nevertheless, with loose stools they may help to spread infection to foods and environment. Examples of food-poisoning outbreaks are given in the next few pages.

Salmonella food poisoning

Salmonella are members of the family *Enterobacteriaceae*. They are Gram-negative bacilli growing aerobically and anaerobically at an optimum temperature of 37°C (98.6°F), readily killed by temperatures above 55°C (131°F). They may be isolated from the intestines of man and animals and from foods of animal origin.

The role of animal and poultry excreters of salmonellae and the spread of the organism to the carcass meat of both animals and poultry and so to the human population is illustrated in many outbreaks.

Salmonella typhimurium in calves and veal

At least 90 persons in 55 separate incidents were ill after eating veal. The phage type of the *Salm. typhimurium* isolated from stools of victims indicated that cases, apparently unrelated, were all part of the same outbreak. Sick calves on a farm many miles away were excreting the same phage type. Calves from this and other farms in the same area were slaughtered in abattoirs providing meat for shops in the districts affected. Calf meat in one form or another appeared to be the vehicle of infection. *Salm. typhimurium* of the same phage type was found in 0.5% of faecal samples from 1000 calves on farms. Faecal samples from calves from the same areas, after they had been herded together in collecting centres for 2–5 days under poor conditions, showed an isolation rate of 36% for *Salm. typhimurium*. Thus stress and cross-infection were important factors in excretion. Many of the strains of *Salm. typhimurium* isolated from such calves were shown to be increasingly multiresistant to antibiotics.

Salmonella typhimurium in pigs and pork

Outbreaks with a similar history have been related to pork products made from carcasses contaminated by the intestinal contents of pigs. Ham, sausages, faggots and pork joints have all been implicated in salmonellosis. Frequently the same types of salmonellae have been isolated from the living animals, the environment of the slaughter house, carcasses, glands and offal. Sometimes sick animals inadvertently killed in the abattoir lead to profuse contamination of the environment and other carcasses. Detention of animals and poultry before slaughter will inevitably spread infection.

The most likely sources of infection of the live animals are the various feedstuffs such as fish, bone and meat meals which, unless specially treated, are likely to contain salmonellae. They are contaminated with small numbers and the animals remain well, but some will retain and excrete the organism. The rate of excretion on the farm is low because the animals are living under normal and usually well cared-for conditions. Exposed to stress by travel, unfamiliar temperatures, deprivation of food and water, overcrowding and possibly fighting, the rate of excretion rises and the infection spreads. The spread of contamination in the environment and amongst livestock will depend on the general hygiene of the establishment. Hazards of continual reinfection would be markedly reduced if feeds were free from salmonellae. (Fig. 6.1)

Salmonella typhimurium phage type 32 in pigs and pork

Hundreds of cases and 12 deaths occurred from pork products contaminated with *Salm. typhimurium* phage type 32. Pork products retailed from shops all over a large city, carcasses and minced meat sent many miles further south were vehicles of the *Salm. typhimurium*. The same phage type was isolated from live pigs in the local abattoir, from the gut room and drains and also from a feedstuff manufactured locally for distribution to farms (Fig. 6.2). It was observed that of two hospitals receiving pork meat from the infected source, one reported cases of salmonellosis due to *Salm. typhimurium* phage type 32, the other appeared to be free from infection. Conditions in the kitchen of the first hospital were poor and morale low, whereas the kitchen of the second hospital was hygienic, well lit and the staff efficient in the use of new stainless steel equipment.

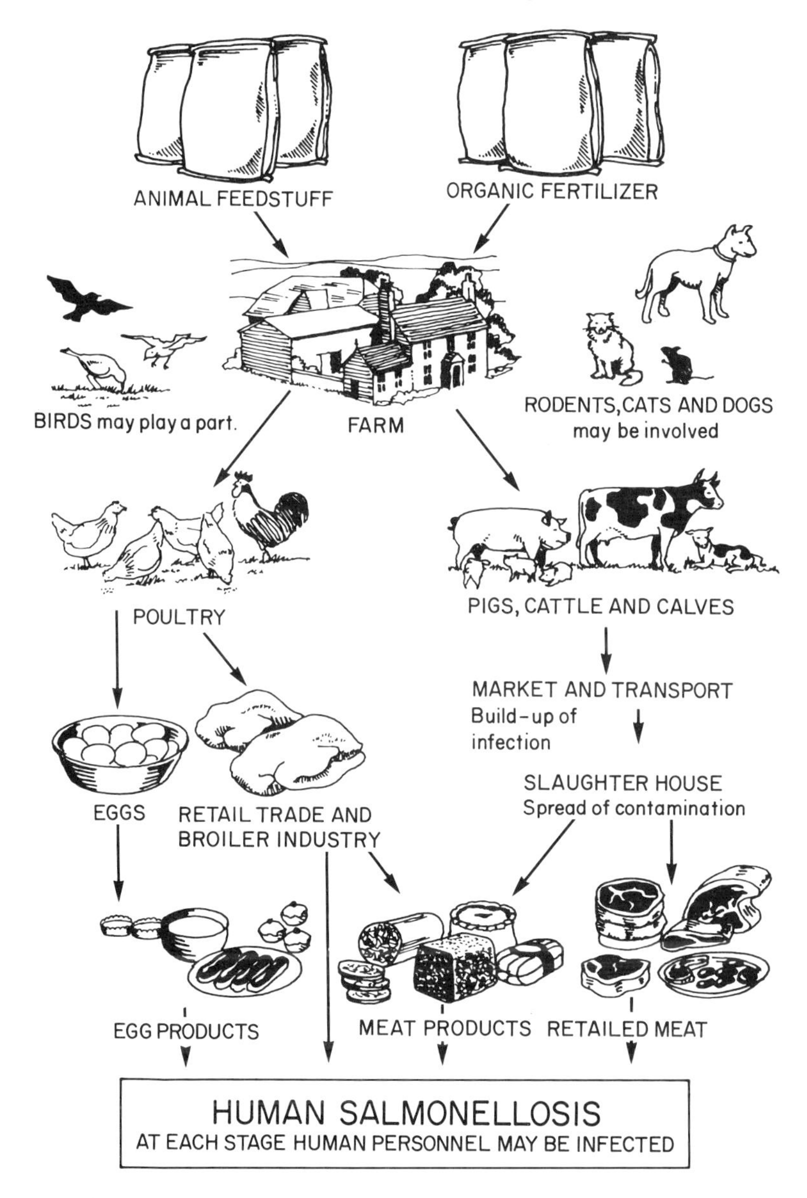

Fig.6.1 The spread of salmonella contamination

Salmonella brandenburg in pigs and sausage meat

The incidence of human cases and symptomless excreters of *Salm. brandenburg* increased 37 times during a period of 5 years and 18 times during a period of 1 year. During this year *Salm. brandenburg* was isolated from 12% of many samples of sausage meat prepared by two factories and retailed in local shops. Investigations led back to a small abattoir with 12–24 pens for pigs held from 2 days to a month pending slaughter. The contamination of carcasses was related to the time spent in the pens. The

(1)

Feeding meals and fertilizers may contain salmonellae which will be eaten and excreted by some animals.

(2)

The number of animals excreting salmonellae increases during marketing, transport and in the abattoirs. Salmonellae may be present on carcass meat as it leaves the abattoir.

Fig. 6.2 The role of animal and poultry excreters in human salmonellosis

(3)

Poultry and meat with salmonellae on them may be sold to the consumer.

(4)

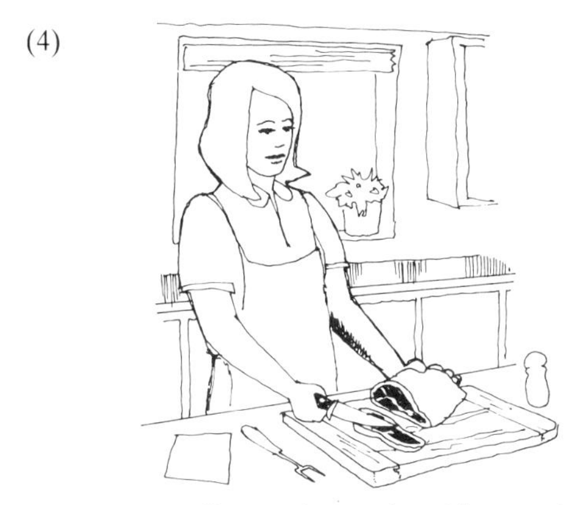

Salmonellae may be transferred from raw to cooked products in the kitchen by hands, surfaces, cloths and equipment.

Fig 6.2 Cont'd

organism could be scraped off the surface of some carcasses, particularly from the area behind the ears. These findings led to a revision of the law relating to period of stay allowed for animals in abattoirs prior to slaughter. *Salm. brandenburg* was subsequently isolated from meat and bone meal.

Salmonella wien in canned ham

Forty-nine residents in two small adjoining towns were ill with diarrhoea and vomiting 4–5 hours after eating canned ham. *Salm. wien* was isolated from samples of stools, the remainder of ham from the 6.3kg (14lb) can and from samples of ham left in households. The canned ham came from a pasteurized batch of 240 imported cans. Such cans are not sterile and are usually marked 'Keep in a cool place' or 'Perishable – keep under refrigeration'. The heat treatment may not have killed the salmonellae already present in the pork or leakage of polluted cooling water into the can may have occurred. Bacteria of various types may be isolated from samples of canned ham. Under conditions of warm atmospheric storage the cans may become blown or spoiled.

Ham, pork and veal in large containers may be lightly cooked to avoid shrinkage and to maintain flavour.

Salmonella manchester in yeast powder

The national outbreak of salmonellosis due to *Salm. manchester* illustrates again the value of the early warning system. This serotype with a plasmid marker is rare. CDSC received 49 reports between 1965 and 1985, a rise to 35 sporadic isolates occurred in 1986, none in 1987 and six in 1988. A rise in the number of identifications to 29 by October 1989 initiated investigation to identify one or more vehicles of infection. The cases were widespread and the median age of those affected was one year, but the age distribution ranged from 4 months to 70 years. Twenty-two cases were 1–5 years of age. Symptoms included diarrhoea, fever, vomiting and blood in the stool (few). An epidemiological questionnaire drew attention to savoury snacks and *Salm. manchester* was isolated from autolysed yeast powder and also cheese and onion flavouring. The organism was found in the roller drying area, drainage (water to cooler) system of the building, in the plant producing yeast powder for the baking industry and in the factory producing yeast in Scotland. Yeast deliveries from distributors were stopped, and the production of yeast at the Scottish factory also stopped pending implementation of control measures. Production was restarted early in 1990.

In two previous yeast-borne outbreaks, 1955 and 1964, both in the USA, a similar vulnerable youthful population was affected; their diet included a high proportion of processed, packaged snack foods. The detection of a small number of cases over several weeks from a wide geographic area indicated low-dose contamination of a nationally-distributed product.

Salmonella in pie meat

Food poisoning due to meat pies occurs from time to time, usually when the cooking is inadequate and the gel not hot enough when poured or injected into the pies. As usual, the severity of the illness depends on the amount eaten. Two outbreaks of salmonella food poisoning from pies illustrate the harmful effect of these faults. In the first, 29 persons in 11 family groups and 21 scouts were taken ill 5–24 hours after eating pies. The scouts were at camp in wet weather, with latrines some distance away. It was their first night. All had eaten pies from a batch of 200 large and small pies prepared with the same batch of meat as that used in 450 sausage rolls. The pies were heated at 232°–246°C (450°–475°F) for 25–30 minutes, and gelatin was poured in from a jug; they were stored overnight at room temperature before distribution.

In this outbreak, most of the affected persons ate portions of the large pies on one day and a few ate small pies 2 days after manufacture. The severity of illness varied according to the amount eaten, and some of the scouts were hospitalized. Salmonellae were isolated from 13 pies, victims and from one symptomless excreter amongst three people who had filled the meat mix into the pastry blocks by hand. This food handler, who liked to eat the meat raw, was regarded as a victim of salmonellae rather than as the original source.

The second outbreak was made up of multiple episodes in hospitals, maternity, orthopaedic, coronary care and baby units, a medical centre and families. Large pork pies appeared to be the vehicles of infection and *Salm. senftenberg* was isolated from 2kg (4.4lb) pies for catering, but not from small pies sold to the public, from eight of nine persons with diarrhoea, 16 persons without symptoms and from three of 17 food handlers. Coliform counts up to 3 million per gram were found in the large pies which were under-cooked, but not in the small pies. All the pies were prepared in one factory and used by catering establishments and hospitals in the region. Illness was prevented in 18 other hospitals by a quick change of menu, but many schoolchildren and teachers were ill after eating sliced pies from the same source at an athletics competition. As well as from patients the organism was isolated from 60 cooks and other food handlers who sliced and nibbled the pies. Persons attending a dinner party were likewise affected.

Many laboratories reported isolations of *Salm. senftenberg* from samples of pie, and two laboratories found the organism in jelly filled into the pies at 59°C (138°F). The machine for injecting jelly was thoroughly examined and 34 removable parts cultured. *Salm. senftenberg* was isolated from one of the injector nozzles. The gelatin accumulated between the hand grip and inner pipe which could not be cleaned because of the spot-welded construction. Thus a culture was established shedding organisms into the pies and contaminating the reservoir of jelly when it cooled. The jelly for the small pies was injected at 71°C (160°F) by an automatic process from a tank of jelly maintained at near-boiling temperature. A stainless steel and glass machine was installed for the injection of the large pies and the cooking time lengthened.

Salmonella typhimurium in salami sticks

In January 1988 a sudden increase in reports of infection from *Salm. typhimurium* DT 124, an unusual phage type were received by the Communicable Disease Surveillance Centre through the weekly laboratory surveillance scheme. Within a week the organism was isolated from salami sticks, the importers and manufacturers informed, the product removed from sale and the public alerted. One hundred and one persons were affected with ages ranging from 7 months to 78 years, the median age 6 years. The incubation period was unusually long, 3–7 days and the duration of illness 1–30 days or more (7 days median). Of the 72 primary cases, 68 had eaten the salami sticks. Eighty-one of 85 cases reported diarrhoea, 35 blood in stools, 38 vomiting and 71 fever. Nineteen cases were admitted to hospital. Of the 72 primary cases 68 had eaten the salami sticks that were imported from Germany and were popular with children. The sticks were 20 cm long, about 1 cm in diameter and each weighed 25 g. Contamination with *Salm. typhimurium* DT 124 was confined to five consecutive batches, which contained one meat component in common. There were small but significant differences in pH and glucose content (high) and acetic acid (low) in the contaminated batches indicating that the fermentative action was less than usual. Refrigerated storage at 10 and 5°C was found to prolong the survival of the salmonellae. The high fat content, thought to protect the organisms in, and delay their passage through the stomach, ingestion of a low dose of organisms and the nature of the serotype may all have contributed to the long incubation periods. *Salm. typhimurium* DT 124 was not found in raw materials, from equipment from the factory environment nor in faecal samples from all workers in the factory. Twenty cases due to the same organism were reported from the Grampian region of Scotland at about the same time.

Salmonella in poultry and eggs

A small prefabricated hospital was closed temporarily because one-third of the staff were excreting *Salm. senftenberg* and the number of sick persons was increasing. There were many excreters in the kitchen and amongst patients, nurses and doctors. The personnel of two departments did not use the canteen and remained unaffected. Turkey was regularly on the menu; the birds came from an area some distance away, where outbreaks of salmonellosis due to *Salm. senftenberg* had occurred amongst turkey flocks. A connection was eventually found between the infected farms and supplies to the hospital. Other hospitals without salmonellosis received turkey from the same source, but the cooking times were lower in the affected hospital and there were many faults in the kitchen which would encourage environmental spread (Fig. 6.3). An additional factor was the proximity of a pig farm where the feed was mixed in a shed windward of the kitchen. *Salm. senftenberg* is a common contaminant of certain feedstuff ingredients for animals.

Salm. virchow was responsible for many episodes of food poisoning, four of which involved 140 people. Spit-roasted chickens were suspected as the vehicles of infection. Warm carcasses pending sale and exposed to non-chilled storage provided an ideal medium for growth of salmonellae. Food handlers excreting *Salm. virchow* were victims of the mass of infected raw birds and responsible also for the mechanical transfer of debris by hands, utensils and cloths to cooked poultry.

Salm. typhimurium from 200 ducks' eggs in Queen's pudding infected 136 hospital staff and patients. The pudding was made with milk and breadcrumbs, and layered with jam and eggs beaten with sugar. Low cooking temperatures for the yolk mix and white topping failed to kill the salmonellae. In another incident, ducks' eggs were used in mousse stored at room temperature for several hours before being eaten by staff at a boy's school; most of the staff were ill, but the boys remained well.

A marked increase in salmonellosis due to *Salm. enteritidis* phage type 4 occurred from 1987 onwards. Vehicles of infection included mayonnaise and other uncooked or lightly cooked egg foods. Warnings, already established for ducks' eggs, were emphasized for the use of hens' eggs as well. As in ducks, the oviduct of the hen is infected; poultry meat also was responsible for infection and inspectors observed overt infection in birds. Breeding stock and also feeding meals were suspected as sources of infection, both home-produced and imported. Imports of eggs and also poultry carcasses have high rates of isolation for salmonellae.

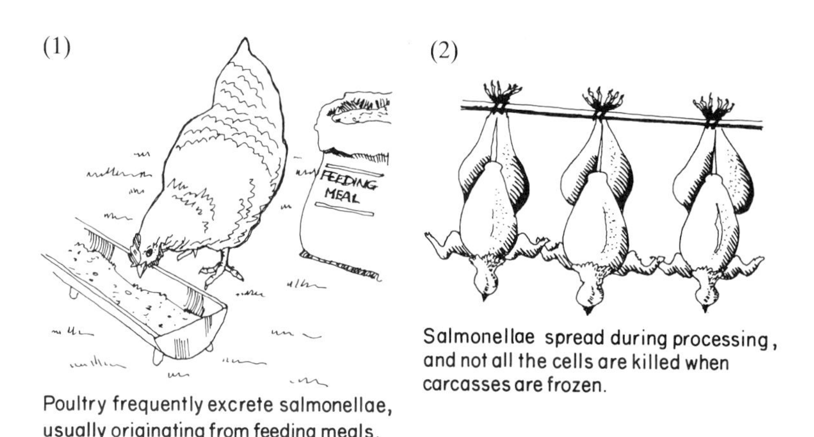

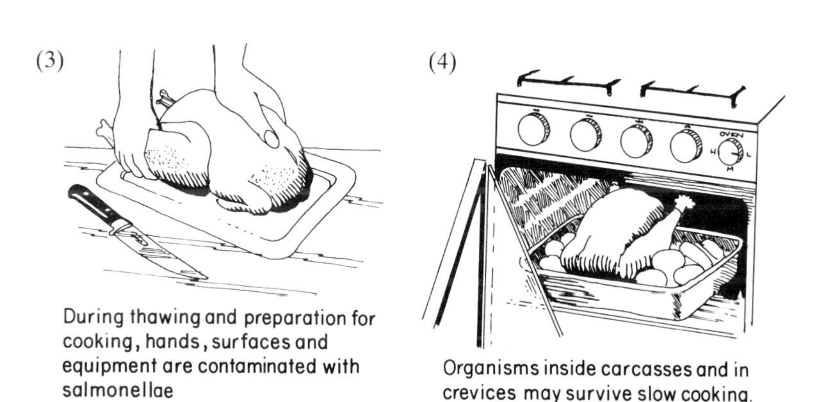

Fig. 6.3 Salmonella contamination from frozen poultry

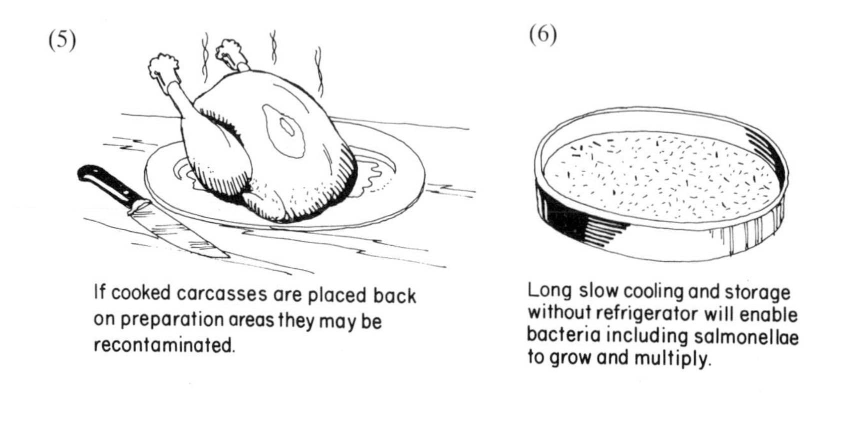

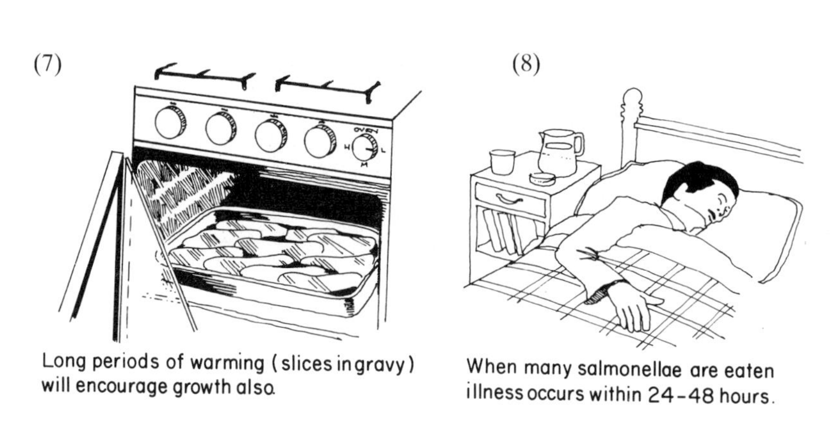

Fig 6.3 Cont'd

Salmonella typhimurium in Baked Alaska

In a home for 35 mentally handicapped young adults 13 of 18 members of staff, but no residents, were ill 18–24 hours after a lunch of mincemeat and vegetables and Baked Alaska. The first course was shared by staff and residents but two Baked Alaskas were made with two different lots of eggs. The first Baked Alaska was served to the residents, the second to the staff. The eggs used for the meringue topping had been bought cheaply from an imported batch and stored in an outhouse for more than a month. *Salm. typhimurium* was isolated from samples of stools from nine of the affected staff and from one husband who became ill one week later. The faeces of one resident who had diarrhoea 2 days earlier was positive for *Salm. typhimurium*. Staff were allowed to return to work when they were well and stools were normal.

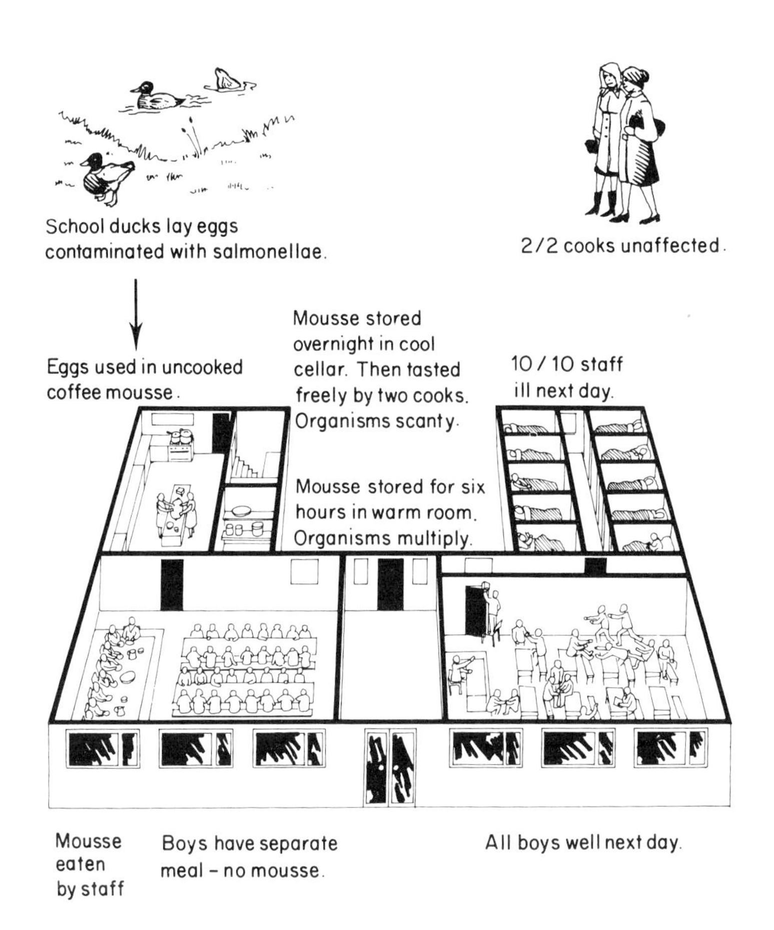

Fig. 6.4 The mousse outbreak

Salmonella enteritidis phage type 4 from home-made ice-cream.

One of the many outbreaks due to this organism occurred in May 1988. Seventy-five guests attended a bridge party, 63 (84%) were interviewed and 16 primary cases identified. Each case either had diarrhoea in the week following the function or *Salm. enteritidis* type 4 was isolated from stool samples and complaints of other gastrointestinal symptoms reported. Thirteen of 27 persons who ate ice-cream were ill compared with none of 25 who did not. The ice-cream mix contained hens' eggs; hens' faeces and dust from the relevant farm were positive for the organism.

Salmonella in confectionery (coconut, chocolate and bakery products)

Salm. typhi, Salm. paratyphi B and other serotypes of salmonellae were isolated from desiccated coconut, eaten as coconut balls, which caused typhoid fever and salmonellosis in Australia. As a result of the reports, samples of desiccated coconut were examined in the UK. *Salm. paratyphi* B and other salmonella serotypes were isolated; the coconut on marsh-mallow sweets was found to be a vehicle of infection. Other uncooked foods garnished with coconut also caused illness. Improved conditions of production enforced by legislation in the country of origin and monitoring of the exports reduced incidents of food poisoning from coconut to a minimum.

Salm. eastbourne in chocolate was reported in the USA and Canada in unusual incidents of salmonella food poisoning mostly in children. Christmas-wrapped chocolate balls were found to be responsible. The source of infection was thought to be the cocoa bean and cocoa bean dust. Various contaminated articles were found in at least three of the moulding plants over a 6-month period. *Salm. eastbourne* was isolated from roast beans cooled in air. Beyond the roaster the chocolate was heated at 60°C (140°F). This heat was not sufficient to kill all salmonellae. Warm chocolate was recirculated between the storage tanks and moulding plants, thus prolonging the period in which microbial growth could occur. The count of salmonellae, 2.5/g in chocolate samples was unusually low suggesting that a dosage of less than 1000 organisms (the number in a 1 lb (454 g) bag) was sufficient to cause illness, 100-to 1000-fold less than the numbers of *Salm. meleagridis* and *Salm. anatum* found to initiate symptoms in adult volunteers. Contributory factors to the apparently low dosage include technical difficulties in the quantitative isolation of salmonellae from chocolate, protection of the organisms by fat against the gastric acid and the age of most patients (young children); the virulence of the organism may be considered also. Regulatory action probably prevented further infections.

Salm. napoli in chocolate bars imported from Italy was responsible for 245 reported cases of infection from the sale of 600 000 bars. The out-break was detected by the surveillance of routine reports of salmonella infections from hospital and public health laboratories. Prompt recognition and rapid identification of the vehicle of infection enabled four-fifths of the consignment of contaminated chocolate to be withdrawn from the market, thus preventing many more infections.

Salm. paratyphi B and other salmonellae were isolated from frozen whole egg mix from China and used in imitation cream cakes and other confectionery products. The condition of the imitation cream as it reached the bakery in cans was good. Accidental contamination of the cream with small numbers of salmonellae from egg products occurred in the bakery and the organism grew in contact with the material of cakes, éclairs and buns. The egg mixes were rarely if ever a constituent of the cream, but mixing bowls, surfaces, utensils and hands served to transfer the organisms.

Cream products from various bakeries were the vehicles of infection in approximately 500 cases of paratyphoid fever. The same phage types of *Salm. paratyphi* B were found in cases and relevant shipments of frozen whole egg. Legislation to safeguard whole egg mix by pasteurization was introduced in 1964 (see Chapter 17). The use of crystalline or pan-dried albumen contaminated with salmonellae has also caused hundreds of cases of salmonellosis. Contaminated powdered products may be even more dangerous than liquids because of air-borne spread reaching dust on ledges and beams. Salmonellae from powdered egg in cake mixes, special diets in hospitals and non-acid mayonnaise have caused many infections.

Salmonella ealing in infant dried milk

A sudden increase in isolations of *Salm. ealing*, mostly from infants, led to a case-control study which showed that 21 cases had been fed on a particular formula of dried milk compared with five of 15 controls; no other food was implicated. There was considerable contamination of the factory with *Salm. ealing*, which confirmed the necessity to withdraw the dried milk from sale.

Staphylococcal enterotoxin food poisoning

Staphylococci are members of the family *Micrococcaceae*. They are Gram-positive cocci, growing in clusters, aerobically and anaerobically at an optimum temperature of 37°C (98.6°F), readily killed by temperatures above 55°C (131°F). They may be isolated from the nose and skin of man and the skin of animals and from skin infections and other septic lesions such as boils, whitlows, cuts and burns. Cold foods, much handled during preparation, are the most common vehicles of infection. Toxin is formed in food.

The first few outbreaks illustrate the ease of contamination of foods by staphylococci from the hands and the time temperature faults which allowed multiplication.

Staphylococcus aureus in ham sandwiches

Sandwiches, prepared at a public house, were eaten in a coach by people on their way to the seaside. On arrival, many were severely ill and taken to hospital. Large numbers of *Staph. aureus* were found in sandwiches left in the coach. The same phage type was isolated from the faeces and vomit of patients and also from the nose of the food handler who prepared the sandwiches; the organism would be frequently on her hands also and readily transferred to the ham. The ham was cooked in the kitchen of the public house, stored in the refrigerator and removed day by day to supply customers. There were three main factors responsible for the outbreak: firstly, the worker used her hands to hold the ham bone and the slices;

Fig. 6.5 The ham sandwich outbreak on a trip to the seaside

secondly, the refrigerator motor failed so that conditions were warm inside the cabinet and thirdly, warm summer weather encouraged growth of the organism on the ham. It is worth noting that the nasal carrier herself was a victim of the toxin in the portion of ham taken home to share with her husband. Two hours after the meal they were violently ill and taken to hospital (Fig. 6.5). Immunity to the toxin formed in food is not acquired while the organism is resident in the nose or throat. It is not always possible to eliminate staphylococci from the nose and they remain on the hands even after thorough washing. In such instances carriers should be transferred to work not involved with food.

Staphylococcus aureus in gelatin glaze on liver sausage

In a widespread outbreak during warm weather the gelatin glaze coating of liver sausage loaves was heavily contaminated with toxin-producing *Staph. aureus* (Fig. 6.6). The foodstuff was prepared on a large scale by a factory with a distribution service over the country. More than 400 cases were reported and many more may have remained unnotified. The examination of samples from various places showed that the glaze contained up to 100 million per gram of staphylococci; the meat part of the loaf was relatively free from organisms. Factory inspection revealed that each ingredient of the glaze was blended by hand in a steam-jacketed container. The temperature of the mix was maintained at 50°C (122°F) or less, too low to kill organisms from the hands. In the lengthy process of glazing, a jug was filled by hand from a small tank and the glaze poured over the meat loaves resting on a grid above the tank, refilled from time to time from the bulk supply in the large jacketed container. The actual glaze during pouring appeared to be at a temperature of 34°C (93.2°F), and the process was continued for several hours. Supplies of glaze remaining at the end of the day were stored refrigerated overnight and added to the fresh batch prepared the following day, thus maintaining a contaminated supply. One week after the outbreak, the food-poisoning staphylococci were found in the nose and on the hands of four employees and in swabs from portions of equipment and from the paper used for packing the meat loaves. Follow-up work indicated that the operative responsible for the preparation of glaze harboured the toxin-producing staphylococcus in the nose. He was transferred to other work. Recommendations were made to minimize the risks by avoiding hand contact with the glaze and ingredients, by bringing the glaze mixture to the boil and thereafter working in a cool room, by maintaining the glaze at approximately 60°C (140°F) during the process and by discarding unused glaze at the end of the day.

Staphylococcus aureus in pressed beef

Another series of outbreaks during a spell of warm weather occurred in various places from distributed pressed beef, contaminated by hand and allowed to cool slowly overnight. There were episodes on an intercity train, at a wedding party and in other areas. Large numbers of toxin-producing staphylococci were found in many samples of the batch of pressed beef, from the hands of the chef responsible for filling the press cans and also from the utensils used by him. Swabs from his hands persistently revealed the same staphylococcus and on one occasion it was isolated from his nose. The organism remained on the hands even after thorough washing and various treatments. Future supplies of the pressed meat were heat treated in sealed cans. (Fig. 6.7)

Staphylococcus aureus in lambs' tongues

Lambs' tongues peeled while warm after cooking were responsible for sickness in a factory canteen (Fig. 6.8). Although refrigerated overnight

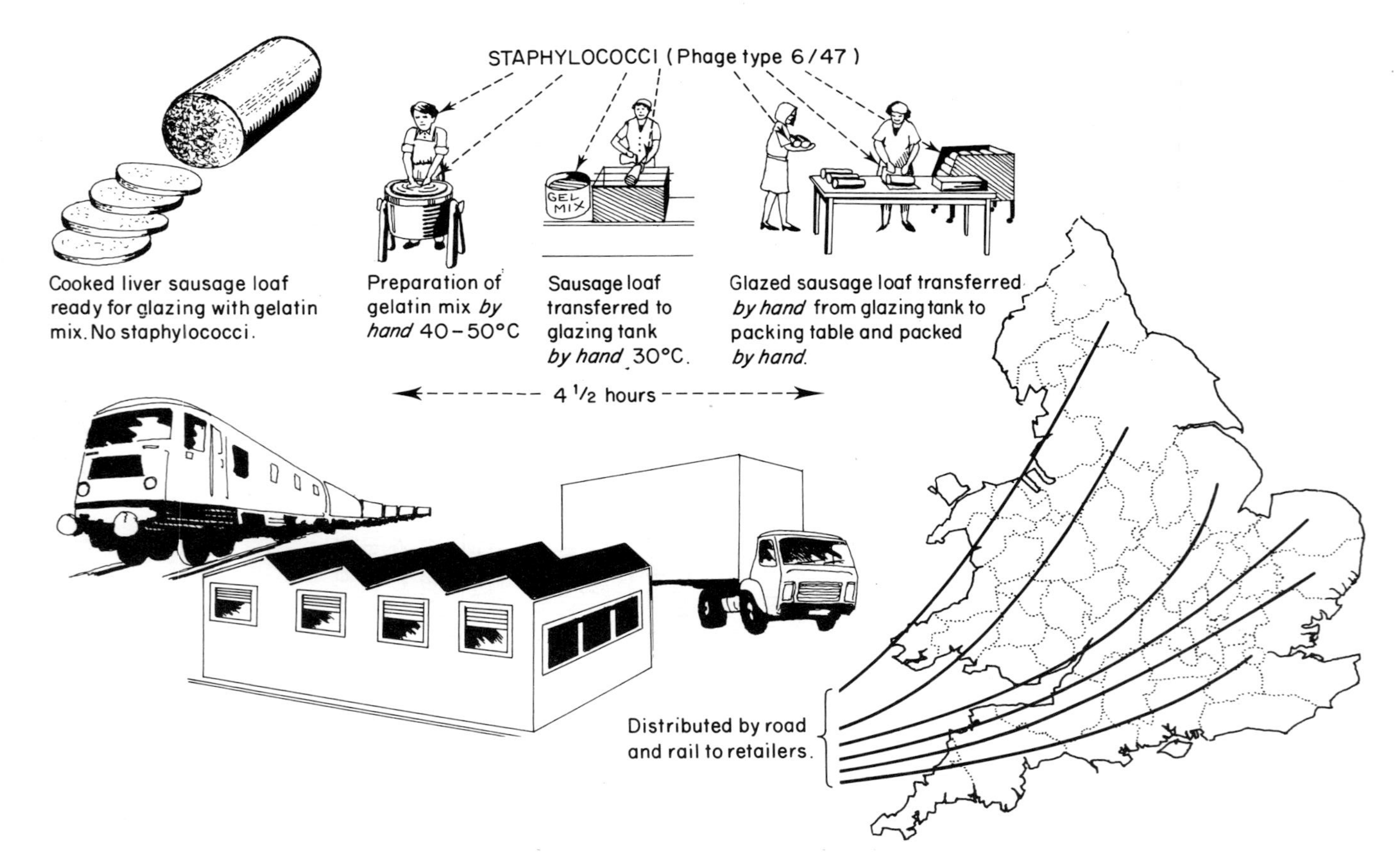

Fig. 6.6 The contaminated glaze outbreak of food poisoning – staphylococcal enterotoxin

they were sliced and kept warm in a hot cupboard before being eaten. Large numbers of toxin-producing *Staph. aureus* were found in the refrigerated and peeled tongues and from wooden table tops and dust, even after instructions had been given to clean all surfaces with detergent and disinfect with hypochlorite. The particular staphylococcus was found on the hands of two employees who skinned the tongues. The organisms were still cultured from their hands months after the outbreak; treatment with disinfectant was required before negative findings were reported. Cooked tongues are not infrequently responsible for staphylococcal food poisoning. Home processing of pressed tongue should include final heat treatment.

Staphylococcus aureus in various cooked meats

Wedding receptions and other large functions are notorious for providing foods prepared too far ahead of requirements and many outbreaks are reported in this country and overseas. For example, 40 of 139 guests at a wedding reception developed typical symptoms of vomiting, abdominal pain and diarrhoea 2 hours after eating turkey and ham contaminated with *Staph. aureus*. The same type was isolated from septic spots on the hand and in the nose of the food handler who prepared the food the previous evening.

Thirty of 120 spectators at a Dog Show were ill 2–3 hours after eating cold chicken left overnight in the tent. The chef had recently returned to work after sick leave with an injured finger and tonsillitis.

Doctors were ill 2–4 hours after attending a luncheon party with a menu including cold chicken and salad. The chickens were cooked the previous day and dismembered by hand while still warm, the portions were piled on trays and eventually refrigerated although the centre portions were still warm. A similar outbreak occurred on an aircraft. The main fault in the two outbreaks appeared to be the portioning of the chickens by hand while still warm. When food is cold, implanted staphylococci will be slow to grow and the numbers will still be low when the food is eaten soon after preparation. In many instances the enterotoxin was demonstrable in *Staph. aureus* cultures and also in the food itself.

Food handlers should be made aware that staphylococci are commonly present on the skin of the hands and various other parts of the body; washing with soap and water is unlikely to eradicate them. Where possible cooked food should be moved using utensils rather than hands. Prompt and persistent cold storage is essential to keep chance contaminants to a low level and thus to prevent the accumulation of toxin from large numbers of proliferating organisms. The larger the amount of toxin, the more severe the illness.

Staphylococcus aureus in ice-cream

Before the ice-cream regulations were enforced in the UK, there were many outbreaks of staphylococcal enterotoxin food poisoning. *Staph. aureus* from

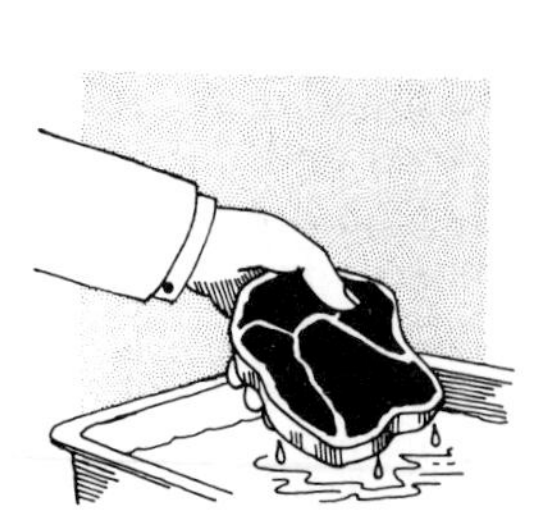

The meat was stored in brine for two days

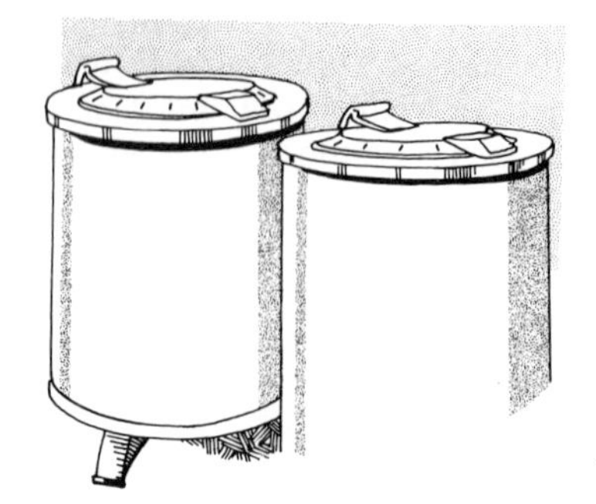

Then boiled in jacketed pans more than 3 hours. By this time nearly all germs had been killed.

BUT it was then cut up and placed in tins by hand and the hands of the Chef carried staphylococci.

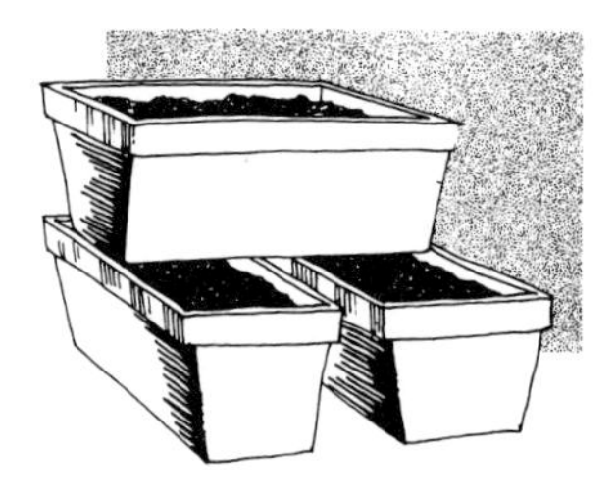

The pressed beef in tins was allowed to cool overnight at room temperature – and consequently the staphylococci multiplied and formed toxin.

Result:

An outbreak in a canteen.

An outbreak at a private party.

An outbreak on a restaurant car.

Fig. 6.7 The spread of staphylococcal infection from pressed beef

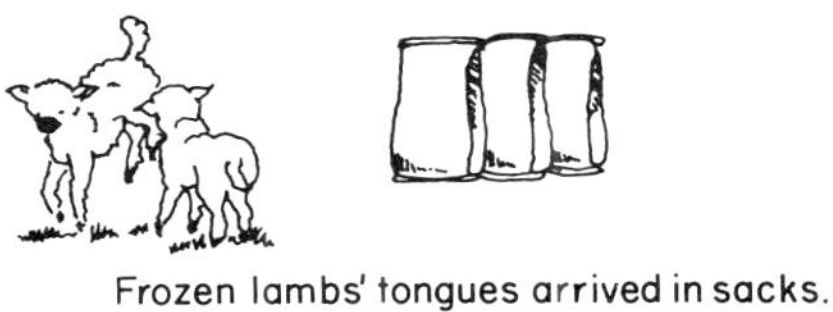

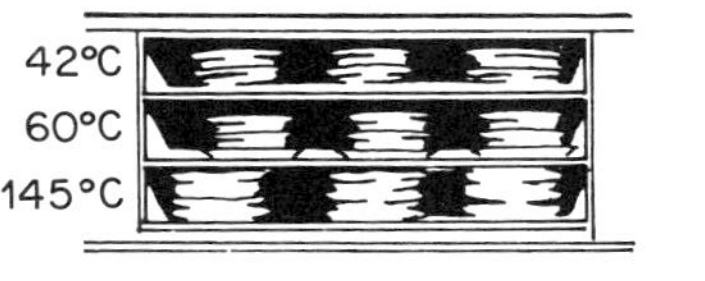
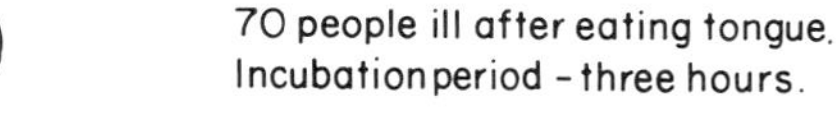
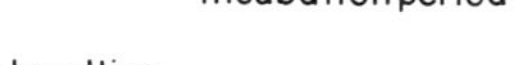

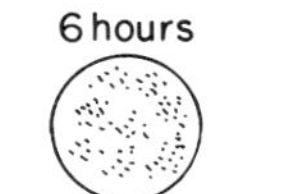
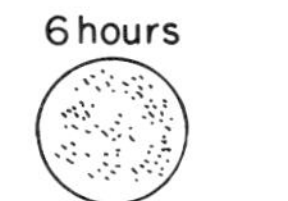

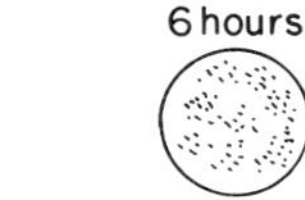
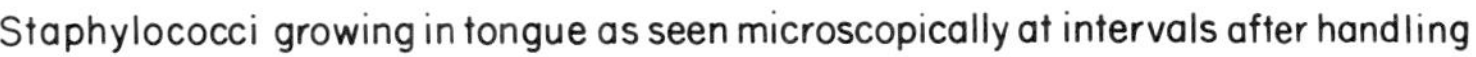
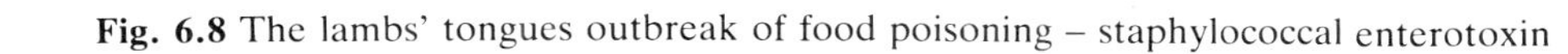

Fig. 6.8 The lambs' tongues outbreak of food poisoning – staphylococcal enterotoxin

nasal and hand carriers contaminated the mix and multiplied before or after heat treatment and before freezing. In one outbreak more than 7000 people were affected by the enterotoxin. The Ice-cream (Heat Treatment etc.) Regulations, 1959, originally 1947, stipulate that ice-cream mix must be pasteurized and immediately cooled to 7.2°C (45°F) within 1.5 hours of heat treatment. In 1984 an outbreak was associated with salmonellae in home-made ice-cream, for which legal requirements of heat treatment do not apply.

Staphylococcus aureus in cheese and milk

Goats' milk cheese containing many millions of staphylococci, presumed to come from the udder of a goat with *Staph. aureus* in her freshly-drawn milk, produced acute toxin-type illness in members of a family.

Cheddar cheese exported as second grade for processing, but distributed for consumption in hospitals, caused *Staph. aureus* food poisoning. Penicillin or other antibiotics administered to cows suffering from mastitis, staphylococcal or streptococcal infections, can inhibit starter cultures in milk for cheese and thus allow staphylococci to grow because the pH is not reduced. It was recommended that all milk intended for cheese should be pasteurized and legislation to that effect was passed in New Zealand. Staphylococcal enterotoxin poisoning has been reported from unheated milk. Milk from tuberculin-tested herds can be contaminated with food-poisoning organisms from the cow and milking equipment.

Staphylococcus aureus in vanilla cakes

Counts of *Staph. aureus* from 18–200 million per gram were found in vanilla cakes eaten by five families who reported sickness, abdominal pain, and in some instances diarrhoea, 3–6 hours later. Staphylococci were grown from a mixing bowl, a healed burn on the hand of one bakery worker and from the eczematous lesions on the right hand and elbow of another. Both had helped to prepare the cakes sold from market stalls in hot weather.

Staphylococcus aureus in canned peas and meat

A series of outbreaks occurred from peas in 2.7 kg (6 lb) cans handled while still wet after water cooling by an employee with a septic lesion on the hands. The enterotoxin-producing strain of staphylococcus was sucked through defective seams with pinhole leaks. Although the organisms had grown and produced toxin the cans were not blown or spoiled either visibly or by flavour.

Similar outbreaks have occurred from *Staph. aureus* in freshly opened cans of meat, such as corned beef contaminated by cooling water or by hands while the sealed cans were still wet.

Large joints of meat arrived on Tuesday

They were boiled on Tuesday afternoon.

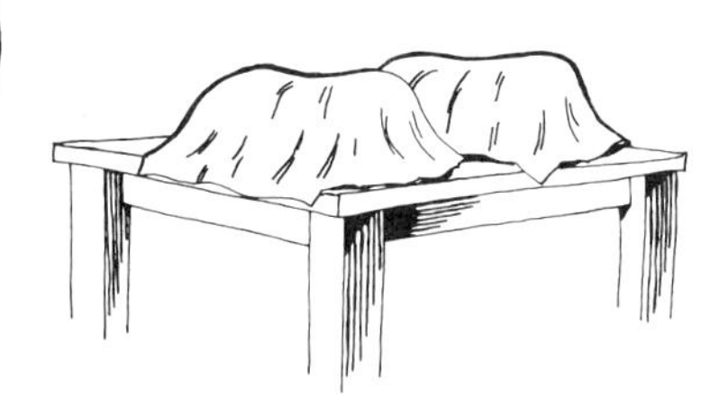

There was no room in the refrigerator, so they were allowed to cool slowly overnight.

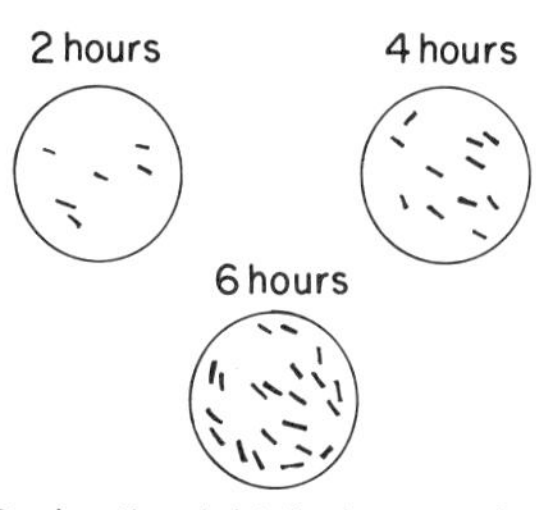

During the night the heat-resistant sporing bacteria grew and multiplied.

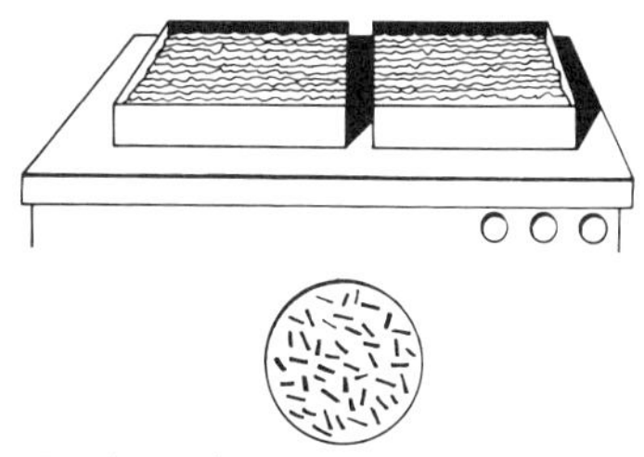

The bacteria multiplied still more when the sliced meat was warmed with gravy on the hot plate.

300 ILL

With pain and diarrhoea after 350 children and staff had eaten meat boiled the previous day.

Fig. 6.9 Outbreak of *Clostridium perfringens* food poisoning in a school canteen

Clostridium perfringens food poisoning

Clostridium perfringens is a member of the family *Bacillaceae*. It is a large square-ended Gram-positive bacillus growing anaerobically at optimum temperatures of 43–47°C (109–117°F). It produces spores some of which can survive boiling. The organism may be isolated from the soil and from the human and animal intestine, as well as from foods of animal and vegetable origin. Toxin is released in the intestine.

C. perfringens gives rise to another well-defined type of food poisoning following storage faults in the kitchen.

Clostridium perfringens in salt beef

A school canteen was responsible for two large outbreaks of food poisoning occurring within a year of each other. Abdominal pain and acute diarrhoea were the predominant symptoms within 9–12 hours of the lunch eaten by the children. In each outbreak cold boiled salt beef was served with potatoes and salad. Joints of beef weighing 1.8–2.7 kg (4–6 lb) were boiled for 2 hours. In the first outbreak the beef was allowed to remain in the liquor overnight. In the second outbreak, rolls of salt beef were lifted from the boiler, drained and placed in enamel dishes, covered with cloths and stored in the larder overnight (Fig. 6.9).

In spite of the warning to boil and eat salt beef on the same day, the second outbreak occurred after faults of long, slow cooling and storage overnight without refrigeration. Spores which had survived boiling germinated and the vegetative cells grew to large numbers of *C. perfringens* mixed with other organisms or almost pure in the rolled meat. Cultures from stool samples showed large numbers of *C. perfringens* able to withstand from a few minutes to hours of boiling.

Clostridium perfringens in brisket

In 1984, 44 of 65 residents and staff at an old people's home were ill after eating boiled brisket which had been left in the liquor to cool overnight. *C. perfringens* was isolated in large numbers from faeces, together with enterotoxin. The same serotypes were found in meat and faecal samples.

Clostridium perfringens in meat meals (hospitals, schools and North Sea oil installations)

Outbreaks of *C. perfringens* food poisoning continue to occur when cooked meat and poultry, soups, stews and gravies are cooked in bulk and stand warm for some hours (Fig. 6.10). Gravy prepared at a central kitchen and distributed next day to schools was responsible for illness amongst many children. North Sea oil installations have suffered *C. perfringens* food poisoning because of the difficulty in off-shore catering; pot roast with gravy, mince, stew and pie have been involved; enterotoxin from faecal samples has provided proof when *C. perfringens* was not isolated from food. Boiled chickens in liquor, cooled overnight and fed to patients caused a series of outbreaks in hospital. Minced meat upset nearly 400 patients and caused one death in hospital, because cuts of meat for mincing were inadequately cooked and cooled, which encouraged the survival and growth of *C. perfringens*. The next day after non-refrigerated storage the meat was warmed through, but the heat treatment was inadequate to kill the *C. perfringens*. A dual-purpose machine was used to mince both raw and cooked meat with scant cleansing in between. Recommendations were given to cool hot food quickly, within 1.5 hours, in a well-ventilated cold room and then to refrigerate. Raw mince meat should be cooked and eaten the same day. When it is necessary to re-heat the mince it should be boiled thoroughly. Many outbreaks with a similar pattern occur in hospitals.

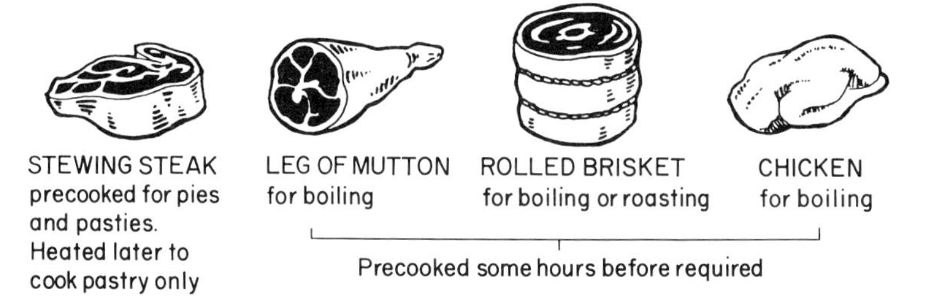

Fig. 6.10 *Clostridium perfringens* food poisoning from meat and poultry

Clostridium perfringens in 'meals-on-wheels'

'Meals-on-wheels' have contributed to *C. perfringens* food poisoning following the consumption of minced beef. Mince in 2.7 kg (6 lb) frozen packs, incompletely thawed, was heated for 2 hours in chicken stock in one large pan. It was 'cooled' for 3 hours in the vegetable room which was inadequately ventilated and found to have a temperature of 21°C (70°F). The mince was reheated for 20 minutes and dispensed with cooked vegetables into foil containers for distribution in electrically-heated cabinets. Where large numbers of meals are prepared, a ventilated cold store, refrigerated with one extractor fan should be available. A blast chiller would hasten the rate of cooling. The cabinets were designed to maintain food at a temperature of 63°C (145°F). The highest temperature achieved was 46°C (115°F) at the beginning and 29°C (85°F) at the end of distribution. Meals should be freshly cooked and dished up rapidly and kept hot whilst delivered.

Clostridium perfringens in curry

Curry, the favourite dish in the Near and Far East, presents constant hazards of *C. perfringens* food poisoning. A picnic at the Bakra Dam in the Punjab, India, illustrated the danger from large quantities of reheated non-refrigerated meat and chicken curries. The coach left at 7.30 a.m. but the chicken curry and rice had been cooked during the night and travelled on the floor of the coach near the engine. At 2 p.m. 50 hungry persons settled on the grass for lunch, and waited 10 minutes for the curry to be warmed on a kerosene stove. Generous helpings of curry and rice were provided, and much enjoyed. Late that evening three people felt unwell in the coach, but at approximately 5 a.m. next day most of the picnickers suffered abdominal pain and diarrhoea. Spores of *C. perfringens* in chicken and spices would survive the boiling process and storage conditions of time and temperature were ideal for bacterial growth. In spite of warnings a picnic to the same place the following year had a similar ending.

Fig. 6.11 illustrates the events which precede a typical outbreak of *C. perfringens* food poisoning.

Clostridium botulinum (botulism)

Botulism from imported canned salmon

Four elderly persons, two couples, took tea together which included canned salmon, salad, fruit and cream. Nine to 11 hours later all four developed nausea, vomiting, a dry mouth and blurred vision. One couple managed to reach hospital with evidence of paralysis, impaired vision, speech, swallowing and breathing. They were placed in the intensive care unit and required respiratory assistance.

Botulism is so rare in the UK that it was a little time before the diagnosis was made and antitoxin therapy administered. Enquiries about others who might be affected prompted a visit to the house of the second couple who were found to be extremely ill and unable to call for help; they were also admitted and given antitoxin but they eventually died whereas the first couple recovered, probably due to the early administration of antitoxin.

C. botulinum type E toxin was demonstrated in the sera of all four patients and in washings of the empty salmon can found in the house of the second couple. Also, spores of *C. botulinum* type E were found in washings from the can.

The batch of over 14 000 cans of salmon, imported from the USA, was recalled and many cans were examined both microbiologically and structurally, but no contamination was found. The rim of the incriminated can was damaged and there was a small hole through which *C. botulinum* type E could have been sucked in as the cans were cooling in the wet environment of the fish cannery.

Prior to this incident there had been no reports of human botulism in the UK for 25 years.

Botulism from hazelnut yoghurt

Hazelnut yoghurt responsible for botulism, *C. botulinum* type B was diagnosed in 27 persons in June 1988; antitoxin was available and only one person (aged 74) died. Ages ranged from 14 months to 74 years (10 males and 17 females) and the incubation period from 2 hours to 5 days, probably depending on the amount eaten, 2 spoonfuls to 3 cartons. The type B toxin was found in 2 opened cartons (from homes) and 15 unopened cartons. The producer had used contaminated hazelnut conserve, a low-acid product insufficiently heated to kill spores. Toxin was found in one blown can and not in 23 other cans. The cans of conserve were recalled, the yoghurts withdrawn from sale and production stopped. The public were advised not to consume hazelnut yoghurt. Communication was good which enabled rapid coordination of action. 1750–3750 MLD (mouse lethal dose) toxin was detected in 125 g yoghurt.

Bacillus cereus food poisoning

Bacillus cereus is a member of the family *Bacillaceae*. It is a large Gram-positive bacillus growing aerobically and anaerobically at an optimum temperature between 28 and 35°C (82 and 95°F). Spores are readily formed, and may survive cooking procedures. The organism is widely distributed in cereals and dried foods, and in dairy products where it is a common spoilage organism. Toxin is formed in food.

Bacillus cereus in cooked rice

Fried rice contaminated with large numbers of *B. cereus* has been responsible for acute vomiting, sometimes followed by diarrhoea, within 2–4 hours of eating food prepared in Chinese restaurants and 'take-away' shops. The spores survive boiling, particularly when protected

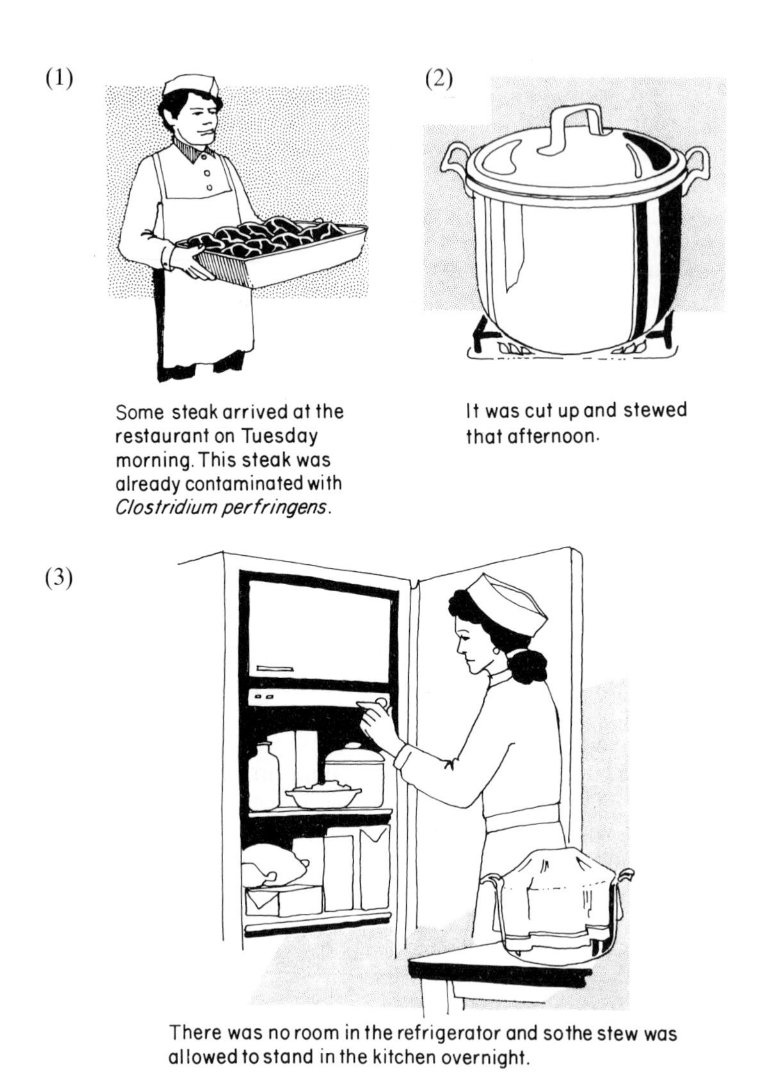

Fig. 6.11 *Clostridium perfringens* food poisoning outbreak from stewing steak

by starch, and the vegetative cells grow at a wide temperature range (Fig. 6.12). A series of five small episodes of food poisoning over a period of 2.5 months affected customers eating in a Chinese restaurant. Incubation varied from 1–6 hours, and the symptoms were characterized by early nausea and vomiting and late diarrhoea. Fried and boiled rice were found to have high counts of *B. cereus* (350 million per gram in one sample of fried rice). Large numbers were found in stool samples also.

Small numbers were found in uncooked rice; all the strains isolated belonged to the same serotype. In the restaurant kitchens rice was boiled

(4)

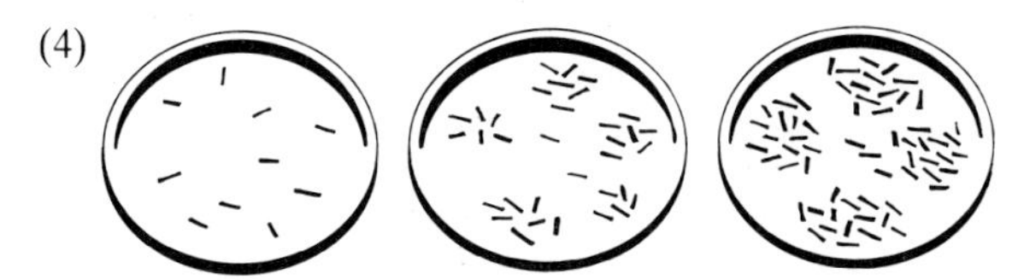

The night was hot and a few heat-resistant sporing germs not killed by cooking started to grow and multiply.

(5)

They multiplied still more when the meat was warmed up for dinner on Wednesday.

Result

52 people were ill with diarrhoea and pain after eating this stew.

Fig 6.11 Cont'd

the evening before it was required for flash frying – a quick turnover in hot oil. In the kitchen overnight opportunity was provided for the spores to germinate. Remains of cooked rice were added to new batches. The restaurant was closed all day on Wednesday and half day on Sunday. It was speculated that rice prepared on Saturday and Tuesday was added to the new batches made on Monday and Thursday when incidents were reported. Many outbreaks of a similar nature occur in the UK, Australia, Canada, Finland, the Netherlands and the USA.

Bacillus cereus in beef stew

Beef stew with creamed potatoes and peas affected 42 persons, mostly residents of a prison, some suffering from nausea and vomiting after 1.5 –3 hours, and some abdominal pain and diarrhoea after 4 hours. No one who ate an alternative meal became ill. Large numbers of *B. cereus* of a single serotype were isolated from samples of beef stew and faeces. The stew was made from raw minced beef together with tomatoes and carrots precooked that morning and left standing in the kitchen. Vegetables left from the previous day may have been added also. The vegetables were thought to be the source of the spores. The cooking time was one hour, and the stew was served directly. Many who selected the dish said that it had a sour smell and did not eat it. There were significant levels of enterotoxin in the stew.

The two syndromes, rapid vomiting and stomach pains with diarrhoea can be recognized in this outbreak; specific serotypes or specific toxins do not appear to be linked exclusively to either form of illness.

Other *Bacillus* species

Other *Bacillus* species of the *B. subtilis-licheniformis* groups have been associated with cases both of diarrhoea with an incubation of 8–12 hours and a rapid (less than 1 hour) onset of vomiting. Minced meat, chicken and rice and various meat and pastry products such as pasties and sausage rolls with high counts of *Bacillus* species have been incriminated. *B. thuringiensis* is also suspected to cause food poisoning. This organism produces a crystal which has a lethal effect inside insects and thus is useful as an insecticide. Common species of bacilli should not be ignored when found in foods in large numbers.

Escherichia coli food poisoning

Escherichia coli is a member of the family *Enterobacteriaceae*. It is a Gram-negative bacillus growing aerobically and anaerobically at an optimum temperature of 37°C (98.6°F), readily killed by temperatures above 55°C (131°F). It is commonly found in the human and animal intestine and may be isolated from foods of animal origin.

Although usually a normal and harmless inhabitant of the intestine of nearly all creatures, certain strains are pathogens in young children, adults and animals. There are five enterovirulent actions as described in Chapter 3. Enterotoxigenic strains are responsible for the incidents and outbreaks of gastroenteritis known as 'traveller's diarrhoea'.

A series of incidents occurred on a ship cruising through the Mediterranean and the Caribbean. Cases of diarrhoea mainly and nausea were reported daily; they increased in numbers and frequency when the climate was warm and humid and when colony counts from food, often served from deck buffets, were high. Catering for many varieties of food in different menus and also for parties imposed heavy demands on food service

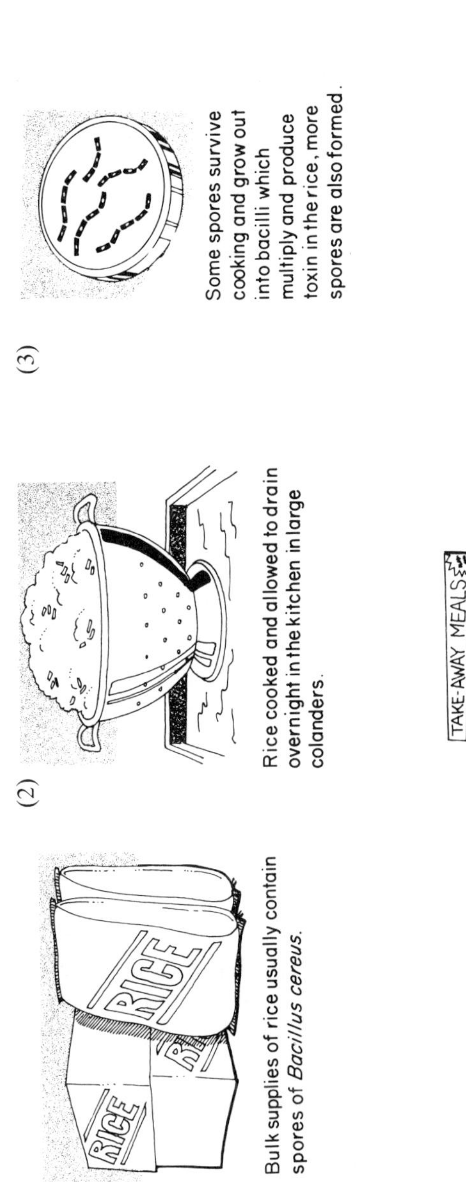

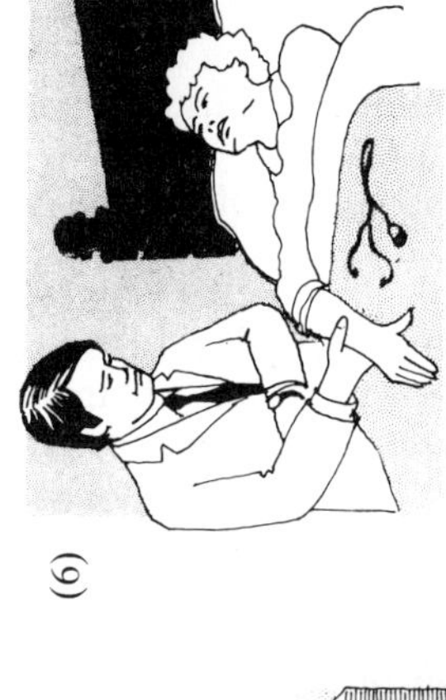

Fig. 6.12 *Bacillus cereus* food poisoning outbreak from rice

and food handlers. Galley staff and foods originated from many different countries; cross-contamination between raw foods and from raw to cooked food was evident in the main preparation areas and overcrowded pantries used for cold meats, seafoods and salads. Prepared dishes were kept warm for some hours pending lunch, dinner and party meals. Various foods for consumption on the ship were monitored bacteriologically both at sea and in port.

Several serotypes of *E. coli* were found in food samples and faeces. Amongst many strains examined the epidemic serotype O27 was prominent in stools; it was also found in cream prepared in the galley and stored in bulk without proper refrigeration. Recommendations on food preparation, storage and cleaning helped to reduce the incidence of illness.

The organism may have originated from raw foods or from excreters; spread and multiplication in food were considered to be the principal faults. Ship-borne outbreaks of gastroenteritis, including salmonellosis and even enteric fever are not infrequent. Courses on food hygiene have been instituted on some shipping lines for chefs and galley staff as well as for other persons responsible for food safety.

Repeated outbreaks of diarrhoea from *E. coli* were reported in newly-arrived personnel in army units in the Middle East. A percentage of persons attending conferences in Mexico City has been affected and stools from control groups examined. A large outbreak in which cheese was the vehicle of infection was described in the USA. One of the original outbreaks involved a can of salmon eaten for a midday school meal. Volunteer tests confirmed *E. coli* as the agent of infection.

Verotoxigenic *E. coli* serotype O157:H7 caused a number of outbreaks of gastroenteritis in both North America and the UK. The illness was often severe causing bloody diarrhoea, which in some cases developed into haemorrhagic colitis. Beef products were implicated in some of the North American incidents, but the particular type of *E. coli* was not isolated from food in the UK.

Campylobacter food poisoning

Campylobacters are curved, Gram-negative bacilli resembling the *Vibrio* and growing under partial anaerobiosis at an optimum temperature of 43°C (109°F). They may be isolated from a wide range of animals and from man, and also from water and foods of animal origin, in particular poultry, meat and milk.

Campylobacter jejuni in milk

The largest reported outbreak of *Campylobacter* enteritis was associated with raw or imperfectly pasteurized milk, when 2500 school children were infected. Two strains were isolated from children, but none from milk; goat's milk was also implicated. Another milk-borne outbreak occurred in a boy's school; 189 boys reported to the medical officer during the week of the outbreak and 102 were admitted to the school infirmary. A questionnaire completed by 775 (99%) of boys revealed that 518 (67%)

had one or more symptoms; 57% had diarrhoea, 30% pain without diarrhoea and 13% symptoms other than pain or diarrhoea. Diarrhoea was correlated with the drinking of milk and was more common with increasing consumption. Older boys were less affected in spite of drinking more milk. *C. jejuni* was isolated from 36 of 41 samples of faeces, and 33 isolates were of the same serotype. Serological evidence of infection was found in 82% of boys with symptoms and 44% without symptoms. It was thought that the milk, supplied from a farm within the school grounds, might have been raw due to an accidental bypass of the pasteurizer; raw milk was habitually supplied to a large processor. Campylobacters of various serotypes, including the epidemic strain, were isolated from rectal swabs from cows and from a sample of milk from one cow.

In a joint medical and veterinary investigation in another school, 29 boys reported with symptoms including headache, fever, abdominal pain and diarrhoea, usually without nausea or vomiting. Eleven strains of *C. jejuni* of the same serotype were isolated from 14 faecal specimens. A questionnaire revealed the incidence of diarrhoea in 32% (34/108) junior boys, 28% (81/290) senior boys and 14% (24/173) of school staff members, monks, nuns and workers on the school farm; there was a significant association between diarrhoea and the consumption of milk which was supplied, unpasteurized, from Friesian cows on the school farm. *C. jejuni* was isolated from 25 animals (21%). Two of six serotypes were also isolated from human cases; faecal contamination of milk was suspected. Serological tests on 451 boys showed that the infection rate, 2 months after the outbreak, was 14–21%. Most of the culture-positive boys had antibodies of at least one sort.

Campylobacter jejuni in chicken

An early prominent and extensive outbreak was reported in the Netherlands among soldiers on a survival exercise. Of 123 cadets given live chicken to prepare for their evening meal, 89 became ill. Abdominal cramps and diarrhoea were predominant, but some patients had fever and a few vomited. The average incubation period was 3 days, but there was a variation from 1–6 and even 8 days; the duration of illness was 4–5 days. *C. jejuni* was isolated from 34 of 104 faecal samples, 27 from sick persons and seven from those who remained well. Neither *Salmonella* nor *Shigella* were found. The chicken meat was consumed in a semi-raw state. Undercooked barbecued chicken was the reported vehicle in another outbreak when 11 of 15 persons were ill.

Campylobacter jejuni in meat

In Japan, 800 of 2500 school children were ill after eating vinegared pork; 54 individuals in a military camp were ill after eating raw hamburgers. The same serotype of *C. jejuni* was isolated from six patients and a food handler.

Campylobacter jejuni in water

Two large water-borne outbreaks in Sweden and the USA are well documented, 4000 persons were involved. The water supplies were incorrectly processed.

Vibrio parahaemolyticus food poisoning

Vibrio parahaemolyticus is a member of the family *Vibrionaceae*. It is a short Gram-negative aerobic bacillus, slightly curved; coccoid and swollen forms may be seen. The optimum temperature for growth is 37°C (98.6°F). It may be isolated from sea water and raw fish, shellfish and other sea foods in many countries.

Illness from *V. parahaemolyticus* in food is common in Japan and other eastern countries where coastal waters are warm. Outbreaks are also reported from the USA, Australia and the UK.

Vibrio parahaemolyticus in crab meat

An outbreak occurred during a flight from Bangkok to London, via Dubai. The number of persons ill was not known, but three of five who reported profuse diarrhoea, vomiting, dehydration and fever were admitted to hospital on arrival at Heathrow. Three of the cabin crew were also ill after leaving the plane at Dubai; thus meals from Bangkok were responsible. Complete frozen meals prepared from the same batches served on the flight were flown to London. *V. parahaemolyticus* isolated from cooked crab in the hors d'oeuvres was the same serotype as that of the vibrio isolated from stool samples of the three hospitalized patients and also from the raw meat in crab claws flown from Bangkok. *V. parahaemolyticus* may have survived boiling or it may have passed from raw to cooked crab meat during preparation and assembly of the hors d'oeuvres. The organism is heat-sensitive and cross-contamination is the most likely explanation. The contamination of cooked crabs by *V. parahaemolyticus* from uncooked crabs stacked in baskets has been described in the USA. This vibrio is rarely responsible for food poisoning in the UK, but holidaymakers were ill after eating crabmeat from crabs caught off the south coast.

V. parahaemolyticus is occasionally isolated from imported frozen cooked prawns which may be responsible for food poisoning when used in seafood cocktails prepared for parties and banquets. Although *V. parahaemolyticus* and also cholera and non-O1 cholera vibrios may travel rapidly from the East to non-epidemic areas, good hygienic control of sewage and water is a safeguard against the spread of infection.

Virus

In a proportion of outbreaks, a bacterial agent is not found in patients' stools nor in the food, if available. Incubation periods are longer than the usual range for bacterial food poisoning and viral infections may be suspected. Viruses were seen in 88% of faecal specimens from outbreaks involving shellfish, but in only 23% of faecal specimens from outbreaks associated with other foods. In such episodes epidemiological data are important and specimens are necessary for virological study.

In 1984, two outbreaks were associated with a Norwalk-like small round structured virus. In the first, 42 of 100 people, mainly doctors attending a meeting, were ill with acute nausea, vomiting and malaise and in some cases fever, headache, fatigue and diarrhoea, 36–48 hours after a buffet meal. The duration of illness was 1–2 days or longer and two caterers had a similar illness 3 days before the meeting; there appeared to be a small secondary wave in families 5 days after the outbreak. A variety of foods was served at the buffet, principally cold meats, salads and sweets. Twenty-two (67%) of 33 who ate the meal were ill, and only one of 11 (9%) who ate no food. Of 41 samples of faeces, 27 from affected persons, no intestinal pathogens were found. Three of five specimens collected from sick persons within 24 hours of onset of symptoms showed small round structured viruses resembling those of the Norwalk group.

In the second viral outbreak 57 (76%) of 75 people were ill with similar symptoms 27–50 hours after attending a funeral dinner. Twenty of those attending also ate a meal of remnants on the following day. For some there appeared to be a short incubation of 10–18 hours, and probably they were infected at the first meal; for others the incubation period was 34–48 hours, indicating infection on the second day. The meal of cold meats, salad and sweet was prepared by the group 2 days earlier. One person was ill the day before the meals and another 48 hours after preparing the food. Four persons who did not prepare the food were ill 1–2 hours after the dinner. The microbiological findings for 60 faecal samples and seven food items gave inconclusive results; one of 15 faecal samples examined by electron microscopy revealed the Norwalk type virus. If the outbreaks were viral in origin it is possible that a number of food items were contaminated with viral particles or that contact at the function and during preparations encouraged person-to-person spread. The effective dose was likely to be low.

An outbreak of gastroenteritis followed an anniversary supper where a wide variety of foods, including seafoods, was served from a buffet. Seventy-one (30%) of the 237 guests and hotel staff were ill with vomiting and more than 50% with diarrhoea and cramps after 3–96 hours (median 36 hours); 179 of 184 questionnaires were analysed. The food history implicated depurated oysters as the vehicle of infection. A parvo-like particle was seen in the stools of six of seven ill persons. Oyster samples were not available for examination.

Scombrotoxin

Gram-negative organisms, mainly *Proteus morganii*, growing on the scombroid fish, tuna and mackerel, are able to synthesize histamine from histidine. The toxic product (scombrotoxin) is believed to consist of histamine and saurine (see p. 38).

Scombrotoxin from tuna fish

Five people developed flushed face, dizziness, nausea and headache within 40–60 minutes after eating tuna fish sandwiches. Catering cans of tuna were used containing 840 g (1 lb 13 oz) of fish; one can filled 20 sandwiches, and five cans were used each week; left-over fish was stored in a plastic bowl until required. A sample from the suspected tuna showed a level of 435 mg histamine per 100 g; one of the highest levels recorded in this country. Tuna from unopened cans showed less than 5–7 mg histamine per 100 g of fish. The level of 7.2 mg/100 g was found in tuna from the same batch of cans. Levels of less than 5 mg/100 g of fish are considered acceptable.

7

Ecology of microorganisms in food

Microbial ecology is the study of the relationship between living organisms and their environment. As regular co-inhabitants of the world, microorganisms must not only survive, but also multiply actively so that replacements are readily available for the inevitable losses in population.

Microorganisms may be divided loosely into two groups, although there is an overlap in some instances. The harmful or pathogenic microorganisms can cause disease because they are able to live on or within humans, animals or plants and in so doing damage the host in varying degrees. They depend on the warmth and nutriment of the human or animal body for growth. If they overcome the host, causing death, they must find and invade other susceptible host tissues. If they are overcome by the immune system of the host they cannot survive, or they may reach a stage of equilibrium with the host. At an intermediate stage, foods are useful for growth when the temperature is suitable; if unsuitable, foods may be used for resting phases.

The harmless or saprophytic organisms live in the outside world; they have less difficulty in maintaining their existence than those which depend on occupation of the human or animal body. The spoilage organisms will proliferate in almost any protein or vegetable matter at a wide range of atmospheric temperatures. Some of these spoilage organisms are pathogenic for plants. Many of them, especially yeasts and moulds and bacteria are harnessed in the manufacture of cheese, butter, alcohol, antibiotics such as penicillin, and solvents for paints and oils, and they are also used to increase soil fertility.

Microbial activity prevents the accumulation of waste matter by disintegration of fats, proteins and carbohydrates, thus dead matter disappears and the earth is kept clean. Broken-down organic matter may be used by plants, which in turn may be eaten by humans and animals. This cycle is operating continuously. Ecosystems are used for the purification of sewage and water supplies.

Of nearly 2000 known kinds of bacteria, about 70 cause disease in humans and only a small proportion are responsible for high fatality rates.

Broad patterns of behaviour can be seen within the families of bacteria which cause gastroenteritis through the consumption of foods in which the organisms have survived and multiplied and sometimes produced toxins.

With regard to the pathogens and in particular those that move between man, food and the environment, each organism can be the subject of an ecological study.

Clostridium perfringens

The clostridia are sporing organisms living without oxygen. The spores enable the organism to survive readily outside the body; they are resistant to dehydration, heat and cold. *Clostridium perfringens* type A flourishes under anaerobic conditions, in the lower intestinal tract and in deep wounds. In the large bowel of man and animals the organism spores readily and the spores will pass out with the faeces into soil or sewage systems. Water and vegetation carry the spores and surviving vegetative cells back into the animal kingdom and into food production systems, including kitchens, in meats, cereals, vegetables, soil and dust. The spores vary in their degree of heat resistance; some can survive for an hour or more at 100°C (212°F) and thus can remain viable throughout cooking procedures. The heat resistance is measured as the decimal reduction time, i.e. the time for the destruction of 90% of spores at a constant temperature.

Another factor of importance is the phenomenon of heat shock or heat activation provided by cooking which encourages spores to germinate, possibly by damaging the outer coat. The anaerobic nature of the organism enables it to multiply in the crevices, rolls and internal cavities of meats and poultry, convenient nooks where the oxygen has been driven off by the heat of cooking.

When cooked meat and poultry or other foods are slowly cooling to a temperature below 50°C (122°F), the freshly germinated cells from the spores will multiply. Another characteristic of this organism is its rapid generation time; cell division occurs every 10–12 minutes at optimum growth temperatures of 43–47°C (109–117°F). A large dose of vegetative cells must be swallowed in order to survive the acid conditions of the stomach and to reach the intestine in sufficient numbers to become established. The natural flora is disturbed as the organism multiplies and sporulates. Symptoms are caused by toxin released during sporulation in the intestine; the vegetative cells break down to release spore and toxin. The organism sporulates rarely if at all in foodstuffs so that there is little or no preformed toxin in the food. In one report low concentrations of enterotoxin were detected in cultures of a non-sporulating mutant of *C. perfringens* growing in a chemically-defined medium. The outflow of spores as well as vegetative cells from those suffering from the acute diarrhoea of *C. perfringens* food poisoning will be profuse and contribute to the survival and spread of the organism. The outflow will stop only when the natural flora of the intestine is restored. Nevertheless, it is possible to find small numbers of *C. perfringens* cells and spores in nearly all stool samples; low titres of anti-enterotoxin found in the serum of many persons may be due to the continual release of small quantities of enterotoxin or to antigenically similar clostridial proteins in the faeces. Conditions such as malnutrition and malabsorption in children will encourage proliferation of the organism and diarrhoea and septicaemic conditions have been described as cause of death.

The enterotoxin can be demonstrated in extracts of faecal material from people with *C. perfringens* food poisoning. The prevention of food poisoning by allowing short intervals only between cooking and eating, and cooking, cooling and cold storage will at least limit the quantity of *C. perfringens* flowing to the outside world.

An example of the adaptation of an organism to a living environment of particular enzymic character is provided by *C. perfringens* type C. The staple ingredients of the diet of the natives of New Guinea are sweet potatoes and bananas which produce a heat-stable trypsin inhibitor. The toxin produced by the organism in the intestine cannot be destroyed because of the low level of digestive proteases (trypsin) in the intestine due to the diet. The situation is aggravated by the feasting habits of the people when there is a sudden intake of pork. There is engorgement with the slow spit-roasted meat contaminated with the vegetative cells and spores of *C. perfringens*. Thus the free proliferation of *C. perfringens* type C and production of toxins is encouraged; acute diarrhoea and necrosis give rise to the often fatal disease of enteritis necroticans reported in New Guinea and also earlier in Germany in association with canned rabbit; it has rarely been isolated or described since. It is probable that type A strains of *C. perfringens* are more adapted to the intestine under the more usual metabolic condition of normal diets. Thus type C is not excreted in sufficient numbers to be of importance in the environment.

Clostridium botulinum

Another member of the same family, *C. botulinum*, is a much more harmful organism. The toxins produced in food are neurotoxins, that is, they attack the nerves. They are likened in action to rattlesnake venom and also to strychnine. The paralytic disease can be rapidly fatal unless antitoxin is given within a few hours of eating the food containing organisms and toxin. As well as the ingestion of toxin in food, *C. botulinum* may grow and produce toxin in deep wounds although rarely so.

A disease of chicken and ducks called 'limber neck' is caused by the growth of *C. botulinum* type C in the intestine. A similar disease, usually due to types A or B, is occasionally found in human babies; the condition is usually mild and is described as the 'floppy' baby syndrome. The toxin is absorbed into the blood stream from the intestine. There are also individual cases of botulism where the cause of the toxicity is unknown.

The organism is strictly anaerobic and most cases and outbreaks are associated with the consumption of preserved or semi-preserved foods in cans or cured products where oxygen has been removed. Faults in canning, such as leaks, under-processing or over-packing, and in the concentration of curing salts are usually responsible for botulism. The spores are exceptionally heat-resistant, but the toxin is destroyed readily by heat.

The botulism from yoghurt (*C. botulinum* type B) episode drew attention to the danger of alterations, however minor, to processes or product formulations. Possible microbiological implications of any change must be considered. (See p. 115).

The concentration of *C. botulinum* in soil and food appears to be low, but it is more common in certain areas than in others. It has been found in the environment of rural dwellings, in dust and water tanks in locations where cases of 'floppy' baby syndrome have been reported; it was not found in an urban environment used as control.

C. botulinum type C is an animal pathogen and the organism survives and multiplies in dead carcasses eaten by scavenger animals and birds. It has been found in worms from soil in the area of outbreaks of 'limber neck' in poultry and also in damp feeds from hoppers. Similarly the cycle of spread for *C. botulinum* type C in fish is through dead fish and birds, coastal waters and pond mud. *C. botulinum* type E occurs in fish and in mud and shore waters, toxin levels may increase so that fish are affected. Intensive growth can occur in over-stocked fish farms where bottom surfaces are earth rather than concrete. Drought encourages proliferation in mud and wild life such as fish, ducks and seagulls. The spores survive cooking, smoking and salting, nitrite is necessary to prevent outgrowth from spores. Salted and air-dried uneviscerated white fish (Kapchunka) from the Great Lakes were responsible for botulism from type E strains in New York (1985, 1987) and California (1981). Storage during shipping and processing would have encouraged growth before salt levels were inhibitory. Other outbreaks described have involved imperfectly home preserved, non-acid foods, cheese and some canned products, for example, over-packed mushrooms.

The gastroenteritis phase of the illness is short and thought to be due to organisms other than *C. botulinum* in profusely contaminated food. Thus, the sparse distribution of the organism may be explained by the rarity of the disease compared with *C. perfringens* food poisoning. The toxin in food is lethal in very small amounts so that the organism is probably not excreted profusely in the stools of patients. Furthermore, patients may die rapidly from the effects of the toxin and it is unnecessary for the organism to be retained in the body. Thus, the period of excretion may be short which may influence the numbers and survival of organisms outside the human and animal host.

Bacillus cereus

Bacillus cereus is another organism which is able to survive indefinitely in the environment by means of spores. It is an aerobic organism not dependent on an oxygen-free environment for growth, and spores are formed freely in almost every cell under good growth conditions. The organism is commonly found in dehydrated cereals and can cause spoilage of milk and dairy products and eggs. Cornflour sauce, milk puddings and rice dishes have been described as vehicles of infection for man. The survival of the spores through the cooking process, germination, proliferation and production of toxins in food are responsible for human sickness. The organism is harmless in small numbers, and the fault lies in multiplication in food prepared in bulk ahead of requirements, for example, rice for the take-away trade.

There is acute vomiting after a short incubation period. The enzyme amylase in saliva may help to break down starch granules containing *B. cereus* so that toxin adhering to the particles is released. The body vomits out the toxin quickly, together with the organism, so that few bacilli reach the intestine and the diarrhoea is usually of minor importance. There is another manifestation of *B. cereus* food poisoning with a longer incubation period and predominantly diarrhoeal character. It is possible that cereal dust and powder are foci for the spores of many different organisms, and that *B. cereus* spores are well adapted to survive cooking, to germinate and to multiply in cereal and meat dishes; the effect of toxin production on consumers draws attention to its presence in large numbers. Short intervals between cooking and eating without storage will prevent *B. cereus* food poisoning.

Staphylococcus

It is doubtful whether the staphylococci, which do not produce spores, have an intermediate environmental phase of significance, unless it is dust, but they do not depend on faecal excretion. Their habitat is assured in the warm, damp and congenial atmosphere of the nose and throat, in the pores and hair follicles of the skin, and on the surface of skin in damp creased areas such as the perineum and axillae. Dermatitis, impetigo, pemphigus, mastitis, upper respiratory tract infections and skin infections such as boils, carbuncles, whitlows and styes comprise a variety of pathological conditions resulting from staphylococcal infection of almost any part of the body. Staphylococci from boils of the nose, hands, arms, legs and face can be transferred to food via the hands. The organism grows readily in non-acid cooked foods, ham, poultry, custards and cream confectionery. Cooking destroys the organisms but not the toxins. It is not easy to rid the skin or the nose of staphylococci of any type. Measures have been suggested, such as antibiotic creams for the nose, which may at least eliminate particular types prevalent in skin lesions or known to produce enterotoxin in food. Freedom from staphylococci may be transient, but the next invader might be less harmful. Disinfectant soaps and lotions for the skin are available, but they require persistent and consistent usage to be effective. Washing the skin with soap and water usually eliminates Gram-negative bacilli, but Gram-positive cocci often remain and sometimes appear in even greater numbers. They rise to the surface of the skin from pores when the hands are soaked, scrubbed or rinsed in hot water; superficial layers of the skin are disturbed by scrubbing and rubbing which may even serve to distribute the organisms. Staphylococci are encouraged to remain on the skin partly due to the high salt content of sweat. Common skin organisms may grow from moist soap passed over agar media.

Breaks in the skin surface will encourage proliferation, and septic lesions forming pus will release enormous numbers of cells which may linger on the skin indefinitely. Growth in food is a poor means of perpetuation, but the enterotoxins which cause illness are produced in the food and although there may be large numbers of cells in stool and vomit when illness is acute, there is no evidence that the numbers remain high. Approximately 30% of persons have small numbers of staphylococci in stools. In staphylococcal enteritis which sometimes follows administration of antibiotics disturbing the natural flora of the intestine, staphylococci may become dominant and be excreted in large numbers; but, unlike the sporing organisms, the time of survival outside the body will be limited.

Compared with the Gram-negative and Gram-positive bacilli, staphylococci are slow in competitive growth. It has been demonstrated that *Staph. aureus* producing enterotoxin type A in foods has a shorter lag phase of growth than strains producing other toxins, which may account for the predominance of A toxin in outbreaks of staphylococcal food poisoning. The short lag phase will enable the organism to compete with the faster growing Gram-negative organisms.

Although staphylococci (including *Staph. aureus*) are often found on live poultry and on the skin of dressed poultry, there is little evidence that these strains cause food poisoning. Phage typing is a useful epidemiological tool and it has enabled the spread of staphylococci to be followed more closely.

Salmonella

The ramifications of the various organisms of the salmonella group are far and wide, the ecology is complex and the damage to animals as well as to man is immense.

The salmonella cycle is important in countries where schemes for the intensive rearing of animals are predominant. The waste products and remains of animals, whether sick or healthy, are not carefully channelled into safe sewage disposal schemes or buried or burned, but are processed into feeding stuffs and fertilizers and placed back in animals and soil in the cycle of infection.

If the methods of rendering gave sterile products or at least ensured freedom from pathogens which infect both man and animals, the economy of the animal-feed-animal-food-man system would be complete. But unless special precautions are taken the finished feeds from rendered animal products are contaminated with salmonellae which are fed back to the animals. It has been demonstrated that the number of animals (or poultry) in any group found to be excreting salmonellae is associated with the level of contamination of feeding stuffs used in the unit.

The animal intestine is as conducive for the survival and multiplication of salmonellae as that of man. The animal system has the added advantages for the organism of rapid spread between animals in overcrowded conditions and of survival and multiplication in rehydrated mash in feeding troughs. In the lairage there may be trauma between animals leading to lowered resistance to infection. Anxiety and stress during transport and in strange surroundings, together with deprivation of food and water before slaughter, predispose the animals to enhanced excretion and infection.

Except under war conditions and during civil unrest, drought and famine, humanity is rarely subject to these circumstances and except for excretion into sewage, the invading organisms may live and die with the host only.

The use of antibiotics will be more effective in man than in animals since both the type of antibiotic and the dose will be controlled more carefully. However, in animals small doses in feeds or for prophylaxis will perpetuate resistant strains of pathogens and non-pathogens and appropriate mechanisms can transfer resistance factors between organisms even in the absence of antibiotics. Plasmids moving in and out of bacterial cells by means of bacteriophage are responsible for the resistance and virulence factors which are gained and lost from time to time. Doses of antibiotics for treatment of scouring animals will, more often than not, be given without knowledge of the organism concerned and therefore of its sensitivity. The administration of antibiotics to animals should be carefully controlled and limited to treatment rather than prophylaxis. Hitherto, statistical records of salmonellosis in animals have been obtained mostly from post-mortem findings from animals brought into veterinary investigation laboratories. In the Zoonoses Order, 1975 (revised 1989) the notification of salmonella excreters in farm animals is required. (See Chapter 17).

Comparative studies have been carried out between countries. In Denmark, dehydrated protein feeding stuffs of animal origin are required to be salmonella-free and there is legislation for efficient heat treatment. Pigs had a lower rate of excretion in Denmark and it is significant that *Salm. typhimurium* was the predominant serotype both in pigs and in the population. In the UK, other serotypes were prominent both in animals and in man, and many of these were found in foodstuffs for both man and animals. The Danes have a pathogen-free programme for poultry which relates not only to feedstuffs but also to all stages in breeding. The success of this system was shown by the low incidence of salmonellae in poultry carcasses and pieces of chicken imported from Denmark and examined in the UK. See Processed Animal Protein Order, 1981, amended 1989 – Chapter 17.

The effect of allowing animals for domestic consumption to eat salmonellae in their food is reflected in the incidence of the organism in raw products – particularly comminuted meats such as sausages and dressed poultry. There are wide differences in the rate of contamination of meat and poultry products between enlightened manufacturers and breeders, including those responsible for slaughtering, and others.

A partial reduction in the incidence of salmonellosis may be brought about by efforts to prevent spread of infection from raw to cooked products at the retail/consumer end of the chain; but without international cooperation between the disciplines responsible for animal care and hygiene, the intestinal disease of salmonellosis will persist and continue to rise. Irradiation with small doses of gamma rays will eliminate salmonellae from raw products and animal feeds and the adoption of such a measure is desirable. (See Chapter 17).

Experimental work carried out up to 1991 indicates that clean intact hens' eggs can be naturally contaminated with more than 100 cells of

Salm. enteritidis per egg. One group of workers strongly suspected that the albumen was the site of contamination, and that the infection could be transmitted vertically from parent birds. The organism could be isolated from the reproductive tissue and the contents of the intact ova. During the passage of the egg through the oviduct the albumen is seeded with a few cells. It was observed also that migration of the yolk towards the air sacs contaminated with *Salm. enteritidis* enhanced multiplication of the organism. Isolation from the ovarian tissue indicates that contamination of the yolk can take place, and thus multiplication can lead to high populations. Although there is much data on the behaviour of the organism within the egg, there seems to be little or no information on the origin of the organism, but it may be assumed that it came from batches of breeding stock.

The adaptation of an organism to an unusual environment was shown by the growth of *Salm. typhi* in canned corned beef. The prolific growth demonstrated under anaerobic conditions and in the presence of salt/ nitrate/nitrite curing salts was not recognized hitherto. Furthermore, the mutagenic effect of the strange environment was shown by a change in the phage type of the strain of *Salm. typhi*. It was suggested that nitratase produced by the organism helped in the breakdown of the curing salts and that a nitrogenous element was responsible for the change in phage type from 34 to A, an ancestral phage type from which all others arose.

Thus fortified in canned corned beef and other meats dispersed in the small supermarket, the organism infected hundreds of people in the well-known 1964 Aberdeen outbreak of typhoid fever. *Salm. typhi* must have poured into the sewage. Nevertheless, the secondary attack rate was minimal, which emphasizes the necessity for good sewage and water systems. Other members of the *Enterobacteriaceae*, for example, *Escherichia coli* and *Klebsiella* were unable to use the same metabolic system. Except for a little growth at the site of inoculation, they failed to proliferate and spread throughout the canned meat in the same way as *Salm. typhi* was able and shown to do.

Campylobacter

Campylobacter jejuni is of worldwide occurrence in patients with diarrhoea. The organism lives as a commensal in the intestinal tract of many warm-blooded animals; both well and sick dogs and cats may be sources of infection. *C. jejuni* is part of the normal intestinal flora of wild birds and migratory wild fowl. Thus the organism may reach the water and feed of domestic animals used for human consumption. Although untreated milk is a common vehicle of *C. jejuni* infection, the organism is seldom detected in milk samples and may not survive well in milk during transport and storage before examination. There were no isolations from 400 samples in the Netherlands, and only 0.9% and 1.5% of samples from bulk tanks in the USA were positive. As *C. jejuni* can cause mastitis in cattle, contamination of raw milk may originate from the udder as well as from faeces.

C. jejuni was isolated from three of 200 packs of fresh mushrooms; epidemiological evidence indicated that the mushrooms were vehicles of infection. *C. jejuni* is a commensal in the intestinal tract of poultry; it has been found in the caecal/faecal contents of flocks at rates of up to 100% of birds sampled and counts have shown 10^4–10^7/g of sample, hence the frequent isolation from carcasses. Numbers are reduced by scald water, but the organisms remain viable in the intestine and recontaminate the carcasses during defeathering and evisceration; isolations have been made from chicken gizzards and livers. Surveys on turkeys gave similar results. Viable cells are reduced during refrigeration, but the cells are not entirely eliminated. The serotypes isolated from poultry are mostly the same as those from human infections. Beef cattle and milk cows are known reservoirs of serotypes corresponding with those from human enteritis. Campylobacters are prevalent at the abattoir level, but the incidence decreases as the meats are distributed to retail shops. Results from red meats gave isolation rates of 5% from 1800 samples in the USA and 1% of 4933 retail samples in the UK; isolation rates from offal were 30.6%, 10.5% and 6.0% for sheep, cattle and pig samples respectively. *C. coli* was found in pork products only and was isolated from 12.5% of freshly slaughtered pig carcasses, but the organism was not found in carcasses stored overnight in the chiller.

Thermophilic campylobacters have been recognized and found to be common organisms in human, environmental and food samples. A 2-year study of their incidence indicated the wide distribution of strains; of 781 environmental samples 529 (67%), and of 2116 food samples 835 (39%) contained campylobacters. Of the food samples both poultry (56%) and offal (47%) were commonly contaminated with the organism and sewage almost always. There was a wide distribution of serotypes, but Penner 2 occurred most often. The potential sources of campylobacters in the sewage system include human faeces, industrial waste and excreta from household pets. The campylobacters are largely removed during sewage treatment such that less than 0.1% remain after full treatment; although this represents a large number discharging into the environment daily. The serotypes in water are less correlated with isolates from human faeces than those from sewage (dilution factor). The surfaces of poultry and offal are moist when sold and campylobacters are protected throughout storage from dehydration. It is thought that these products are the most likely items to transmit campylobacters to humans. Cases of campylobacter infection have occurred following the consumption of milk from milk bottles where the cap has been pecked by birds, in particular the corvids (magpies and crows). Scavenging birds have often been found to excrete campylobacters.

Aeromonas

The aeromonads are ubiquitous with a seasonal incidence higher in summer. They are water- and food-borne, and survive in fresh and salt water and also in chlorinated as well as non-chlorinated supplies; they multiply in domestic waste water drainage systems. Aeromonads are essentially aquatic organisms found in all surface waters; blue green algae and waterweed (*Elodea canendensis*) have a negative influence on their growth possibly because of the excessive changes in pH and oxygen content of such water. The aeromonads are found in wild fish, pool, cultivated edible and ornamental fish, retail fin fish and shellfish, particularly oysters.

Massive production occurs in waste water and other places where there is intense degradation of certain high molecular weight compounds for example, proteins, fats and starch. The ability of aeromonads to degrade simple compounds for use as carbon sources is limited compared with other aquatic organisms such as the pseudomonads. Utilization of high molecular weight substances like starch in a poor medium such as water (tap) is possible only when degradable low molecular weight substances, for example glucose, are present. When provided with both high and low molecular weight degradable substances in waste water there is proliferation to high numbers. Temperature is also important in competitive growth, at 15°C the aeromonads were ten times as numerous as the coliforms. Most strains of *A. hydrophila* from environmental sources do not grow at 37°C in contrast to strains from diarrhoeal cases which can grow at 41°C. Many rivers and lakes receive untreated waste water and even when treated the aeromonads may still remain at levels of 10^3–10^5/ml. It is assumed that most aeromonads in surface water originate from domestic and other waste water. Twenty five per cent of *A.hydrophila* and *A.sobria* are aerogenic, the bulk of *A. caviae* are anaerogenic. *A.hydrophila*, *A.punctata* and *A. sobria* have a surface water habitat. *A. caviae* is found in waste water, activated sludge and percolating filters. The clinical manifestations of *Aeromonas* toxaemia may vary according to the counts in fresh and salt water where people have had prolonged water contact. Infection may be asymptomatic. Many isolates from aquatic sources do not possess adhesive and virulence factors to enable them to colonize the human gut.

As *A. hydrophila* and *A. sobria* are potential pathogens in human gastroenteritis they should be considered in suspected food poisoning. In the manufacture, distribution and preparation of food, high levels should be regarded as unsafe, particularly for infants, elderly and immunocompromized persons.

The public health importance of aeromonads in drinking water is not well understood. Evaluation must take into consideration local conditions, well construction and ground water protection.

Listeria

Listeria monocytogenes occurs widely in the environment. It can be isolated from soil and sludge fertilizer and also from many different species of animal, both wild and domesticated, distributed over many continents. The different serotypes are found in both man and animals. *L. monocytogenes* is commonly found in wet situations in food factories, and the maintenance of a dry environment is necessary to discourage survival and spread; efforts are required to reduce the general contamination of the environment. Sampling and testing of plant surroundings should consider the overall microbial evaluation of maintenance and sanitation practices without due attention to pathogens and specifically *Listeria*. Hygiene measures will reduce the total load of contamination and the *Listeria* will be reduced also. Organisms in both the *Listeria* and *Yersinia* groups are able to grow slowly at low temperatures, which gives them an added advantage for survival in chilled foods stored under lowered temperature conditions. Every means used to destroy microorganisms such as chlorination of water, pasteurization of milk and other heat treatments of milk and other foods will help to reduce the overall distribution of *Listeria*. Soft cheese, implicated in outbreaks, appears to provide a convenient substrate for growth; the market for soft cheese has increased and thus the potential hazard for the increased numbers that eat it. The clinical niche for an organism must suit its metabolic requirements and it is presumed that the location of the unborn child is a suitable area/medium for the proliferation of *Listeria*. As usual for food-borne pathogens the numbers eaten must influence the speed of onset and severity of symptoms. The survival of this ubiquitous organism in the environment appears to be assured.

Yersinia enterocolitica

Yersinia enterocolitica can be isolated from a variety of living reservoirs: birds; frogs; fish; flies; fleas; snails; crabs; oysters and; a wide array of mammals. The organism is found in lakes, streams, well water, soil and vegetables. Swine provide a major reservoir of O3 and O9 types from tongue, throat, tonsils, caecal contents and faeces. The organism has been isolated from pork and ham, beef, lamb, poultry and cutting boards in butcher's shops, dairy products such as raw milk, whipped cream and ice-cream. Biogroups and serogroups of strains from human sources differ from those isolated from the environment. Water, bean sprouts growing in well water, chocolate flavoured milk and bean curd have been reported as causative agents in cases and outbreaks. *Y. enterocolitica* in pasteurized milk was thought to be post-treatment contamination. Apparent transmission to humans from cats, dogs and swine has been reported. The high incidence of infection in young children in Belgium was thought to be due to the practice of feeding raw ground pork as a weaning food. The organism grows at a pH range of 5–9 with an optimum of 7–8.

Pathogenic vibrios

The pattern of spread is likely to be similar for *Vibrio cholerae*, the non-cholera vibrios (NCV or non-O1 *V. cholerae*) and *V. parahaemolyticus*; excreters, sewage, water, sea and river foods all playing a part. Other foods may be contaminated in the kitchen directly from water, human sewage and from the raw foods by the usual routes of hands, equipment and surfaces. None of these organisms survives well in a dry environment. Where sewage systems and water hygiene are efficient, outbreaks are unlikely to occur even though the organisms may be introduced in cases and symptomless excreters. The sudden influx of vibrios in sewage from rare incidents is unlikely to pollute controlled water systems. Shellfish infected from sewage outflows may be responsible for local incidents of short duration.

Outbreaks of cholera in Italy (1973) and in Portugal (1974) were initiated by polluted water and shellfish harvested from this water. The vibrio was consumed in shellfish which harboured the organism; raw or poorly cooked cockles were vehicles in Portugal. In Portugal also, spring water both commercially bottled and consumed direct from the spring was found to be a vehicle of transmission of cholera. The cholera epidemic in South America in 1991, which began in Peru, is also thought to have originated from contaminated fish. Subsequent spread throughout Peru and into neighbouring countries was due to poor water supplies and inadequate sewage disposal.

The so-called non-cholera vibrios (non-O1 *V. cholerae*) are also found in choleraic and milder diarrhoeas: they have been isolated together with *V. parahaemolyticus* from frozen cooked shrimps and prawns harvested and processed in eastern countries. They are also isolated from stools in association with *V. cholerae*. One reported outbreak from non-O1 *V. cholerae* in airline passengers followed 6 months after a food-borne outbreak of cholera on a similar flight. The food originated from the same flight kitchen and many airline passengers were ill in both outbreaks. Nevertheless, in neither instance were there repercussions related to secondary outbreaks from carriers.

It is interesting to note the adaptation of *V. cholerae* biotype eltor to the environment and the gradual decline of the classical biotype. It is suggested that the physiological and/or biochemical characteristics of the eltor strains enable them to compete and survive better than the classical type. In laboratory experiments the lag phase of the eltor strains growing in some foods was observed to be considerably shorter than that of the classical strains in the same foods, giving a competitive advantage to the eltor strains.

The ecology of *V. parahaemolyticus* appears to be closely connected with a water to sea creature cycle. In its natural condition, *V. parahaemolyticus* seems to alternate between a free-living phase of existence in sea water and a phase in sea creatures. Outbreaks, described frequently in Japan, are associated with the consumption of fish and other seafood both raw and cooked. *V. parahaemolyticus* has also been isolated from the stools of persons with cholera-like symptoms when the cholera vibrio was not found. There are no reports of water-borne outbreaks of *V. parahaemolyticus*, in contrast with the cholera vibrio, and this fact emphasizes the role of polluted water supplies and sanitary incompetence in the rapid spread of cholera. When sewage becomes overloaded with *V. cholerae* the local water will be invaded by the organism and where there is no purification system it will reach the people. Whereas food is of only secondary importance to the cholera vibrio, *V. parahaemolyticus* appears to be dependent on food for transmission to the human body. An important factor in determining the vehicle of infection for these organisms may be the differing doses required to initiate symptoms.

In Japan, food poisoning due to *V. parahaemolyticus* and the isolation of the organism from coastal waters are summertime occurrences; warmth and bacterial growth are correlated. Although *V. parahaemolyticus* has been isolated from coastal waters and sea creatures in England, Germany and other countries, the numbers will be small compared with those which accumulate in the environment in epidemic periods in warm countries; infected sewage from sporadic cases and outbreaks will contribute large numbers of organisms. The sudden influx of vibrios in sewage from rare incidents is unlikely to pollute controlled water systems, but shellfish could be infected from sewage outflows.

There are other vibrios, for example, *V. fluvialis* and *V. vulnificans*. *V. fluvialis* has been isolated from diarrhoeal stools in Bangladesh and other parts of the world; blood in stools has been reported and enterotoxins. The organism is present in shellfish and estuarine waters and patients have reported the consumption of seafood prior to illness. The organism may be under-reported because of misidentification with A. *hydrophila*.

V. vulnificans is associated with wound infection and septicaemia which may originate from the gastrointestinal tract or traumatized epithelium surfaces. Prior consumption of raw seafood has been described. The organism has been isolated from sea water, sediment and shellfish.

Efficient water supplies will keep the threat of most water-borne diseases under control, but the use or disposal of contaminated food is another matter.

Virus

The life cycle of the virus suggests that viral particles made up of genetic material and protein capsid may lie dormant in food and water. When they enter certain living cells, the protein capsid may be removed enabling the viral genetic material to participate in cell function. Thus viral particles can survive and spread until the genetic material becomes active in living cells. Free particles and viral shapes seen in electron microscopy studies of stools may be active or passive.

Shellfish harvested from sewage-polluted water and inadequately depurated or further processed have been implicated in many incidents of viral gastroenteritis and hepatitis. Shellfish are filter feeders and viruses in the water may be concentrated in the flesh and gills and initiate infection in human cells when consumed. It is unlikely that present day methods of cleansing oysters, for example, can wholly remove the viral particles.

8

Spoilage and preservation

Living plants and animals are used as raw materials in the food industry. If freshly produced they are usually sterile inside and have a mixed microflora on the outside; the numbers and types of organisms depend on contact with the environment and the storage time and temperature. Food spoilage, and possibly food poisoning, may be expected when the organisms grow to unreasonable numbers during transportation after harvesting and during storage before and after processing. Foods may be regarded as perishable when their constituents and moisture content encourage the growth of microorganisms causing spoilage or food poisoning, and/or they are subject to oxidative changes. Non-perishable foods are unable to support the growth of microorganisms because they are too dry, too acidic or they have a high sugar or salt content. Also they are slow, or unable, to show chemical changes. Semi-perishable foods are slow to exhibit microbiological changes, for example, those protected by curing salts; oxidative changes may occur slowly. The level at which there are organoleptic changes will depend on the kind of food and the types of predominant organisms. Spoilage is generally recognized by changes in odour, colour, texture and flavour and even frank decomposition. At this stage bacterial counts will have risen to more than a million, often 10–100 or more million per gram. There comes a time when nutrients are exhausted and growth and cell division are held in balance. Although a known pathogen may not be present such profuse mixed growths may be dangerous. There will be metabolic products from growths of various organisms and the possibility of non-specific toxins; the foods are usually grossly unpalatable. There are exceptions with regard to fermented products. It has been suggested that fermentation was discovered accidentally when some fruits and vegetables and milk became acidic or alcoholic after storage at room temperature. Pleasant or desirable flavours developed and the keeping quality was enhanced. Certain organisms such as yeasts, streptococci and lactobacilli were later cultivated and used to manufacture the fermented products now in daily use, such as cheese, yoghurt, dahi, alcohol, vinegar, pickles, sausages, sauerkraut and others.

It may be assumed that before microbiological techniques permitted isolation, identification and maintenance of culture stocks of known harmless organisms suitable for fermentation purposes, accidents occurred due to the use of toxin-producing organisms, or accidental contamination with pathogens. Even today there are outbreaks of food poisoning due to enteropathogenic *Escherichia coli, Staphylococcus aureus* and even *Clostridium botulinum* growing in cheese, for example, faults in processing which allow such growth include failure to lower the pH to a level inhibitory to organisms other than those used in fermentation, and attack by bacteriophage on the starter culture used for fermentation.

Food technology includes safety factors in the production of foods of good keeping quality. Most of the methods are concerned with controlling the growth of bacteria and moulds on and in the food. Methods of preservation and fermentation have developed as fast as knowledge and the engineering of equipment have allowed. The last few generations have seen great advances in scientific principles and products.

Temperature control

Temperature control ranks first in the list of preservation methods. Sterilization in metal or foil containers, pasteurization, originally applied by Pasteur to wines to delay or prevent spoilage, and cooking methods in homes and catering establishments aim to produce wholesome and palatable food.

Use of high temperature

(1) *Sterilization*: temperatures and time can be calculated precisely to destroy not only vegetative cells but spores also, for example, canning and bottling procedures of non-acid foods.
(2) *Pasteurization*: heat processes at lower times and temperatures will destroy most vegetative cells but not spores. Spoilage is delayed, but cold storage is necessary to delay growth even if curing salts are involved, as in pasteurized canned hams. The pasteurization of milk, originally designed to destroy tubercle bacilli, eliminates the vegetative cells of other pathogenic organisms also. Since the days of paratyphoid fever and salmonella food poisoning from frozen whole and dehydrated egg products, the pasteurization of liquid egg mix has been a safeguard. The recommendation to pasteurize milk for cheese-making followed staphylococcal enterotoxin food poisoning from imperfectly processed cheese.

Organisms vary in their ability to withstand heat and cold. Some, the thermophiles, grow at temperatures above the common optimum temperatures for growth and may spoil canned food stored in hot climates and in the holds of ships. Others in the opposite range, the psychrophiles, will survive and grow at average refrigerator temperatures. Freezing limits growth, but does not kill spores or even all vegetative cells. Abnormal temperature conditions are likely to cause injury and gradual death.

(3) *Blanching*: heat may be applied to foods for reasons other than for destruction of organisms, such as for the blanching of vegetables before freezing to slow oxidative changes. Also, to loosen the shells of crustaceans. It is a quick heat-treatment.

In the kitchen, cooks have a responsibility to cook food so that it is both palatable and safe. In principle, pressure cooking, frying and roasting will destroy most spores. Cooks must ensure that delay between cooking and eating and storage conditions will not allow growth of vegetative cells from recently germinated spores. *C. perfringens* food poisoning is caused by the consumption of large numbers of vegetative cells growing in foods, particularly meats. The spores which survive cooking germinate into vegetative cells able to multiply under conditions of long slow cooling at ambient temperature, or even in the refrigerator in the centre of large masses of meat in bulk or in piled slices.

Use of low temperature

Spoilage organisms able to grow at cold temperatures are called psychrophiles; they have an optimum for growth of 15°C (59°F) or lower and the maximum is 20°C (68°F) or lower. Psychrotrophs grow at 0°C (32°F), but they do not meet the optimum and maximum requirements for psychrophiles; they are found in many genera. In chilled food held below 7°C (44.6°F) only psychrotrophs will cause spoilage. Many millions can grow in a few days, giving objectionable odours, flavours and texture. Fluctuations in temperature for storage should be avoided as the growth rate will increase as the temperature rises.

(1) *Chilling temperatures*.These are close to but above the freezing point of fresh foods, usually −10°C (14°F) to +7°C (44.6°F). As the temperature is lowered from the optimum, growth slows and eventually stops. In a mixed flora of psychrotrophs and mesophiles (moderate temperatures with optimum growth between 30°C (86°F) and 45°C (113°F) and a minimum range from 5°C (41°F)–10°C (50°F)), low temperatures exert a selective action and lead to a change in flora, notably in meat and milk.

Rapid cooling may injure mesophilic organisms. Low temperatures also cause morphological and physiological changes, such as increase in cell size, formation of filaments, mesosome deterioration and double cell wall. Slowing of enzyme action may change metabolic pathways and end products. There may be an increase in unsaturated fatty acids.

Most pathogens are mesophiles and unable to grow below 7°C (44.6°F). Staphylococcal enterotoxin has been detected at 10°C (50°F), but the production is slow with no growth below 4–5°C (39.2–41°F), and *Yersinia enterocolitica* has been isolated from vacuum packed meat stored at 1–3°C (33.8–37.4°F). The *Listeria* and *Aeromonas* can also grow at chill temperatures of 4°C; the generation time for *Listeria* at 3°C is 1–2 days. *C. botulinum* type E and the non-proteolytic strains B and F grow and produce toxin at 3.5–5°C (38.3–41°F). Indicators of faecal pollution, *E. coli* and *Streptococcus faecalis* can grow at 8–10°C (46.4–50°F). Strains of *Klebsiella, Enterobacter* and *Hafnia* are reported to grow at temperatures down to 0°C (32°F).

The correct use of refrigeration and freezing greatly reduces spoilage and health hazards (Fig. 8.1). The length of time foods may be refrigerated depends on their bacterial and fungal content before storage. Fresh raw food and freshly cooked foods with low bacterial counts can usually be stored for 3–4 days or longer before psychrophilic organisms grow to spoilage numbers. If already contaminated with large numbers of organisms foods may spoil in 1–2 days. Frozen foods should be placed in freezing cabinets as soon as possible after purchase. Foods to be frozen in the home should be freshly picked or prepared from fresh, clean ingredients, cooked or blanched, cooled rapidly and frozen.

(2) *Freezing*. Some microorganisms, such as most spores and some vegetative cells survive freezing unharmed, some survive but are damaged, others are inactivated. Parasites such as protozoa, cestodes and nematodes are more sensitive to freezing, and they are destroyed by frozen storage. High freezing temperatures (−10 to −20°C) (14 to −4°F) are more lethal than lower temperatures (−15 to −30°C) (5 to −22°F), but the food may be harmed by change of structure. Many enzymes and some toxins remain active during frozen storage so that foods may slowly deteriorate in quality. The microbial flora after

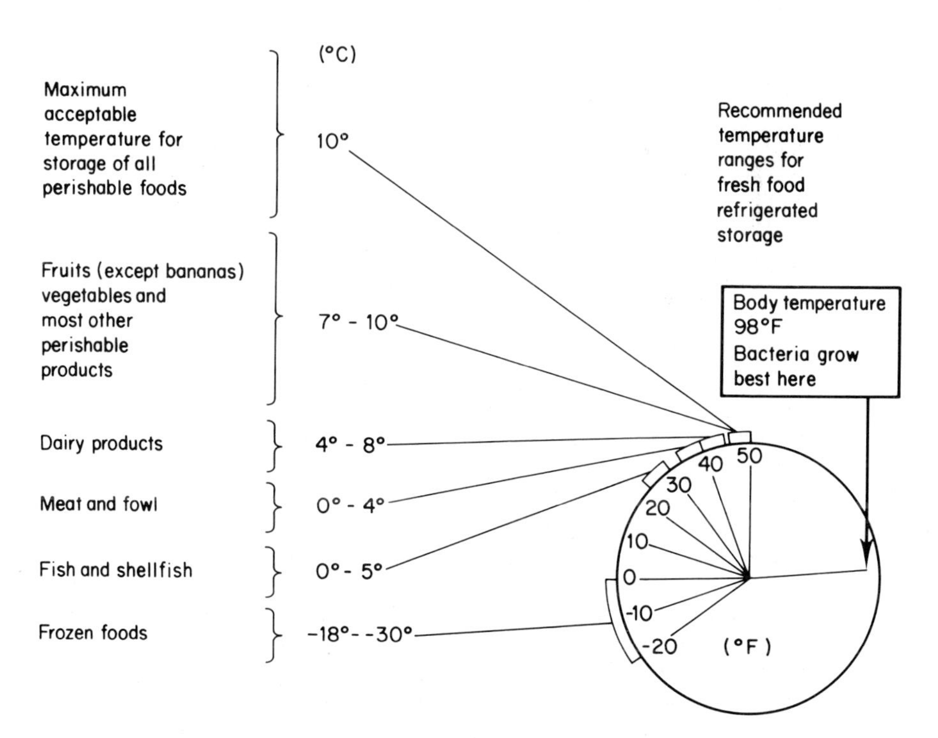

Fig. 8.1 Cold storage for foods

thawing will depend on the numbers and types of organisms in the food before freezing, although some will be destroyed. The conditions of thawing and the time/temperature range after thawing are important with regard to the growth of surviving organisms. Organisms may grow on the surface of blocks of frozen food, while the centre is still frozen; 10°C (50°F) is the most satisfactory temperature for thawing. Condensation and the accumulation of moisture in packs during thawing may encourage growth. Thawed materials spoil in the same way as unfrozen materials and they should be kept chilled.

Dehydration

Dehydration to powder reduces the moisture content to levels below which microorganisms cannot grow. Dehydrated foods, although not sterile, may be stored indefinitely because of their low water content. The bacteriological state of a product depends on the degree of contamination before dehydration. The addition of water will stimulate growth if the food is suitable. The growth and metabolism of microorganisms require water in an available form. The availability of water or water activity (a_w) in a food or solution is the ratio of the water vapour pressure of the food to that of pure water at the same temperature.

The a_w range of 0.995–0.980 is best for most bacteria and spoilage is encouraged in meat, fruit and vegetables. At a_w 0.98–0.93 spoilage by Gram-negative bacteria gives way to spoilage by certain Gram-positive organisms; below 0.93–0.85 micrococci, yeasts and moulds can grow. Below 0.85–0.60 certain fungi and yeasts predominate in spoilage and below 0.60 there is no growth.

The function of so-called preservatives, such as salt and sugar, is to reduce the a_w; heat may also remove water. Most canned, cured meats have an a_w near 0.93. The reduced a_w is important in combination with salt and nitrite to prevent outgrowth of spores.

Dehydration, or the removal of water, may be carried out by utilization of solar energy, or by tunnel and belt, fluid bed, pan, foam mat, drum or roller, spray, freeze drying and also by concentration. Concentration implies the reduction of water content in liquid foods without reaching a dry state, for example, meat extract, tomato paste, fruit juice concentrate and condensed milk. Microorganisms are more resistant to destruction by heat in dry foods than in moist foods. The action of preservatives is enhanced when microbial growth is retarded by reduced a_w.

Acidity (pH)

Acidity may be the primary factor in the preservation of fermented food, such as yoghurt, sauerkraut and pickles or together with reduced a_w, heat or chemical preservative it may be part of a combined process. Normal fermentation and acid formation may be prevented, in yoghurt for example, by an excess of sugar added to the milk. This fault was responsible for an outbreak of staphylococcal enterotoxin food poisoning in a military hospital. Yeasts may spoil yoghurt, utilizing the lactic acid and producing 'off' flavours with gas production. There are many psychrotrophic strains among yeasts and moulds able to spoil foods at low temperatures. Interference with the starter culture in cheese by bacteriophage or other contaminants may reduce acidity and give rise to hazards from pathogens and spoilage organisms. Fluorescent pseudomonads are active in spoilage, particularly of egg and meat products; they too can grow at low temperatures.

Organic acids occur widely in nature in fruits, animal tissues, leaves and spices. Also they appear as end products of fermentation. The short-chain organic acids such as acetic, benzoic, citric, propionic and sorbic acids are most commonly used as preservatives for reasons of solubility, taste and low toxicity. The antimicrobial activities of organic acids generally increase with chain length, but their use is limited because of low solubility.

Acetic acid (vinegar) and its salts are extensively used as acidulants and preservatives. Some lactic acid bacilli, *Acetobacter*, yeasts and moulds are resistant. The presence of undissociated acid (1–2%) in meat, fish or vegetable products will usually inhibit or kill many microorganisms. The level may be reduced for refrigerated products or for those with high salt or sugar content. The growth of most food-poisoning organisms and spore-forming bacteria is inhibited by 0.1% and mycotoxigenic moulds by 0.3% of the undissociated acid.

Benzoic acid (160–2000 ppm) may be used for soft drinks, fruit juices and pulps, flavourings and liquid colouring matter, liquid coffee and tea extracts and liquid rennet. It is an antimycotic agent, and most yeasts and moulds are inhibited by 0.05–0.1%, food-poisoning organisms by 0.01–0.02% undissociated acid; many spoilage organisms are more resistant.

Propionic acid may be used in bread (3000 ppm calculated on the weight of flour) and in Christmas pudding (1000 ppm). This acid and its salts are effective mould inhibitors, but at concentrations used in foods they have little effect against yeast. They inhibit many bacterial species at concentrations of 0.05–0.1% undissociated acid.

Sorbic acid (1000 ppm) is useful against mould growth in flour, confectionery, cheese, marzipan, prunes and solutions of food colouring. It is sometimes included in wrapping papers, although restricted to papers used for products in which this preservative is permitted. It has a wide spectrum of activity against some microbes including yeasts, moulds and bacteria. It is a useful inhibitor of aerobic contamination in fermented and acid foods (0.01 and 0.3% undissociated acid) depending on a number of factors and pH up to about 6.0. It has been shown to be of some value as an inhibitor of *C. botulinum* in cured meat products, but not as effective as nitrite. The action of sorbic acid is more protective than acidic.

Thiabendazole (3–10 ppm) may be used on bananas and citrus fruit; *biphenyl* may also be used on fruit.

Citric and lactic acids have moderate antimicrobial activity and inhibit microbial growth only at low pH values. A concentration of 0.001% undissociated citric acid is said to inhibit the growth of *Staph. aureus* under anaerobic conditions. Both acids inhibit the formation of the mycotoxins aflatoxin and sterigmatocystin.

β-*hydroxybenzoic acid* as the propylester inhibits the growth of food-poisoning and spore-forming bacteria, it also prevents the formation of staphylococcal enterotoxin at concentrations that do not prevent growth of the organisms (0.02–0.03% undissociated acid).

Organic acids must be of some direct or indirect benefit to the consumer with regard to nutrition and keeping quality and the amounts permitted depend on toxicological considerations.

Alkalinity

Mixtures of solid salt and alkali were used for storing eggs in an Asian process known as 'alkali eggs'. The alkali slowly penetrates the egg and the albumen becomes gelatinous and brown. The yolk solidifies and the egg resembles soft cheese. The yolks, however, are not excessively alkaline.

Use of chemicals

(1) *Cures*: salt/nitrite have been known and used for generations, starting with crude salt containing nitrate. Today sodium chloride as brine and sodium chloride together with sodium or potassium nitrite or nitrate are used as curing salts. A suggested carcinogenic hazard from nitrite in food has resulted in reductions in concentrations of nitrite added to cures. Sodium nitrite (50–200 ppm) may be used in specified types of cured meat and in a small concentration (5 ppm) in some types of cheese. It is now prohibited to add either sodium nitrate or sodium nitrite to foods specially prepared for babies or young children. Attempts to find an alternative substance effective against *C. botulinum* outgrowth from spores have not been very successful so far, although the addition of sorbic acid to low concentrations of nitrite has had some initial success.

Additives known as adjuvants are used in many cured meat products. They include ascorbates for colour and phosphates, glucono-δ-lactone and sugar for pH, texture and flavour. All these substances will alter susceptibility to bacterial growth. Sugar in high concentration has a preservative effect by withdrawing water from the food (osmosis). Curing usually changes the food to prevent spoilage or to change the type of spoilage.

The preservation and safety of a cured product are the result of interactions of many delicately-balanced factors in the system. These factors include curing salts, microbial content of the raw product, time and temperature of processing, types and numbers of organisms that survive or gain access after processing, and the conditions which determine growth, such as pH, Eh (redox potential), a_w, the decreasing nitrite concentration, packaging and time and temperature of storage; even the labelling is important and must state clearly the expiry date and the required temperature of storage.

Non-cured meat and fish spoil with offensive odours due to the growth of *Pseudomonas* and other Gram-negative psychrotolerant species. Some cured products can spoil without offensive odours, and pathogens may reach dangerous levels before obvious spoilage is evident.

(2) *Smoking* is an ancient process now used for a few products only mostly to give flavour and colour to meats and fish. Smoke contains a wide variety of organic compounds, tarry fractions and formaldehyde, also antoxides and oxides of nitrogen. Natural smoke may be absorbed directly onto surfaces. Hot smoking takes place at 60–80°C (140–176°F) and cold smoking at 25–30°C (77–86°F). There is a preservative effect throughout small sausages and on the surface of large pieces of meat, but not internally.

Compounds of smoke may be dissolved in water and the liquid smoke used as a spray or dip; there is little or no antibacterial activity. Smoking is generally effective against Gram-negative rods, micrococci and staphylococci. Smoke and spice may increase the period at risk for spoilage by masking odours.

Gases

Gases may be used to inhibit spoilage organisms and thus to prolong storage life. Carbon dioxide, ethylene oxide, propylene oxide, sulphur dioxide and ozone, are all used to kill or inhibit Gram-negative bacteria, moulds and yeasts. The ethylene and propylene oxides and ozone have some effect on bacterial spores. Safety precautions are necessary in factories where they are used.

Carbon dioxide (CO_2) at high concentrations and low temperatures is useful against a wide range of Gram-negative spoilage organisms. There will be little effect against pathogens except perhaps staphylococci because of low temperature storage. Refrigerated fresh meats vacuum packed in gas-impermeable plastic film keep much longer than meat stored in air. There is a rapid increase in CO_2 in the gas phase of the package to 10–20% in 4 hours to a maximum of 30%; there is a reduction in oxygen content to 1–3% due to enzymic activity in the meat. For atmospheric storage the best concentration is about 20%; an atmosphere of 35–75% CO_2, 21–28% oxygen and the remainder nitrogen has been used for transportation. CO_2 inhibits moulds and yeasts in refrigerated fruits and vegetables, and it helps to control ripening in fruits. The concentration used varies according to the product, for example, 5–10% for apples. The gas has an inhibitory effect on microbial growth, it also helps to maintain the physiological health of plant tissue. Nevertheless, CO_2 damage may produce 'off' flavours, discolouration and tissue breakdown. Solid CO_2 is used as a refrigerant in the storage and transport of unfrozen eggs, meat and poultry, and of frozen food such as ice-cream. The gas from dry ice helps to prevent the growth of psychrophilic spoilage organisms. CO_2 gas allowed to accumulate naturally or added directly to a contained environment extends the storage life of millions of tons of respiring fresh plant and animal produce.

The carbonation of soda and fruit drinks and of mineral waters to levels of 3–5 atmospheres of CO_2, kills or inhibits the growth of spoilage and pathogenic bacteria. The higher the CO_2 pressure and the lower the sugar content the faster the death rate. There is thought to be a combined effect of high CO_2 pressure and low pH.

Sulphur dioxide (SO_2) (50–2500 ppm) is applied to foods and beverages as the liquefied gas or as the salts, sulphite, bisulphite, or metabisulphite. It is used extensively to control spoilage organisms, mostly moulds and yeasts, in soft fruits, fruit juices and pulps, syrups, desiccated coconut, jam, fruit yoghurt, wines, sausages and hamburgers, fresh shrimp, acid pickles and in the starch extraction process. In fruit juice and the starch process it prevents fermentation, in grape juice and wines it inhibits moulds, bacteria and undesirable yeasts, although it has no effect on wine yeasts. As sulphite or bisulphite in sausages it delays growth and it controls 'black spot' in shrimp. SO_2 is also used in many foods as an antoxidant or reducing agent to inhibit enzymic and non-enzymic action or browning. Its use is subject to statutory regulations and it is not permitted in food sources of thiamine (vitamin B_1).

Sulphur dioxide and the salts produce a pH-dependent equilibrium in solution. As the pH falls, sulphite ions decrease and the proportion of SO_2 increases at the expense of the bisulphite ions. The death rate increases with decrease in pH, with greatest effect at pH less than 4.0.

Ethylene oxide has been used in the past to reduce microbial contamination and to kill insects in dried foods, such as gums, spices, dried fruits, cereals and potato flour. Regulations were introduced because of the toxic products of hydrolysis, ethylene glycol and ethylene chlorhydrin. Now ethylene oxide is permitted only for whole or ground spices without salt. The residue of the gas in the product must not exceed 50 ppm. It is unlikely that protein foods would be treated with ethylene oxide because certain vitamins and amino acids are destroyed.

Propylene oxide is similar to ethylene oxide, but less active biologically and less volatile; about double the concentration is needed compared with ethylene oxide. Bacteria including streptococci and staphylococci are more resistant than yeasts and moulds, and bacterial spores more resistant than the vegetative cells. Both ethylene oxide and propylene oxide are thought to have the same mechanism for killing cells (alkylating agents). The breakdown product, propylene glycol is non-toxic. In the USA propylene oxide is used as a fumigant for bulks of cocoa, gum, processed spices and nutmeats (except peanuts) and starch when the foods require processing. One treatment is allowed at 51.7°C (125.1°F) or below and the residue limit must not be more than 30 ppm.

Ozone has been used mainly for the treatment of water. It decomposes rapidly in the water phase of foods so that antimicrobial action takes place mostly at the surface. Ozone is a powerful oxidizing agent and gives rise to rancid flavours from fats, lowers pH, coagulates protein and inactivates enzymes. It is produced by ultraviolet (UV) lamps and is partly responsible for the killing effect of UV light in cold storage rooms. Bacteria are more susceptible than yeasts and moulds, but bacterial spores are resistant. Gram-positive cocci are more sensitive than Gram-negative bacteria and less so than Gram-positive bacilli. The threshold for killing and the rate depends on the amount of organic matter. In clean water less than 10 ppm are bactericidal, much larger concentrations – greater than 100 ppm – are required to repress spoilage organisms from food surfaces.

Ozone may be used as a maturing agent in ciders and wine and to disinfect the interiors of soft drink and mineral water bottles before filling. It has been described as a means to preserve eggs at high relative humidity, and to destroy botulinum toxin in raw sewage.

Oxidation-reduction potential (redox potential) Eh

The changes in value between oxidizing and reducing agents determine the Eh of a bacterial culture. The lowering of the redox potential in a food is attributed to the liberation of gaseous hydrogen and reducing metabolites by food, enzymes and growing organisms. Redox potential is not recognized as a processing parameter, but there will be interactions between pH and gaseous atmospheres to determine spoilage microflora; it may influence the development of pH, pleasant and unpleasant flavours. The ability of organisms to alter Eh in a food and to tolerate changes is important in microbial competition. The strict aerobes, such as members of the *Bacillus* group, micrococci, pseudomonads and other Gram-negative organisms are active at the surface of food; where oxygen is readily available, they cause spoilage such as slime on meat and ropy bread. The facultative organisms can compete in a wide range of foods and are generally responsible for spoilage of food of low Eh.

The strict obligate anaerobes grow as contaminants in the inner part of non-processed foods and together with facultative anaerobes are the main spoilage flora. Many anaerobes cannot grow below about 10°C (50°F), so that rapid chilling is a deterrent. Many clostridia are able to grow in the presence of oxygen and at high potentials in the absence of oxygen. The non-sporing anaerobes such as the *Bacteroides* cannot tolerate high redox potentials. Environmental Eh is an important determinant of microbial growth; it acts as a selective agent and influences the metabolic products of microbes.

Irradiation

Ultra violet light has limiting penetration, thus it is of value for thin layers of fluid only. It is used, for example, in the purification of water for the depuration of shellfish, and for the surfaces of food to prevent spoilage by moulds. Clean surfaces given adequate exposure can be disinfected. UV light has been used for the decontamination of air in filling and sterility testing and inoculating areas of microbiology laboratories and in bakeries where it helps to control the occurrence of mould spores on bread and custard pies, for example.

UV radiation is unsuitable for certain foods containing fat since it accelerates lipid oxidation and rancidity, for example, in butter. Also it produces spots of discolouration on the leaves of green vegetables.

Ionizing irradiation with gamma rays has advantages over other methods used to destroy bacteria in food and it can be used for the preservation of food. It has a high energy content, great penetration and lethality due to action at the cellular level. The penetration is instantaneous, uniform and deep. At low levels there are no organoleptic changes in the food product, and even at high levels chemical changes are small. Foods may be processed in any state, frozen or liquid and in their packages, whatever they may be.

Process control is precise and more effective than for heat processing, and regulatory control is likely to be more strict.

The few disadvantages include the continued activity of enzymes in irradiated food during storage. Some chemical changes may take place, such as rancidity in sensitive foods. Strict safety measures are necessary to protect operators and food handlers.

There are three levels of ionizing irradiation according to the type of food and the various spoilage organisms:

(1) *Radurization* aims to prolong storage life. It is used for the low acidity foods spoiled by bacteria and for the acid or dry foods spoiled by fungi. The nature of the package and the temperature of storage influence the dosage required from 1–5 KGy. Typical spoilage organisms, such as *Pseudomonas* and *Acinetobacter-Moraxella* will be destroyed by the lower dosage. At 5 KGy the more resistant spoilage organisms such as the faecal streptococci and yeasts will survive; spore formers will remain. Radurization extends the shelf life of cereals, bread, fruit and vegetables commonly spoiled by fungi. Liquids such as fruit juice may require light heat treatment in addition to irradiation.

(2) *Radicidation* is irradiation with small, but higher than radurization, doses of gamma rays. It is recommended for the destruction of salmonellae in foods such as uncooked chicken, meat, animal feeding meals (bone, fish, meat and cereal), frozen and dried egg products. Either 2.5 KGy or more will reduce the content of vegetative cells of bacterial pathogens to negligible levels; thus salmonellae can be destroyed in meat and chicken and vibrios in fish. Twice the dose is needed for frozen and dried food. Four KGy can reduce the numbers of conidia of *Aspergillus*. The use of gamma irradiation for this purpose has been pronounced safe, while changes in taste and smell are barely detected. It is hoped that the method will come into common use and help to reduce considerably the distribution of salmonellae to animals and man in feeds and foodstuffs.

There are many advantages for the use of radicidation for foods of normal hygienic quality, unspoiled and prepared with good quality raw materials; no procedure is adequate when raw materials are heavily contaminated. Spoilage after irradiation may take unusual forms such as souring rather than putrefaction, but there is a reduced public health risk as the organisms likely to cause food poisoning, the salmonellae for example and also *Listeria*, will be destroyed. The use of radicidation for foods known to have a high rate of contamination with salmonellae, such as uncooked poultry and some meats, has been recommended for some years.

(3) *Radappertization* is the name given to the use of high-dose treatments to destroy or render inactive resistant spores. The spores of *C. botulinum* are the most radiation resistant so that sterilization doses of gamma rays must be used, about 45 KGy with strict process control, or lower doses of the order of 20 KGy combined with acidity or curing salts; even so such high dosage would damage many fruit products. The spoilage organisms of concern in acid foods are *Bacillus coagulans, B. thermoacidurans* and yeasts. Cured foods such as canned ham are safe from *C. botulinum* and stable with irradiation doses of 25 KGy. As with heat processing, cells are damaged and unable to grow in the presence of sodium chloride (2–4.5%) and sodium nitrite (about 50–150 mg/Kg.) The amount of nitrate normally required (200 ppm) can be reduced also.

Antibiotics

Antibiotics used in food processing have a limited value because resistant spoilage organisms as well as pathogens, such as *Salm. typhimurium*, may be selected. Experiments showed that when spoilage organisms were repressed by chlortetracycline, resistant *Salm. typhimurium* grew more readily on dressed broiler chickens fed with the organism during life. The proportion of antibiotic resistant strains of *Salm. typhimurium* rose sharply when antibiotics were added to animal feeds in order to increase growth rate, and also for prophylactic measures as well as for treatment of animals; resistant organisms could monopolize the gut flora.

Recommendations to limit the use of certain antibiotics, except in treatment, to those not normally employed for treating man are given in the *Report of the Joint Committee on the use of Antibiotics in Animal Husbandry and Veterinary Medicine*, (Swann) 1969. They were aimed to reduce the spread of antibiotic-resistant strains of salmonellae. The three antibiotics still in use are natamycin, nisin and tylosin. None of these is of value therapeutically. Natamycin is an antifungal agent; it is used in some countries to protect raw ground peanuts, the surface of sausages and the rinds of soft and hard cheese against mould growth. If mycelium has developed, for example *Aspergillus flavus*, enzymes can inactivate natamycin. Nisin is a naturally occurring antibiotic produced by strains of *Strep. lactis*. It affects Gram-positive bacteria only and may be used to prevent 'blowing' of cans and to preserve cheese, chocolate, milk and clotted cream. Also it is used in the preservation of green peas, beans, sauces and ketchup. Tylosin is sometimes used as a feedstuff additive.

Packaging

Packaging is designed to preserve the quality of the food. Its various functions include protection against water vapour, oxygen and other gases, light, dust and other dirt, weight loss, mechanical damage and to prevent the entry of microorganisms and insects. It may affect the mode and rate of spoilage and the survival and growth of pathogens. Care must be taken that the contents do not damage the pack. Packaging material, if not sterile, should introduce few organisms. In the USA, paper for food containers should have counts of not more than 250 organisms per gram, containers and seals for milk not more than $1/cm^2$, plastic foils and tubes for packaging, less than $10/1000\ cm^2$ and plastic beakers even less. The UK recommendation for counts on milk bottles is less than 600 per bottle. Containers include, cans, paper, cardboard, glass and plastic (rigid) and plastics and foil (flexible).

The most important factor is the permeability of the packaging material to oxygen, carbon dioxide and water vapour, particularly if preservative gases are used for the pack and for perishable products such as meat, poultry and fish. Film may be highly permeable so that conditions in the pack are similar to those in the unpacked product, and permit the growth of pseudomonads which are inhibited in gas-impermeable film. Polythene may allow the passage of moisture, but not oxygen, and surface spoilage occurs. Within a pack the usual factors govern the growth and activity of microorganisms such as the food medium, temperature, a_w, pH, gases and competitive flora. In gas-impermeable, hermetically-sealed evacuated or unevacuated packages, oxygen is used up and CO_2 forms, the pH drops and lactic acid organisms become active; the typical aerobic spoilage flora is retarded and altered so that shelf life is increased by 50%. *Enterobacteriaceae* including salmonellae survive and sometimes increase on the surface of packaged fresh meats even when large numbers of lactic bacteria are present. Anaerobes rarely cause trouble in packaged fresh meat even under vacuum. In cooked or lightly cured meat with few competitors *C. perfringens* and *C. botulinum* can grow even in the presence of gaseous oxygen. Botulinum toxin has been demonstrated in cooked products. CO_2 alone or mixed with nitrogen may be used to replace the atmosphere in hermetically sealed packages. A mixture of CO_2 and air may increase time to spoilage.

Aseptic filling is used to fill sterile food into sterile containers, for example UHT milk, so that shelf life is increased. All packaging requires careful control of the process of production.

Details about preservatives including antibiotics are contained in the Preservatives and Food Regulations, 1979 as amended 1980 and 1982 (SI 1979, No. 752, SI 1980, No. 931, SI 1982, No. 15).

The distinction between the terms 'spoilage' and 'pathogenic' for bacteria is not always clear, some organisms fall into both classes. In general food-poisoning organisms give no indication of their presence in food by the common spoilage signs, such as off-odours and tastes, or changes in appearance, for example, colour, consistency or gas production. A manifestation such as gas production by *C. perfringens* may come too late to prevent food poisoning, when the organism has already reached numbers initially required to invade the intestine where they sporulate and release enterotoxin.

B. cereus is common in the environment and in uncooked cereals such as rice and flour. Counts of *B. cereus* in cooked rice and other foods which have caused food poisoning rise to greater than $10^6/g$, often 10^8–$10^9/g$, without visual or flavour evidence.

B. cereus is a common cause of spoilage in cream and milk, shown by the appearance of particles, 'bitty' cream or milk. The counts at this stage are said to be too low to cause food poisoning through toxin production and the products are unlikely to be consumed because of their appearance; occasional milk-borne outbreaks occur.

E. coli strains can be enterovirulent and produce gas and acid, thus they may be active in both roles, pathogenicity and spoilage. Yet there is little published evidence to indicate that foods are frequent vehicles for the transmission of infection. There are probably two explanations for the lack of information, firstly the difficulty of diagnosing the disease, and secondly the difficulty in detection, enumeration and proof of pathogenicity. Nevertheless, there have been outbreaks of *E. coli* gastroenteritis from canned salmon, camembert cheese and cream, and also of the enterohaemorrhagic *E. coli* O157:H7 from ground beef in fast-food establishments. Coliforms including *E. coli* are often present in cheeses, sometimes in large numbers. In the making of camembert cheese with impaired starter activity populations of *E. coli* reached 10^9 in 24 hours and then declined. Semi-soft surface-ripened cheeses are susceptible to abnormal fermentations because of slow acid development or lack of acidity. Thus they are considered to be a potential hazard from *E. coli*. Adequate heat treatment of milk, active starter cultures and strict hygiene measures are recommended. Temperatures should be as low as possible, 10–11°C (50–52°F) during the development of the surface mould matt.

Medical and science workers interested in food microbiology should be familiar with the organisms responsible both for food poisoning and food spoilage, and also with their metabolic activities.

Some of the factors leading to spoilage include:-

(1) Storage under conditions of ambient temperature, which will encourage growth of bacteria and moulds.
(2) Storage of moist food in cellophane or plastic containers without means to eliminate moisture. Vegetables, fruit and meat should be removed as quickly as possible from synthetic temporary packaging materials which are closed and do not allow exchange of air.
(3) Storage of fruit and vegetables wet after washing, at ambient temperature.
(4) Washing of eggs not required for immediate consumption.
(5) Storage of meat, cheese and other foods in hermetically sealed synthetic containers, particularly if the meat is not properly cooked.

Damp, dark conditions of storage encourage spoilage and rot, particularly inside synthetic wrappings.

Reference should be made to the ICMSF series of books *Microorganisms in Food 3* (1980). *Microbial Ecology of Foods*: Volume I, *Factors affecting Life and Death of Microorganisms*. Factors affecting spoilage organisms and pathogens, control measures and interrelationships. Volume II, *Food Commodities* (2nd edn in preparation) describes the microbiology of various commodities in detail. The properties of the foods affecting the microbial content, the initial microbial flora, the effects of harvesting, transport, the process and storage on the microbial content and means of control. Typical spoilage associated with the foods is described together with the application of preservatives and other methods of control to the foods.

Those interested in the ecology of organisms in food and the nature of spoilage and types of spoilage organisms, preservation and safety measures should read both books; they provide excellent reference material.

Part II

Food hygiene in the prevention of food poisoning

9

Factors contributing to outbreaks of food poisoning

The previous chapters have described the bacterial causes of food poisoning, characteristics of the agents responsible, the sources or reservoirs of these germs and the conditions which encourage them to grow and reach numbers that are dangerous to the health of those who eat them.

Food hygiene may be defined as the sanitary science which aims to produce food that is safe for the consumer and of good keeping quality. It covers a wide field and includes the rearing, feeding, marketing and slaughter of animals as well as the sanitation procedures designed to prevent bacteria of human origin reaching foodstuffs. The veterinary and manufacturing aspects are beyond the scope of this book and although mentioned they are not considered in detail. With an acceptance of the fact that raw materials may contain food-poisoning organisms, what can the food handler do to prevent food poisoning?

From the statistics on outbreaks and cases of food poisoning (Chapters 3 and 4) it can be seen that rather than abating, food-borne illness has increased. Improvement in methods of food preparation and education of those responsible for the provision of food, particularly in mass catering situations, would undoubtedly reduce the incidence of food poisoning. To accomplish this it is essential to know not only the food vehicles of infection, bacterial agents, the places where the incidents occurred and where food was prepared, but also factors which have contributed to the incidents.

Various characteristics of the food itself may allow or inhibit bacterial multiplication. These include factors such as pH, water activity (a_w), redox potential (Eh) and level of organisms competing with or inhibiting the growth of potential pathogens.

Similarly, certain characteristics of food-poisoning bacteria, such as production of heat-resistant spores, ability to grow at relatively high or low temperatures and tolerance of high salt or sugar levels also contribute to incidents of food poisoning and must be taken into consideration when attempting to improve food-handling practices.

It is mentioned in earlier chapters that for persons to succumb to bacterial food poisoning the dose of organisms in a food must usually increase from an initially harmless level to a harmful one – millions per gram or more. For multiplication to take place there must be sufficient nutrients, moisture and warmth together with a time lapse between preparation and consumption of the food. Many malpractices take place in food preparation which permit contamination with, survival and growth of food-poisoning bacteria. A study of these faults or factors will produce data which can be used to educate those involved in the production of food at all levels.

A summary of the results of a study made of almost 1500 general and family outbreaks of food poisoning which occurred in England and Wales between 1970 and 1982 is given in Table 9.1. Most of these outbreaks are of bacterial origin but some episodes of scombrotoxic fish poisoning, red kidney bean poisoning and viral gastroenteritis are included. In some outbreaks as many as five or six contributory factors may be noted, whereas in others only one may be recorded; in such instances, it is likely that other factors were involved, but were overlooked by those responsible for the investigation. The 1479 outbreaks represent only about 20% of the incidents reported in that period. The figure is low because epidemiological data were lacking in many reports of outbreaks, particularly for salmonella food poisoning. As usual meat and poultry were responsible for about three-quarters of the incidents, raw milk was associated with salmonellosis and rice with *Bacillus cereus* food poisoning. In the late 1980s eggs, either eaten raw or in dishes incorporating uncooked or lightly cooked hens' eggs figure more prominently as a vehicle of *Salmonella* infection.

The largest proportion (58%) of incidents studied (851 of 1479) occurred from food prepared in restaurants, hotels, clubs, hospitals, institutions, schools and canteens; 15% only were associated with food prepared in family homes – mainly for home consumption, but occasionally for external catering. Although some 60% of food-poisoning incidents occur in the home, only a small proportion were included in the study due to the incomplete epidemiological data for such incidents. Situations where food is prepared in quantity for a large number of people are most likely to give rise to most food poisoning.

Table 9.1 shows that for all types of food poisoning the factors which are recorded as most commonly contributing to outbreaks include:

(i) preparation of food more than half a day in advance of needs;
(ii) storage at ambient temperature;
(iii) inadequate cooling;
(iv) inadequate reheating;
(v) use of contaminated processed food (cooked meats and poultry, pies and take-away meals prepared in premises other than those in which the food was consumed);
(vi) undercooking;
(vii) cross-contamination from raw to cooked food. This appears ninth on the table because, although important, it tends to be overlooked.

The study was carried out some time ago, but there is no evidence to suggest that the factors contributing to food-poisoning incidents have changed markedly. The factors associated with *Salm. enteritidis* and shell

Table 9.1 Factors contributing to 1479 outbreaks of food poisoning, England and Wales, 1970–82

	Number of outbreaks in which factors recorded (%)					
Contributing factors	*Salmonella*	*C. perfringens*	*Staph. aureus*	*B. cereus*	Other	Total
(i) Preparation too far in advance	240 (42)	464 (88)	80 (48)	54 (86)	6 (4)	844 (57)
(ii) Storage at ambient temperature	172 (30)	276 (53)	75 (45)	39 (62)	4 (3)	566 (38)
(iii) Inadequate cooling	125 (22)	313 (60)	12 (7)	17 (27)	1 (<1)	468 (30)
(iv) Inadequate reheating	76 (13)	275 (52)	5 (3)	33 (52)	2 (1)	391 (26)
(v) Contaminated processed food	100 (19)	19 (4)	27 (16)	4 (6)	86 (54)	246 (17)
(vi) Undercooking	139 (25)	74 (14)	2 (1)	1 (2)	7 (4)	223 (15)
(vii) Contaminated canned food	2 (<1)	4 (<1)	42 (25)	1 (2)	55 (35)	104 (7)
(viii) Inadequate thawing	61 (11)	34 (6)				95 (6)
(ix) Cross-contamination	84 (15)	8 (2)	2 (1)			94 (6)
(x) Raw food consumed	84 (15)		1 (<1)		8 (5)	93 (6)
(xi) Improper warm holding	15 (3)	52 (10)		8 (13)	2 (1)	77 (5)
(xii) Infected food handlers	13 (2)		50 (30)		2 (1)	65 (4)
(xiii) Use of leftovers	25 (4)	25 (5)	11 (7)	1 (2)		62 (4)
(xiv) Extra large quantities prepared	29 (5)	17 (3)	2 (1)			48 (3)
Total	566	525	166	63	159	1479

eggs would have increased the number of outbreaks associated with raw food, but they would also have figured under temperature control and cross contamination.

The first two factors are involved in outbreaks associated with all the main bacterial agents of food poisoning, but with other factors some association can be seen with particular organisms. In outbreaks due to salmonellae, undercooking, inadequate cooling and the use of contaminated processed food are the next most commonly reported factors. Consumption of raw foods such as shell eggs would also be associated mainly with salmonellae. The hazard of cross-contamination from raw to cooked food is a basic fault. In outbreaks due to *Clostridium perfringens* multiple factors are frequently recorded with inadequate cooling and reheating playing important roles. Food handlers and contaminated processed and canned food are the chief factors in staphylococcal food poisoning. Factors contributing to *B. cereus* food poisoning are similar to those for *C. perfringens*.

Preparation in advance and storage at ambient temperature

In the domestic kitchen it is fairly easy to time preparation and cooking so that a meal can be served hot. The timing in mass catering is not as straightforward. Large numbers of people may require to be fed in a short space of time or a meal service may be required outside the normal working day, for example, with shift workers. Thus it is often necessary to prepare food hours before it is needed and to hold the food, under refrigeration, in a hot-holding apparatus, or even at atmospheric temperature. If the procedures are strictly controlled and the storage temperatures are at levels which will not permit bacterial growth, then the hazards are reduced. Food to be held hot should be placed in a hot-holding apparatus already at a temperature of at least 62.8°C (145°F) and maintained at that temperature until required. Foods left for long periods at ambient temperature, or placed in hot-holding equipment not preheated or set at too low a temperature, or 'reheated' by the addition of hot gravy or sauce, are most likely to be involved in outbreaks of food poisoning. The legislation relating to holding temperatures for certain food groups has been updated (see Chapter 17). A two tier system has been introduced. A wide range of high-risk foods must be kept at the revised lower storage temperature of 8°C, and in 1993 foods of greatest risk will be held at 5°C. The upper storage temperature has not changed.

Cooking and cooling

When foods are undercooked or inadequately thawed, particularly large joints of meat or poultry, and especially large frozen turkeys, with cooking times too short and temperatures too low, salmonellae and other organisms may survive. Subsequent poor storage will permit multiplication.

Poultry is particularly troublesome in relation to both salmonella and campylobacter food poisoning. Many oven-ready chickens and turkeys enter the kitchen aready contaminated with both organisms. The procedures carried out during the preparation of poultry for the table offer many opportunities for the spread of contamination and for the survival and multiplication of the organisms. Thus in poultry-associated outbreaks, factors such as inadequate thawing, undercooking and cross-contamination are important; the larger the birds the greater the hazards. Undercooking or very light cooking have been important factors in the large number of outbreaks reported in the late 1980s associated with dishes prepared from fresh shell eggs such as meringue, souffle, quiche and various sauces. Temperatures achieved during the short or low temperature cooking have been insufficient to eliminate the salmonellae present inside the egg.

Heat-resistant spores of both *C. perfringens* and *B. cereus* and other *Bacillus* species survive most conventional cooking procedures. The organisms are widespread in nature and *C. perfringens* in particular in the stools of man and animals. Cooks should be aware that these organisms are present in most raw foods and may remain alive after cooking. Unless precautions are taken to cool meat and poultry and cereal dishes quickly and to cold store those not to be eaten freshly cooked, outbreaks of *C. perfringens* and *Bacillus* food poisoning will continue.

Contaminated processed food

With the increase in the number and variety of premises distributing food in a cooked and ready-to-serve state, there has been a rise in the number of incidents of food poisoning associated with cooked meats and poultry, pies and take-away meals. The purchaser of the food has no control over the preparation of the dish and may purchase food already contaminated with organisms such as *Salmonella, Staphylococcus* and *Listeria*. It is often impossible to determine the time and cause of the contamination, but at some stage during production there must have been faults due to undercooking, poor storage and/or cross-contamination. Further storage at the wrong temperature by the purchaser will enhance bacterial multiplication to levels resulting in food poisoning.

Contaminated canned foods

The number of incidents of food poisoning reported from canned food is small in relation to the large amount of canned food consumed. Under-processed cans or cans with pinhole leaks or poorly made seams often show visible signs of spoilage and would not be consumed. Nevertheless, in some the food may look and taste normal. Cans may be contaminated after processing and during cooling. Many of the incidents of staphylococcal food poisoning reported come from both freshly opened cans, indicating post-processing contamination, and from cans which have been opened and stored, when contamination probably comes from hands when opening or after opening the can. There have been two well-documented outbreaks of botulism in recent years from canned food in the UK. One in 1978 involved canned salmon, which was recontaminated after the canning process when water containing *C. botulinum* type E was sucked in through a break in the can during cooling. The other, in 1989, resulted from the use of an underprocessed nut purée in a batch of hazelnut yoghurt. The heat treatment applied during the canning process for the purée was insufficient to kill the spores of *C. botulinum* type B naturally occurring on the hazelnuts.

A number of incidents of scombrotoxic fish poisoning have been attributed to canned fish (mackerel, tuna, pilchards). Here the raw material already contained the scombrotoxin due to poor pre-processing storage; the toxin is not inactivated by the canning process.

Thawing of frozen foods

Some distributors recommend cooking frozen food from the frozen state. This may be safe for small items such as chops and steaks where the depth of the food is shallow and heat penetration will be rapid. It is important that large joints of meat and poultry are thawed out before cooking. Foods cooked from the frozen state may still be cold in the centre when the outside looks cooked. Bacteria in the centre of such food can survive and subsequently multiply if stored at warm temperatures. Many incidents of salmonella and *C. perfringens* food poisoning have occurred following the consumption of meat and poultry, liver and offal inadequately thawed prior to cooking.

Cross-contamination

If raw and cooked foods are prepared on the same surfaces using the same equipment and by the same food personnel, or if they are stored in close proximity, organisms may spread from raw ingredients to foods which will receive no further heat treatment before consumption. Separate surfaces, equipment and personnel for raw and cooked foods, regular hand washing, particularly after handling raw foods, and good cleaning schedules regularly enforced are essential to reduce cross-contamination from raw to cooked foods.

Raw food

Outbreaks of food poisoning following the consumption of uncooked foods are occasionally reported. In some areas untreated milk, previously (see Chapter 17) distributed by milk roundsmen, has been implicated in both *Salmonella* and *Campylobacter* enteritis. Raw oysters have led to incidents of both viral gastroenteritis and hepatitis. The red kidney bean causes illness only if eaten raw or undercooked; the toxic component is inactivated by thorough boiling. Lightly cooked dishes, especially those with eggs, have also caused illness, the cooking temperatures being sufficient to coagulate the protein, but not to kill the bacteria.

Improper warm holding

Warm holding equipment such as hot cabinets and *bains marie*, if held at the correct temperature, should prevent bacterial multiplication, but they may serve as incubators and even encourage growth. Such equipment must be preheated and used to hold only hot, cooked food, it should never be used to warm food from the cold unless a reheating cycle is built into the mechanism.

Food handlers

Food handlers as carriers do not play a significant role except in *Staphylococcus aureus* food poisoning. This organism is frequently carried in the nose, less commonly on the skin and – most important of all – in septic lesions on the hands and other parts of the body. In many incidents the same phage and enterotoxin-producing types of *Staph. aureus* are isolated from both food handler and implicated food. Foods which act as vehicles of intoxication are usually those which have received much handling during preparation and which have been subjected to poor storage conditions, such as cold meats and desserts which are consumed without further heat treatment. In a small proportion of salmonella outbreaks food handlers excreting salmonellae may have been the source of the agent; more often, the food handlers are victims, not sources, becoming infected either from frequent contact with contaminated raw food, from tasting during preparation or from eating left-over contaminated food. After touching raw food their hands can contaminate cooked foods and also implements or surfaces used for preparation. Where food handlers as excreters have been implicated in outbreaks they have usually just returned from visits overseas or they may continue to prepare food while actively suffering from symptoms of gastroenteritis.

Leftover food

A number of outbreaks have been traced to the consumption of food left over from meals served earlier the same day or even a day or more previously. When left-over foods are correctly stored and eaten cold from the refrigerator or thoroughly reheated there is no danger, but they are often left at ambient temperature and served with minimal or no further heat treatment.

Extra large quantities of food

The preparation of food for large numbers of people is difficult when cooking and cooling and storage facilities are inadequate. Catering for dinners, receptions and parties in domestic kitchens or the small kitchens provided in church and other community halls is awkward, because food must be prepared in batches to fit the available oven space. The cooked food must then be stored until required and as domestic refrigerators are too small for large bulks of food, they will be left at ambient temperatures until transported to the venue of the event. The risks inherent in this type of catering arise from inadequate cooking, cooling and storage and also from cross-contamination when preparation areas are restricted.

The final line of defence in the prevention of most types of bacterial food poisoning is good kitchen practice and requires knowledge of safety measures through education of those involved in the preparation, processing and service of food on both the commercial and domestic scale. Food handlers must be taught that raw foodstuffs are frequently sources of food-poisoning organisms and that precautions must be taken to check their spread and multiplication in cooked food. Factors related to temperature control, i.e. storage at ambient temperature, inadequate cooling or reheating, warm holding and undercooking are frequently associated with outbreaks. Emphasis placed on improving these aspects together with strict cleanliness of surfaces and equipment, would significantly reduce the number of outbreaks of food poisoning.

The following chapters describe in more detail the measures which will lessen the risk of food poisoning; these factors have been recognized while studying the spread of infection both in the field of outbreaks and in the laboratory.

There are other factors, sometimes magnified in the public mind, which may be relatively unimportant in themselves and which may serve merely as an indication that the conditions in a particular establishment are not all that may be desired. The presence of lipstick on a cup is not necessarily a sign that dangerous germs must be present also, but that the washing-up has not been carried out with due care and attention. Varnish on the nails of a waitress or kitchen employee is not in itself a harbour for bacteria, but perhaps an indication that the hands will not be washed or the nails scrubbed as often as required, for fear of damaging the cosmetic effect. The flakes of whitewash which may fall from the ceiling into food during its preparation may not constitute an immediate danger, but indicate that the general care of the kitchen is poor. The much-handled unwrapped loaf will not give rise to food poisoning or even food-borne infection except in quite unusual circumstances, yet it indicates a lack of care and respect for the food which others have to eat.

There are many such examples which arouse feelings of apprehension in the minds of people and which perhaps divert attention from more serious lapses in personal habits, such as fingering the nose and mouth with hands used for the preparation of foods, failure to wash the hands between and after jobs, which include work with raw foodstuffs, as well as the after-care of cooked foods cut up for the table or to be cooled and stored.

The remaining chapters on prevention are concerned with the food handler, the food itself, the environment and equipment of food premises, and the parts which can be played by education and legislation.

10

Personal hygiene of the food handler

The food handler has an important part to play in the prevention of food poisoning and other food-borne disease. The common concern is with the passage of organisms from persons to food, from the nose, skin of hands and other surfaces and from the bowel. More important still is the transmission of organisms from raw to cooked foods with the hands as means of transport as well as surfaces, utensils and cloths. The last factor which predisposes foods as vehicles of food poisoning is the slow cooling of cooked food and the time at room temperature at which the food stands before it is eaten or eventually refrigerated. Furthermore, without some means of rapid cooling, food in bulk may be placed in the refrigerator while still warm so that organisms may continue to multiply and produce toxin until the temperature of the food is low enough to stop growth.

The words 'food hygiene' are usually associated with personal cleanliness which is often limited to the care of the hands. Hands are rarely free from bacteria, which may be transient or semi-permanent in or on the skin. The commensal flora of the hands usually consists of staphylococci. They cling to the skin surface and persist in hair follicles, pores, crevices and lesions caused by breaks in the skin, and they are not easily removed. Many strains of staphylococci are harmless in food, but when strains producing enterotoxin join the skin flora, probably from the nasal mucosa, there is a continual hazard from the hands of those who prepare cooked food for the table. Semi-preserved meats and cooked poultry eaten some hours after preparation and not refrigerated are frequently proved to be vehicles of staphylococcal enterotoxin food poisoning. A few staphylococcal cells newly implanted on cooked food can be eaten without harm. When the number of organisms increases to millions per gram of food during unchilled storage, sufficient toxin can be formed in the food to cause vomiting.

It is impossible to sterilize the hands, disinfection by heat is impracticable, but chemical disinfection may be used. Many transient bacteria picked up from raw foods, excreta and the environment can be removed from the hands by washing with soap and water, but much of the resident flora of staphylococci will remain, however carefully the hands are washed and scrubbed.

The hands should be washed with plenty of soap and water, preferably warm, and rinsed in running water. Nails should be short, unvarnished and scrupulously clean. Nail brushes should have nylon bristles set in a plastic back, so that they may be effectively and regularly cleaned and disinfected by heat or in a hypochlorite solution; a nail file should also be available. Intestinal, nose and skin food-poisoning bacteria have been isolated from swabs rubbed under the nails.

The hands must be washed carefully before touching food of any sort and particularly after handling raw food ingredients, which will introduce bacteria daily to the kitchen, and before continuing with other cooking preparations. They must be washed after handling waste food and refuse and after visits to the toilet.

In critical situations, such as the preparation for cooking of poultry carcasses known or suspected to be contaminated with salmonellae, the use of a disinfectant hand dip might be considered.

Bacterial flora of the hands

Investigations of the bacterial flora of the hands before and after washing with soap and water, alone or with temporary antiseptic treatment, showed that the plain soap and water wash was effective for the removal, or at least the reduction in numbers, of coliform and other Gram-negative intestinal organisms picked up from foods. In spite of washing the hands in soap and water, organisms known as *Pseudomonas* have been isolated from the skin under wedding rings worn by nurses. *Pseudomonas aeruginosa* is responsible for much infection in hospitals. Although it is not an agent of food poisoning, other organisms such as salmonellae could survive on moist skin surfaces beneath rings, for example.

The resident population of staphylococci, although usually reduced, could still be cultured from the hands after washing; sometimes the cultures were more profuse than before. The staphylococci that lodge in the hair follicles and cracks in the skin surface come to the surface after scrubbing in hot water. Because it is difficult to alter the resident flora of organisms on the hands, washed hands do not necessarily mean safe hands. Thus, foods which readily support the growth of staphylococci, such as cured and uncured cooked meats, creams, cooked seafoods and other delicacies which will be eaten without further heat treatment, should ideally not be touched with the hands; care must be taken that such foods be kept cold after preparation and before consumption.

Fingerprint cultures may be used to demonstrate the fluctuation in the numbers and types of organisms on the hands. Dry hands, without immediate washing, have fewer organisms than wet hands. After touching raw materials, such as meat and poultry, and also after handling wet cleaning cloths, sponges and rags, enormous numbers of organisms can be isolated from the hands, as shown by the fingerprint cultures (Fig. 4.1, p. 62). Cooks and other kitchen workers are astonished to see the extent of the bacterial contamination on their hands after cutting up raw foods, and also to see that the surfaces and utensils in contact with raw materials are likewise profusely contaminated with the organisms from the foods. Such demonstrations serve to emphasize the necessity to wash the hands with care before transferring attention from raw to cooked foods, even though the cooked foods are required for immediate consumption. If it is considered advisable to use a disinfectant dip, the hands should be washed with soap and water before the application or submersion in the chemical solution, just as a work surface should be cleaned before the application of a chemical disinfectant. Chemical disinfection should not be used as an alternative to washing with soap and water.

Facilities for washing, drying and care of the hands

Washing facilities should be clean and in good order, and should include a hand wash-basin in the kitchen separate from the sinks intended for food preparation. Wall dispensers can be used for liquid, finely flaked or powdered soap. They must be cleaned with hot water when empty and before refilling; they should not be topped up because the residue is likely to be contaminated. Newer types of dispenser contain sealed disposable units of liquid soap each with their own nozzle. Thus there is no opportunity for a build-up of soap residues on any part of the dispenser. If tablet soap is used it should be stored dry and soap dishes washed and dried daily. Soap tablets can be suspended from metal holders by means of a magnet pressed into the soap surface. When soap is passed from hand to hand, bacteria may be trapped in scum or curd and remain on the surface. Soaps both liquid, dried and in tablet form can include a disinfectant such as hexachlorophene. The action of the disinfectant is slow and attributed to a film left on the hands after washing with the medicated soap. It is necessary to use the soap in cake or liquid form exclusively and frequently for a prolonged period of time; it may help to reduce the bacterial flora of the hands and the distribution of contaminants. A few drops of alcohol rubbed well into the hands is effective.

Disposable paper towels and electrically operated hot air dryers are the most commonly used means to dry the hands. The continuous roller towel provides a portion of clean towel for each person and it is a vast improvement over the communal roller towel which can transfer infection from person to person. When the continuous roller system is used, provision must be made for spare rolls of towel to be available.

Some food establishments dispense creams and lotions for the hands. They can be plain barrier creams to decrease the sensitivity of skin to certain food substances which are known to cause dermatitis. They also serve to keep the hands soft and supple and to reduce roughness and the cracks which harbour bacteria. Such creams may incorporate a disinfectant to prevent the growth of contaminants in the cream itself. It has been suggested that in some instances, allergens (lanoline) might be absorbed onto the film of barrier cream and sensitivity reactions increased.

A simple and inexpensive hand lotion which can easily be made up contains tragacanth and glycerin in the proportion of one part of tragacanth to two parts of glycerin in water with the addition of a few drops each of the oils of lemon and lavender. There are also commercially available hand creams which are effective in keeping the skin of the hands soft and healthy.

Cuts, burns and other raw surfaces, however small and apparently healthy, can harbour staphylococci. The Food Hygiene (General) Regulations, 1970 (Chapter 17), impose a measure that in the kitchen all lesions must be covered. Waterproof dressings help to prevent the passage of bacteria outwards in serous fluid and inward from fluids in the environment. When lesions or sores on hands or other parts of the body are obviously infected, as shown by inflammation and pus formation, the person concerned should not handle foods. Even after healing, agents such as staphylococci may still linger on the skin, and it may be difficult to eliminate them from the hands. Various skin diseases are colonized by staphylococci, such as dermatitis from handling certain food ingredients. Care should be taken to avoid touching irritant substances, unnecessary contact with dirt, chapping, or exposure to very hot water.

Rubber gloves worn by food handlers may not necessarily help to improve the bacteriological condition of foods, unless the gloves retain a smooth unbroken surface and are washed frequently. They should be washed inside as well as outside to prevent soiling the hands after continual use. Gloves are recommended for procedures with frozen foods and also when there is prolonged immersion of the hands in hot detergent solutions; higher temperatures can be used for washing up. Thin disposable gloves are available for the assembly of salads and sandwiches particularly when cold cooked meats and poultry are involved (Fig. 10.1). They aim to restrict the passage of staphylococci from hands to foods. It is unlikely that they would stop the spread of salmonellae and other intestinal pathogens from carcass to carcass of poultry, or from raw to cooked food. Gloves should not be worn for too long, but they may prevent the skin of the hands becoming dry and cracked. Heat-disinfected implements should be used to manipulate foods where practicable; tongs, metal meat slices and other metal gadgets are available (Fig. 10.2).

To summarize, the hands may be responsible for: (a) the transference of intestinal pathogens such as salmonellae, campylobacters and *E. coli* from raw food and utensils, as well as faecal matter, to cooked food, and (b) the transference of nose and skin organisms such as staphylococci from the person of the food handler to cooked food. Recommendations to overcome these hazards include:

(i) hands to be washed not only after using the WC, but also with care between handling different items of food and especially after touching raw food;
(ii) as far as possible, cooked food should not be touched with the bare hands, because even washed hands may not be free from staphylococci;
(iii) no worker with septic lesions anywhere on the body should work with susceptible foods. Clean, non-suppurative cuts and abrasions should be covered with a dressing.

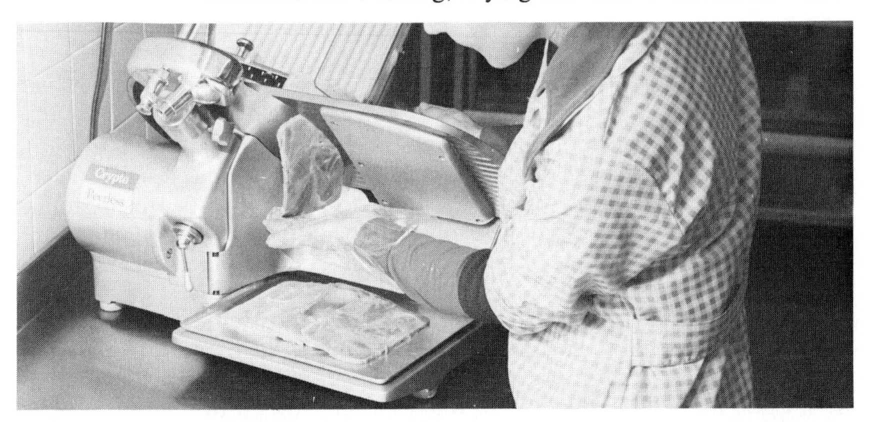

Fig. 10.1 Hygienic method of slicing meat

Fig. 10.2 Pastry tongs

Nasal and throat carriage

The usual sites for harbouring staphylococci are the nasal mucosa, hands and other skin surfaces, cuts, burns, abrasions and pustular lesions such as boils, carbuncles, whitlows and styes. More rarely, staphylococci are found in the throat. Streptococci may be found in the nose and throat and in some skin infections. The habits of fingering the nose and spots will increase the hazard of passing staphylococci from hands to foods.

Clean cloth and paper handkerchiefs are almost free from bacteria, but when dirty they may harbour many millions of organisms including staphylococci. Paper is more hygienic than cloth; paper handkerchiefs may be flushed individually down the water closet, burned in a sanitary incinerator or collected in a pedal-operated bin and subsequently incinerated. Cloth handkerchiefs should be washed and boiled frequently.

The use of tobacco in any form by kitchen staff is prohibited under the Food Hygiene (General) Regulations, 1970, so that smoking while on duty must be avoided. Although smoke and ash are harmless they are aesthetically distasteful and fingers will be in contact with the mouth. Licking the fingers to lift paper or turn pages is unhygienic, particularly if the paper contaminated with saliva is to be used for wrapping food. Other habits such as chewing gum or taking snuff while on duty should be discouraged for fear of contaminating the hands and the environment, and thus food.

The hair and head are less likely to be touched by the hands if a cap, net or headscarf is worn, and loose hair will be prevented from falling onto food. The hair should be clean and tidy and the scalp free from dandruff. Abrasions on the scalp may spread staphylococci in dry scales of skin. Beards should not be encouraged unless short, well-trimmed and clean and should preferably be covered by a beard guard.

Intestinal secretion

Intestinal bacteria are more likely to spread from the fluid stools excreted by those suffering with diarrhoea than from a well-formed stool. Aerosol sprays are formed when the WC is flushed, and general toilet cleanliness is more difficult with fluid excreta. People with diarrhoea should not work in the kitchen, whatever job they are doing. The Food Hygiene (General) Regulations, 1970 require that food handlers suffering from food poisoning or certain other diseases, or carrying the causative organisms, shall not be allowed to work with food until permission has been given by the proper officer of the local authority, usually the Medical Officer for Environmental Health or Consultant in Communicable Disease Control (CCDC).

Regulations

The Salmonella Subcommittee of the Public Health Laboratory Service (UK) recommends the exclusion of cases until free from symptoms. Recovered cases and contacts with well-formed stools could return to work even though they may still be excreting salmonellae other than the enteric fever organisms. Advice should be given with regard to personal hygiene and also food hygiene. For cases and excreters of typhoid and paratyphoid bacilli it is suggested that 12 consecutive specimens taken over a period of 6 months, eight in the first 2 months and four thereafter (two after purges) should be negative. Urine samples should be examined at the same time. After cases of cholera and bacillary dysentery, three consecutive stools at intervals of not less than 24 hours should be negative before handlers return to work.

With regard to *Staph. aureus*, food handlers with septic lesions should be excluded from work until successfully treated. Nasal carriers of *Staph. aureus* need to be excluded if they are implicated as the source of an outbreak. Nasal carriers can be treated with an antistaphylococcal cream.

The treatment of healthy salmonella excreters with antibiotics is not recommended because such treatment may prolong the period of excretion. Where it is suspected, but not proven, that there may be healthy excreters with well-formed stools in the kitchen, instructions about care in washing hands after visits to the toilet should be given.

The spread of intestinal pathogens, such as salmonellae in hospital wards – particularly paediatric and geriatric wards – is difficult to check. Nurses administering bedpans and other needs of the patient, the disposal of excreta in sluice rooms not far from the ward, and the service of meals by both nurses and patients, all raise peculiar difficulties in control. The use of iodophors for hands, or even better alcohol or spirit rubbed in, for those attending patients is recommended.

Clothes

Protective clothing should be light coloured and light in weight; it should be changed frequently. Drip-dry fabrics ease the work of daily laundering. All large establishments should be provided with adequate changing rooms, and means to store clothes and other personal belongings; equipment for drying clothes such as hot air racks or tumble driers, should be available. The provision of shower baths in changing rooms is also recommended. In addition to changing rooms, areas for rest and relaxation are essential together with canteen facilities.

Health

Occupational health services are available in industry. Most large departmental stores and factories possess well-equipped medical departments including a surgery and waiting room. When an identified outbreak of food poisoning or other disease occurs in a factory or in the community arising from factory procedure, the first people to be involved are the on-site doctor and nurse. It is then the responsibility of the medical staff to inform the management and the medical environmental health departments. Furthermore, the occupational health nurse can be a valuable health educator in relation to personal hygiene. She has a unique position by means of clinical involvement with both employers and employees. Education on good standards of health, including hygiene and nutrition, should be provided on a regular basis.

Illnesses, however mild, must be reported at once to the medical department in a large establishment or to the person in charge in smaller places. Clinical conditions which are likely to spread bacteria known to cause food poisoning or sepsis are required to be notified to the medical officer for environmental health or CCDC. Small shops and kitchens should keep a first-aid box, and train at least one member of the staff in first-aid treatment.

Medical examination

Physical examination is usually required at the beginning of employment to establish freedom from tubercle infection, intestinal pathogens and skin infection. Also a medical certificate should state freedom from past typhoid-like illnesses and skin diseases. Periodic examination of stools from healthy food handlers is not recommended. Amongst meat and poultry workers salmonellae may be excreted for a short time only, the serotypes changing according to the pattern of serotypes in the animals and birds and in the carcass and poultry meat.

Typhoid carriers are in a different category; any discovered should not be permitted to work in food premises, and it is advisable to give specific treatment. Those who travel abroad, especially to middle and far eastern countries, may bring back intestinal pathogens infrequently found in their home countries. The examination of their stool samples might be advisable and even essential if there has been an episode of diarrhoea while away.

During and after an outbreak of food poisoning, it is necessary to trace the origin of the agent of infection or intoxication. Swabs from food handlers may reveal enterotoxigenic staphylococci from nose and skin; it must be established that the phage type is the same as that isolated from the suspected food vehicle.

The examination of stools for salmonellae, not recommended except following an outbreak of salmonellosis, may reveal excreters amongst the food handlers in kitchen, shop, or factory. However, the infection may have come from the foods handled and the excreters regarded as victims of the environment. Within this context the food handler's work could be listed amongst those described as occupational hazards. Nevertheless, the presence of many excreters at any one time in kitchens or other food establishments will help to spread infection and extra care will be needed with hand-washing, cleaning and sanitizing operations. When there are many incidents of salmonellosis, efforts should be made to find the food source of the relevant serotype or serotypes and their mode of spread. Similar batches of imported or home-produced food should be banned.

There are other diseases which can follow the handling of contaminated food, for example, streptococcal and staphylococcal infections, leptospirosis, brucellosis, anthrax and viral infections. The pathways of spread of enteropathogenic *E. coli* are likely to be similar to those of salmonellae.

Special care should be taken by persons handling the more susceptible foods, those which encourage the growth of microorganisms, such as cream and custard dishes, cold cooked meats, gelatin and salads. Compulsory food hygiene training of food handlers is under consideration for the future (Food Safety Act, 1990, see Chapter 17).

A special chapter is devoted to 'Food Hygiene in the Tropics'. The high ambient temperatures and the lack of sophisticated facilities give rise to hazards which are far greater than in western countries. There are many opportunities in the West for the education of food workers in matters of food, environmental and personal hygiene; the public can also participate in the teaching programmes. Such opportunities are minimal in the tropics.

It is impossible to compare statistics for the incidence of illness and predominant intestinal pathogens in the East and West. There is no compulsory reporting system for food poisoning or other food-borne disease, with the possible exception of cholera, in eastern countries. Diarrhoea is usually accepted as part of daily life. Notification of all infectious disease including food poisoning and other food-borne disease is essential for assessing the importance of the agents most involved and for finding the means of control.

11

Preparation, cooking, cooling and storage

Preparation and cooking

General methods of cooking

Fresh food cooked and eaten immediately while hot should never be responsible for food poisoning due to microorganisms. Nevertheless, heat-resistant bacterial spores frequently survive cooking and give rise to large numbers of bacterial cells when cooling is slow and the storage time of cooked foods in the kitchen is too long.

Short, high-temperature cooking is best for the safety of the foodstuff and thereafter there must be an understanding of times and temperatures of storage in relation to bacterial growth.

Steam under pressure, thorough roasting of small solid chunks of meat, grilling and frying are considered to be the safest methods of cooking. Infra-red rays and high-frequency waves have good penetration but may be patchy in action. The results of comparative experiments with chicken inoculated with sporing and non-sporing organisms showed that some spores survived inside and on the outer surfaces of the carcasses after cooking in a microwave oven and also after conventional roasting. Both methods of cooking destroyed vegetative cells. The rapid reheating of chilled or thawed frozen meals in a microwave oven has advantages in some circumstances and for certain foods, but the bacteriological condition of the cooked food before reheating is important. All the benefit of heat destruction of microorganisms in raw food by cooking may be undone if manipulation and handling after heating allows recontamination of the cooked food.

Convection ovens

A fan in a forced air convection oven recirculates air in the oven, which improves heat transfer, and the cooking is faster, more even, and more effective (Fig. 11.1). The oven shelves can be filled with food without affecting efficiency, and similar articles of food will be cooked to the same degree. Cooking times generally can be reduced by one-third compared with those of the conventional oven. Forced air convection ovens are available in three types according to the speed of air movement and the input of heat. Care is needed to select the correct oven. As with conventional ovens vegetative cells of bacteria are killed but not all spores.

Microwave ovens

The microwave oven (high frequency) uses little electricity and it reheats cooked food quickly: the oven can be used in conjunction with a deep freeze for fast food service. Microwaves boil food by the agitation of molecules, especially water molecules, the waves penetrate into the food, the distance depending on the type of food, and they gradually lose energy. The energy is absorbed and converted into heat which then spreads inwards. Most foods need to be left to stand at the end of the cooking process to allow the centre to reach the same temperature as the surface of the food. Metal containers will slow down microwave cooking by reflection and may even cause a breakdown in the system. There is a tendency towards uneven cooking which is overcome to some extent by the inclusion of a turntable which rotates the food during cooking, also wave stirrers or rotating antennae deflect the waves off the metal walls of the oven to give a more even wave distribution. Turning solid

Fig. 11.1 Forced air convection oven

foods and stirring liquids at intervals during the cooking period will aid distribution of heat. As the microwaves heat only the food and not the container, care must be taken to keep the oven and the containers clean. It has been shown that organisms can survive in dried films on the walls of the oven.

The drawbacks of microwave cooking are irregular heat penetration and lack of browning; they are overcome by the Mealstream Micro-Aire ovens, which are designed for fast-food menus. These cook by continuous microwave energy or pulsed microwave at 7-second intervals, in combination with forced air convection making use of the adjacent infra-red wave band as the source of heat. They can be used to roast, bake, fry, grill, braise and boil and save labour, food, power, space and time compared with traditional cooking methods. There is less dirt and the hygiene is thereby improved; ventilation requirements are also reduced. A Code of Practice for hygiene in microwave cooking is available, DHSS No. 9 (see Appendix B).

Pressure cooking

Pressure steamers enable food to be cooked to order within minutes and in conjunction with bratt pans they can replace the boiling top and its battery of pans. Modern pressure steamers are of two types:

(i) dynamic or high pressure, where steam jets will rapidly defrost and cook frozen food at 12–15 lb per square inch;
(ii) low pressure, operating at about 6 lb per square inch when there is greater flexibility and control of the cooking process.

In a pressure cooker, as in an autoclave, there will be sterilization if the conditions are hot enough for long enough. At a pressure of 15 lb per square inch, the temperature will probably reach 121°C (249.8°F) in 15 minutes. At pressures below 15 lb., sterilization cannot be assured although most bacteria will be destroyed and some spores also.
Pressure cooking is similar in principle to the hospital autoclave; all bacteria and spores are destroyed by the combination of pressure and heat. As well as a steam generator, a water treatment plant is often an essential prerequisite for trouble-free pressure cooking.

Slow cooking

Slow cooking dates back to ancient times when stone pots were first used for this purpose. With the increasing costs of fuel and food there has been renewed interest in this method of cooking as a means of household economy; maximum tenderness can be obtained from less expensive cuts of meat when they are cooked slowly at low temperatures. Electric cooking casseroles have been popular in the USA for a number of years and they are also available in Britain.

A typical slow cooking vessel consists of a glazed earthenware bowl and lid set into an outer aluminium casing. The cooking heat comes from an element wound around the sides between the earthenware bowl and the outer casing. The heat used amounts to approximately 60 watts on low and 120 watts on the high setting – as little electricity as a light bulb.

Some concern was expressed about the possible bacteriological hazards of cooking food at low temperatures for long periods of time, particularly with respect to *Clostridium perfringens*. Experiments have shown that the hazards of using an electric casserole are no greater than those for conventional cooking methods providing that the manufacturer's instructions are followed. It is important that the food be eaten hot immediately, or removed from the casserole, cooled rapidly, and refrigerated, although these instructions should be followed when foods are cooked by other methods.

When a dish with red kidney beans is cooked in a slow cooking vessel the beans should be precooked by boiling for at least 10 minutes before they are added to the other ingredients. Food poisoning due to the red kidney bean haemagglutinins has occurred when the beans were cooked from the raw state in an electric casserole. The temperatures are sufficient to cook food and kill vegetative cells of bacteria, but not to inactivate the toxic component of the bean.

Electronic catering size ovens are now available on the market. They cook meats at low temperatures to a predetermined centre temperature detected by means of a probe. When the required temperature is reached in the centre the oven switches to a non-cooking cycle which holds the food at 62.8°C (145°F) for hours, even overnight, until required. It is claimed that shrinkage is minimal and that natural enzyme action during the holding cycle makes the meat more tender.

It was thought that there might be hazards with this type of oven due to the survival of organisms in the centre of the meat and poultry during slow, low-temperature cooking, and that growth would occur during the prolonged holding. Tests have shown that if the oven is operated correctly there is no danger. Bacterial spores will survive, but there is no outgrowth during the holding cycle.

Cook-freeze

Cook-freeze can be defined as a system whereby high quality food is prepared and cooked in economic quantities, retained in a state of 'suspended freshness' by rapid freezing and freezer storage, and served when and where required from finishing kitchens with low capital investment and minimal staffing. Food is prepared, cooked, and frozen centrally with distribution of the frozen food to service areas for final heating and service (Fig. 11.2).

The many advantages of the cook-freeze system include: centralization of stores, saving on staff, better portion control, and a better balance of the day's work, thus avoiding the traditional peaks of production around mealtimes. From a microbiological point of view greater control can be exercised over preparation and storage as urgency to present meals at set times is eliminated. The dangers of cross-contamination from raw materials to other foods can be reduced by the preparation of single items at a time using equipment which will not be used for other foods on that occasion. If necessary, samples taken from large batches can be checked bacteriologically before they are eaten.

However, mistakes in batch preparation could lead to a far greater outbreak of food poisoning than small-scale cooking on a day-to-day basis. In dealing with large quantities of food, there is an increased hazard because it is difficult to ensure adequate heat penetration and rapid cooling after cooking.

The main emphasis in cook-freeze, as in any other catering system, must be placed on correct storage between cooking and service to prevent multiplication of organisms which may have survived the heat process, or which have reached the food after cooking.

Cook-chill

Cook-chill is a catering system based on the cooking of food followed by fast chilling, storage under controlled low temperatures above freezing point, 0°C (32°F)–3°C (37°F), and subsequent reheating immediately before consumption. The advantages of this system are similar to those described for cook-freeze, but abuse of temperature control in the preparation, storage, distribution or reheating of food is more likely to result in hazards to health in a cook-chill system than by the cook-freeze method.

The chilling process should commence within 30 minutes of the food leaving the cooker and be complete within a further period of 1.5 hours. The chilled food should be distributed under controlled conditions and the temperature not allowed to exceed 10°C (50°F). It should be reheated to

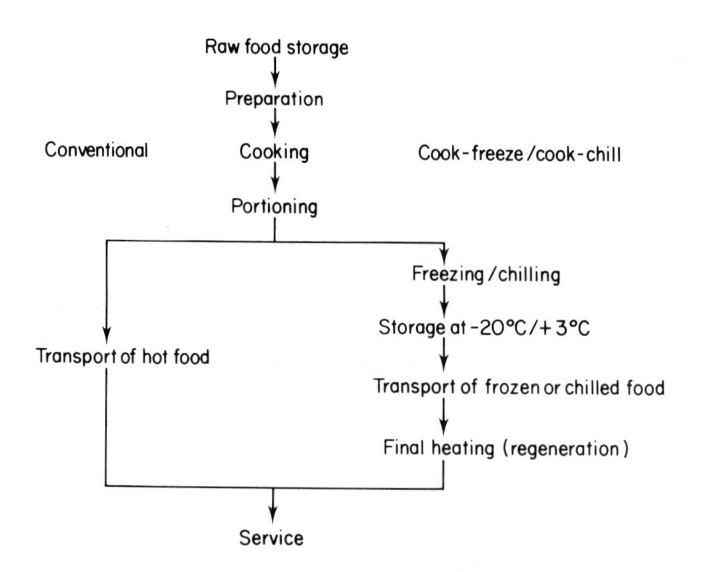

Fig. 11.2 Operation flow – conventional cooking versus cook-freeze or cook-chill

at least 70°C (158°F) for consumption. The maximum life of the cooked products when held at or below 3°C (37°F) should not exceed 5 days including both the day of cooking and the day of consumption. If the temperature exceeds 5°C (41°F), but does not rise above 10°C (50°F) the food should be consumed within 12 hours; if 10°C (50°F) is exceeded the food should be regarded as unsuitable for use.

A Code of Practice has been issued by the Department of Health for both the cook-freeze and the cook-chill systems (see Appendix B). However, this guidance does not apply to chilled foods prepared under special conditions of processing and packaging, which provide a product with an expected shelf life of more than 5 days, for example, '*sous vide*'. In the basic '*sous vide*' process the raw or parcooked food is heated gently under vacuum, in high-barrier plastic bags or pouches, at low temperatures, often only pasteurization. The food may be served immediately, or more usually, chilled to 0–3°C, stored for up to 21 days, reheated and the bag opened for service of the meal.

The process does not kill all anaerobes, some spores will survive, so for subsequent preservation of the food the bag may require flushing with an inert gas (modified atmosphere packaging) before sealing and refrigeration. The system improves both the nutritional value and the palatability of the foods and the longer potential storage life gives greater flexibility in catering and retail operations. There are drawbacks with the system that include a greater risk of spoilage or even food poisoning, if production controls are not carefully enforced. Storage under vacuum may mask signs of spoilage and surviving anaerobic organisms may multiply and produce toxins. Of particular concern are the psychrotrophic *C. botulinum* (types B and E). Regulations governing the production of hermetically sealed foods including '*sous vide*' are currently in preparation.

Cooking frozen food

With the rapid expansion of the home freezer market there has been much discussion on the hazards of cooking food, particularly joints of meat and poultry, from the frozen state.

It has long been accepted practice to cook small items such as vegetables, fish fingers, beefburgers, pies, chops and ready meals without preliminary thawing. These items are relatively small so there is little delay in heat penetration and adequate cooking is ensured in fairly short periods of time.

The main difficulty arises in the cooking of joints of meat and poultry. Although some frozen food companies and authors of books on home freezers and cookery advocate cooking most meat from the frozen state and give times and temperatures to use, it is safer to recommend proper thawing of frozen meats before cooking.

When a frozen joint or poultry carcass is cooked, the outside may appear to be well done while the centre remains quite raw. Organisms present in the centre of a rolled joint or in the cavity of poultry, for example, may survive and if the cooked food is stored incorrectly, may multiply to numbers able to cause food poisoning.

Even if recommendations for cooking times and temperatures are laid down there will be occasions when rules are forgotten. The housewife tends to judge 'doneness' by the external appearance of meat and by probing with a sharp implement – how many use or would purchase a meat thermometer?

Joints for the average family will be small and consumed quickly; the number of organisms surviving after cooking will most likely be small, probably too small to cause illness. Catering size joints are a much greater hazard as they are more likely to be cooked in advance of needs and may be improperly stored before use due to lack of cooling facilities. The cooking of large joints from the frozen state requires longer cooking times and hence more fuel; the practice should be discouraged.

Many food-poisoning outbreaks have been attributed to the large 'banqueting' size frozen turkeys, often 14–18 kg (30–40 lb) in weight. There are difficulties with the size at all stages during preparation, cooking and storage. The time to thaw may be several days; thorough cooking to the centre without overcooking the exterior is difficult and cooling prior to refrigerated storage requires many hours without mechanical coolers such as cold rooms with fans or blast chillers. Thus there are many opportunities for survival and growth of food-poisoning organisms, in particular, salmonellae and *C. perfringens*.

Meat and poultry dishes

Heat penetrates slowly into joints, poultry carcasses, and made-up dishes such as pies, and adequate cooking times and temperatures should be allowed for the centre to reach boiling point. For instance, a 2.9 kg (6.5 lb) meat pie requires a temperature of 177–204°C (350–400°F) for 2.5–3 hours for the centre to approach boiling point at 100°C (212°F). An outbreak of food poisoning has already been described in which an organism of the salmonella group was isolated from several pies of two sizes: 453 and 113 g (1 lb and 4 oz). The minced meat, probably already contaminated with salmonellae, was hand-filled raw into the pastry cases and it was thought that the cooking, at 232–246°C (450–475°F) for 25–30 minutes, was either inadequate to destroy the contaminants or that there were temperature fluctuations. It was pointed out that pies made with raw meat should be cooked to the point of sterility even though the pastry may become a little over-brown in the process. It is worthwhile to note that for the red muscle colouring matter of meat to change to the familiar grey colour the temperature must be at least 73°C (163.4°F). Heat penetrates meat slowly and the English roast beef which remains red inside has not reached an internal temperature of more than 63–65°C (145.4–149°F). Slowly cooked rare beef used for sandwiches prepared for customers at their request has caused *C. perfringens* food poisoning.

Precooked meat is often used for making pies and pasties so that a final cooking time sufficient only to bake the pastry is given. The temperature reached in the centre of the mass of meat, sometimes mixed with vegetables, would not necessarily destroy sporing or even non-sporing organisms, which would multiply actively while the meat was still warm.

Between 1988 and 1991, 146 of 611 (24%) general outbreaks of food poisoning in England and Wales in which the vehicle of infection was traced, were due to *C. perfringens*; the foods most commonly mentioned were cold and reheated meat and poultry, and foods such as, stews, casseroles, curries and minced meat dishes. Thus a change in cooking methods and a greater awareness of the necessity for rapid cooling and cold storage would help to eliminate this type of food poisoning.

Part-cooking of meat in hot weather or at any other time is frowned on by the bacteriologist. Some organisms will be killed but others may be encouraged to multiply. The final cooking may be too light either to kill bacteria or to destroy toxins.

It is safer to keep meat in the raw state overnight, preferably in the cold, and to cook it thoroughly on the day when it is required. If, from motives of economy or for some other reason, it is essential to cook meat the day before it is eaten, it should be cooked thoroughly, cooled rapidly and refrigerated overnight. When there is insufficient refrigerator space, it should be a strict rule that all dishes of stewed or boiled meat, whether as stews, pies, or joints, must be cooked and eaten the same day, preferably with no delay between cooking and eating and certainly not more than 1–2 hours. The danger of eating meat and poultry cooked a day or two earlier seems to be far greater in the communal canteen dealing with large masses of meat than in the home, although in no instance should stewed or boiled meat be allowed to stand at a warm temperature for several hours. Well-roasted solid meats should be safe, but the rolled roast joint frequently gives rise to trouble because the contaminated outside is folded into the centre. The size and shape of joints of meat are important in relation to heat penetration and heat loss. Temperature recordings from thermocouples showed that during cooking the temperature at any point within the meat was dependent on the distance of that point from the outer surface of the meat. The centres of large bulky cuts of meat will, therefore, take longer to heat and cool than those of long slim portions of like weight. It was also shown experimentally that the centres of large portions of meat were slower to heat up in an hot-air oven than in a moist-air oven.

The degree of heat reached inside a grilled or fried sausage or a sausage roll will depend on the preference of the cook for a well-browned or lightly cooked article. Rissoles and fishcakes made with precooked meat or fish or potatoes, may be placed in the frying-pan or grilled for a few minutes only. Boiled rice may be quickly turned over in a little hot fat for frying. Raw foods such as fish, sliced or minced meat, and bacon are likely to be grilled or fried thoroughly on both sides in fat, which should ensure sterility.

The installation and use of pressure cookers for quick, high-temperature cooking might solve many of the difficulties associated with the preparation of foodstuffs on the day they are required to be eaten hot. Even heat-resistant spores would be killed by this method of cooking, given the correct time and temperature.

Canteen-prepared tongue has been a frequent vehicle of *C. perfringens* and staphylococcal food poisoning. *C. perfringens* from the organisms already present in the raw tongue, and staphylococci from the hands of the cooks, will multiply in the cooked product. The following method is recommended for safety. The cured tongue should be washed and allowed to soak in water for 8 hours in a cold room overnight. It should be washed well and boiled quickly in a lidded saucepan or steamer. The skin should be removed while the tongue is hot, and the small bones cut away from the thick end. The tongue is then ready to be pressed in a container which is scrupulously clean and either sterilized or disinfected in hot water; at this stage it is allowed to cool, but it is cooked again in the press, cooled rapidly, and refrigerated until required.

Outbreaks of food poisoning frequently occur from cold or warmed poultry meat. Salmonellae and *C. perfringens* are commonly present in the frozen or chilled birds as they arrive. The spores of *C. perfringens* may not be destroyed by cooking and grow out later during long slow cooling and kitchen storage. Even a few salmonellae may survive in underthawed and undercooked carcasses, although they are most often picked up after cooking from traces of the raw product left on surfaces and utensils. Left unchilled, the organisms grow in the meat.

Outbreaks of salmonella food poisoning from spit-roasted chickens have been described in Chapter 6. Careful instructions are needed for thawing, cooking and, most important of all, for handling and storage of cooked poultry. Staphylococci are usually implanted by hands after cooking. Cooked carcasses may be cut and torn apart while still warm and the portions piled high on trays, so that, even if refrigerated, the cooling rate is far too slow. When removed from the refrigerator or cold room the next day the organisms continue to grow and produce toxin.

Milkfoods

The hazards associated with the preparation of food a day or so ahead of it being required apply also to custards, trifles, blancmange and other milk puddings; these dishes should be freshly cooked unless they can be stored in a refrigerator.

By heating (or cooking) milk gently, as in pasteurization, all the vegetative cells of organisms which are dangerous are destroyed so that the milk is made safe unless recontaminated in the home. The term 'sterilized milk' means that a temperature is reached at which all or nearly all bacteria are destroyed, and the milk will remain close to sterility for an indefinite period whilst it is kept in the sealed container. Ultra Heat Treated (UHT) milk is subjected to a high temperature by indirect or direct application of steam; not all spores are destroyed, but the chilled shelf life of the milk in unopened cartons may be as long as 3 months. There have been some noteworthy episodes of food poisoning from dehydrated milk. Several outbreaks are described in the UK due to staphyloccal enterotoxin and *Salm. ealing* and in the USA due to *Salm. new brunswick* (see Chapter 6).

The susceptibility to bacterial contamination of certain confectionery creams has already been emphasized, and their preparation and storage should be considered with care. All ingredients used in the vicinity should be free from pathogenic organisms and gross bacterial contamination. Equipment should be scrupulously clean and disinfected frequently, preferably by heat. Savoy bags for dispensing cream and other toppings and also mashed potato should be made of materials which can be washed and boiled after use, such as cotton, linen or nylon. After boiling they should be dried thoroughly and stored in a special place protected from dust-borne or other contamination. Disposable paper bags may also be used for this purpose.

Although it may be considered necessary to ensure freedom from salmonellae of persons handling cream and milk, the presence of these organisms in the cow herself may be of even greater importance. Little space has been devoted to *Brucella abortus* in cows and *Brucella melitensis* in goats, but these organisms and others like the tubercle bacilli are scourges to cattle and to those who drink raw milk in countries without eradication schemes and without pasteurization. The restriction of sale of raw milk in England and Wales (see Chapter 17) should help to prevent outbreaks of salmonellosis and campylobacter infection attributable to milk.

Infant bottle feeds

Careful precautions are necessary in the preparation and storage of bottle feeds for infants. Bacteria may be introduced into prepared feeds by constituents of the feed, bottles, teats, utensils and also from hands used in the preparation. Investigations have shown that high counts may be due to bacterial growth during cooling and storage at atmospheric temperatures; unless care is taken at every stage of preparation unsafe feeds, contaminated with pathogenic bacteria including *Bacillus cereus* and other aerobic sporing bacilli, may be given to infants. There are hazards in the use of breast-milk substitutes in developing countries, including over-dilution of the powder, preparation with unboiled contaminated water, and inadequate washing of bottles and teats.

There is a growing interest in the procedures of hospital kitchens and in the preparation of milk and safe infant feeds. Hospitals in the UK, in the USA and in other countries are using terminal sterilization techniques. In one large hospital in India the incidence of *Escherichia coli* enteritis in infants has been markedly reduced by this procedure. The cooked feed is poured into sterilized bottles; a sterilized or disinfected teat is placed on each bottle and covered with a paper cap; the complete bottled feeds with teats are placed in small crates, ready for feeding each child for a day, and given heat treatment in a pressure cooker or steamer. Special equipment is available for cleaning bottles and teats. Milk remaining in infant feeding tubes used for premature babies may be a source of infection.

Much care, attention and legislation have been given to the hygienic production of milk, both liquid and dried, for the general public; similar care is needed to ensure a safe clean product for the infant consumer.

Until recently, little attention was given to the contamination of nasogastric tubes for babies unable to suck because they are premature or ill. The small capsule through which the milk or other fluid is injected into the small bore plastic tube must be handled; the liquid may overflow and the lumen of the tube becomes coated in coagulated milk affording a good medium for bacterial growth during the time, from 24–48 hours, that it may be in place. Bacteria from the hands of the nurses, the nasopharynx of the baby, the environment of the cot and even from the mother may multiply in the tube. Various pathogens have been isolated in large numbers from tube washings, including *Staphylococcus aureus*, virulent enteropathogenic *E. coli*, salmonellae, *Pseudomonas* and *Klebsiella*. Little is known of the effect on the infant and so far no alternative method of feeding has been suggested. The nasal feeding of adults carries the same risks. It is important that there is strict microbiological quality control of the ingredients and stringent hygienic precautions during the preparation and handling of all enteral feeds. It has been shown that bacteria can survive and multiply in feeds with low pH and high osmolarity, as well as neutral, isotonic feeds.

Likewise, care must be taken in the administration of all enteral and nasogastric feeds. Although the complete feed is supplied in a sterile sealed pack, either home-produced or from commercial sources, it may be given to the patient over a long period of time – 24 hours or more. Ward temperatures are often high and ideal for bacterial multiplication. All associated tubes and attachments must be sterile and care must be taken to avoid contamination while the feeding set is attached to the patient; so far as possible the backtrack of the bacteria from the patient should be avoided.

The Parenteral and Enteral Nutrition Group of the British Dietetic Association have produced a guidance document on microbiological control in enteral feeding. It proposes microbiological limits for raw materials used as enteral feed ingredients and the finished product (in the nutrient container prior to administration) and also gives guidance on administration, sources of contamination and sterilization and disinfection. They recommend that dried ingredients with an aerobic plate count of greater than 200 per gram should be rejected and that liquid ingredients should not exceed ten organisms per ml. The products should be free from *Salmonella*, *E. coli*, *Klebsiella* spp, *Pseudomonas* spp, *Clostridium* spp, *Staph aureus* and *B. cereus*.

The administration time of non-sterile complete feeds should be limited to 4 hours in order to ensure that microbial numbers will not exceed 1000/ml at the end of administration.

Other infant feeds

Bone or meat and vegetable broths and purées are common weaning food for infants during the second 6 months of life. The care of these foods, which may be canned or prepared in large quantities in the home, is important because they are excellent media for the growth of bacteria. The sterile contents of small bottles or cans are intended to be eaten for one meal. There are closure methods for bottles designed to prevent opening by curious mothers anxious to know the smell of the product before purchase. The preparation in the home of small quantities for individual meals may be uneconomical; large quantities may be stored for a few days only in the refrigerator or distributed into meal-sized portions and frozen for future use.

Contaminants can be introduced from hands or from utensils, and they will grow rapidly in warm weather without refrigeration. Infants are more sensitive than adults to food infections and are more likely to succumb to the dehydration caused by diarrhoea and vomiting.

All foods and liquids intended for infants should be stored in covered containers in a refrigerator or, if canned, the portion left in the can should be refrigerated in the can and covered. Three days should be the maximum time of storage. Supplies prepared for 2 days should be separated into two portions to avoid further touching. Where there is no refrigeration the covered container should be stored in the coolest place and the food boiled immediately before the meal. When the indoor temperature is greater than 21°C (70°F) and there is no refrigeration, bone and meat meals ought not to be used unless the contents of a freshly opened can are consumed; anything left over should be abandoned.

All containers and utensils used for the preparation of infant feeds should be boiled after cleansing and allowed to drain dry.

Egg products

Small numbers of organisms of the salmonella group may be present in unpasteurized egg products, including liquid and dried whole egg, white, and yolk. It is recommended that liquid egg mix and rehydrated powders should be cooked within 2 hours of preparation unless refrigerated in small amounts. Such egg products should be used only in recipes requiring thorough heat treatment. Prepared in bulk such egg products should be pasteurized.

An outbreak of food poisoning which occurred in an army camp illustrates a number of these points. Reconstituted spray-dried whole egg was prepared early one morning; part of it was scrambled for breakfast and eaten by 70 men, only one of whom was affected. The portion which remained was allowed to stay in the cookhouse until tea-time when the original bacterial population, including salmonella organisms, would have increased enormously. The mixture was lightly scrambled with insufficient heat treatment to kill all the salmonellae, and of 20 men who ate it 16 were taken ill.

Infection from contaminated ducks' and hens' eggs can be avoided by boiling until hard, frying them well on both sides, or confining their use to baked products, such as cakes and puddings, which require cooking temperatures high enough to destroy the organisms. But care should be taken that all equipment used in the preparation of mixes containing ducks' eggs is thoroughly cleaned and disinfected. Outbreaks of food poisoning have been caused by the contamination of imitation cream mixed in a bowl previously used for the sponge or cake mixture containing ducks' eggs. Lightly cooked or uncooked dishes, such as scrambles, omelettes, meringues, mayonnaise, mousse, or similar foods, should not be made with ducks' eggs. Following the recent trouble associated with the contamination of hens' eggs with salmonellae, *Salm. enteritidis* phage type 4 particularly, the Chief Medical Officer advised that raw hens' eggs should not be used in such lightly cooked or uncooked dishes. Other outbreaks from eggs and egg products are described in Chapter 6.

Gelatin

Powdered gelatin is another substance that requires particular care in preparation and addition to foodstuffs, because it may contain a large and varied flora of bacteria. Melted gelatin in water for use in cooked meat pies and also for other purposes such as for glazing meat loaves, cold meats and pâtés should be nearly boiled, and used as rapidly as possible with the temperature maintained above 60°C (140°F). This procedure may involve the use of a higher concentration of gelatin than formerly needed at low temperatures to produce an effective gel. Microbiological standards for the various forms of gelatin used for cooking and also for feeding animals will reduce the hazards associated with this product. Gelatin has been responsible for outbreaks of food poisoning as glaze (see p. 77) and as jelly for pies; when poorly designed filling equipment was at fault (see p. 97), or prepared in bulk with the re-use of glaze which dripped off foods after coating.

Coconut

Methods for the production of desiccated coconut are so much improved that contamination with salmonellae has been reduced to a minimum. When purchases of desiccated coconut are made from reputable firms, who know safe sources of supply, there is little danger from the use of this popular garnish. Coconut milk is sterile unless the outer husk of the nut or seed is damaged and bacteria pass through into the white flesh. Coconut milk is used as a refreshing drink in hot countries where the palm is common.

Rice

Dried polished rice has low counts of bacteria but *B. cereus* is often present; the spores of this organism can survive cooking. The spores germinate into bacilli which multiply in cooked and moist cereal products left unrefrigerated, and a toxic substance is formed. Boiled rice intended for frying should not be stored in the kitchen, but should be prepared in small amounts and fried for customers as required. Outbreaks of food poisoning caused by the growth of *B. cereus* in boiled and fried rice are described on p. 115.

Cake mixes

Packaged mixtures for cakes and sponges have given rise to many cases and outbreaks of salmonella food poisoning in the USA and Canada due to the inclusion of salmonella-contaminated egg products. It is recommended that such mixes should not be used unless there is an assurance either that they do not include egg products, or that those included have been subjected to pasteurization before being mixed with the other ingredients of the powder. Clostridial spores in flour may germinate in undercooked cakes (see p. 77).

Salad vegetables and dessert fruit

In large-scale catering it is recommended that vegetables and fruits with thin skins used for salads and desserts should be washed in water containing a solution of sodium hypochlorite. There are opportunities for salmonella contamination from fertilizers and soil as well as from food handlers. To facilitate the correct use of hypochlorite a marked sink should indicate a known volume of water and a small measure used to add the required concentration of solution; 60 to 80 ppm is considered to be satisfactory for this purpose. The household use of hypochlorites for salad vegetables (including watercress) and fruits is also advocated; exposure to solutions should not be less than 30 seconds.

The poor sanitary conditions and polluted water in many countries, particularly in the Middle and Far East, emphasize the need for care of vegetables and fruits purchased in shops and local markets. As well as the direct transfer of bacilli and amoebae (causal agents of dysentery), from product to consumer, cross-contamination from raw to cooked vegetables may take place in the kitchen.

Cooling

Domestic and small scale catering

Domestic refrigerators are intended to keep food cold, and not to cool hot food which may damage the cooling coils and cause moisture to condense on adjacent cold foods: this may encourage the growth of slime bacteria and moulds. Cooked foods should therefore be cooled before they are placed in the domestic refrigerator. The cooling time should be short – within 2 hours of cooking.

The penetration of heat into large joints of meat and the loss of heat after cooking are slow, so that cuts should be limited in size to 2.7 kg (6 lb) unless special precautions are taken for cooling in a chilled atmosphere with a good circulation of air. In homes and small catering establishments when cool rooms may not be available, hot meat should be left in a cool and draughty place for not longer than 1.5 hours before refrigeration.

Large bulks of food should be portioned into smaller lots to accelerate cooling, and liquids decanted into shallow containers for the same reason. Hot foods can be placed directly in large refrigerators with proper ventilation where they will cool more rapidly. For small catering establishments, a simple circulating fan installed in a well-ventilated cooling room or larder situated on the north side of the building will provide a satisfactory cooling system. A cabinet with shelves and containing a fan and air filter has been designed for quick cooling. Shallow rather than deep containers provide a larger cooling area for stews, gravies, and other liquid foods prepared in bulk. Household refrigerators could be adapted to provide space for shallow trays stacked one above the other for the cold storage of liquids, although once cool the food can be transferred to a clean deep container again.

The provision of proper facilities for the thawing, cooking, cooking and storage of large turkey caracasses is essential. Many outbreaks of salmonella and *C. perfringens* food poisoning from turkey meat have arisen because of faults in preparation and storage.

Storage of large cuts of meat cooked by boiling or by other methods not requiring temperatures above 100°C (212°F) should be discouraged.

Large scale catering and food production

The rapid cooling of bulks of cooked food intended to be eaten hours or even days later is difficult. The results of experiments to estimate the cooling rate of large cuts of meat showed that immediate storage in a well-ventilated cold room was the most effective method: 2.3 kg (5 lb) cuts of meat lost approximately 60°C of heat, cooling from 70 to 10°C (158 to 50°F) in 1.5–2 hours.

In commercial premises, blast chillers may be used to increase the speed of cooling. A cooling rate survey made in the USA showed that rapid chill refrigeration through 'blast chillers' cooled food at a consistent rate, while with walk-in coolers it was impossible to ensure steady cooling rates.

Some food factories cool by wind-tunnels measuring about 1.5 by 1 by 24 m long (5 by 3 by 80 feet), provided with iron rails, and, while the food on metal trays passes through, it is subjected to the wind from a fan 1.5–2 metres (60–80 inches) in diameter; cupboards with descending cold air-streams are also used.

For large catering establishments compelled to cool masses of food required for frozen or chilled meals, the temperature of the cooling rooms should be maintained at 10°C (50°F). Some new kitchens for school canteens include this facility; it is hoped that in time, accommodation for efficient cooling will be provided in all kitchens cooking for large numbers of people.

Ways in which the food handler can help to prevent food poisoning have been discussed in previous chapters. Yet without facilities to aid rapid cooling and to store foods under cold conditions, bacteria will multiply to dangerous levels and there may be an accumulation of toxic substances.

Improvised cooling

Older houses and flats possess a cool cupboard or larder fitted with slate or stone slabs and with a ventilator protected against flies. Where the air flow is limited a small fan would be helpful. Foods should be covered, but all materials must allow an exchange of air otherwise there will be an increase in humidity, which will encourage mould and bacterial growth. Where space for cool storage is not available a box, louvered to allow a free flow of air, but protected from rain, should be hung on a wall away from direct sunlight.

There are various improvised methods for keeping food cool in the home. Porous earthenware vessels cooled with water help to keep foods such as milk and butter cold by evaporation. Muslin covering a bottle of milk or other container may be kept damp by dipping the four corners in cold water held in an outer receptacle; the container should be placed in a draught and out of the sun. A small louvered cupboard standing in water contained in a large basin or bath may be used for more bulky foods. Milk delivered after a family leaves for work should be placed in an insulated box, such as a polystyrene container, kept by the door in the shade.

Frozen packs of food allowed to thaw should be cooked and eaten within a short time; a vacuum flask will keep frozen foods solid for an extra day and delay the growth of bacteria for 3–4 days. Polystyrene containers designed to keep ice-cream solid are available; they will delay the growth of bacteria for at least 24 hours. Frozen meals for elderly incapacitated folk living alone may be stored in either of these containers for 2–3 days after delivery.

A wide variety of insulated containers, cool boxes or bags are now available which can be used to keep food cool during transport, for example, to picnic sites or while travelling. Such containers are also useful for keeping frozen foods in the frozen state after shopping at supermarkets or freezer centres.

Storage

Cold storage

In the UK the summer temperatures vary between 18 and 20°C (64 and 69°F) with occasional heatwaves with temperatures up to 32°C (90°C). Winter temperatures are usually 1–5°C (35–41°F) with occasional cold spells, and warm days when temperatures of 10°C (50°F) may be recorded. Kitchen and shop temperatures will be considerably higher.

The incidence of food poisoning is highest in the summer, but cases and outbreaks occur throughout the year. *C. perfringens* grows rapidly in food left to cool in the warmth of kitchens, and outbreaks due to this organism occur in any season. Adequate cold storage facilities in homes, shops and canteens would reduce the incidence of food poisoning; there is no other method which can effectively replace cold storage as a preventive measure.

Cold affects microorganisms in different ways depending on its intensity. As the temperature falls, bacterial activity declines; therefore foods which support bacterial growth should be stored at low temperatures to prolong their life and maintain safety. When foods are 'chilled' or stored at temperatures near but above freezing point, some bacteria will grow slowly, but in the frozen or solid state, many microorganisms will be killed directly in the process of freezing; the remainder will not multiply and the numbers tend gradually to diminish. Hence freezing preserves foods for a long time, while chilling merely delays the growth of organisms and extends the shelf life of the food.

Chilled storage

In the refrigeration trade, the term 'chilling' is used to cover any reduction in the normal temperature of the article concerned. For example, the ripening of tropical fruits is delayed during transit by storage at a temperature not far below that of the atmosphere. whereas the decomposition of imported meat is delayed by storage at −3 to 1°C (26.6–33.8°F) on ships.

Some foods cannot be chilled at too low a temperature because there may be harmful changes; for instance, the inside flesh of apples turns brown if chilled below 3.5°C (38.3°F) and the resistance of some fruits to moulds may be destroyed by chilling, so that the rate of spoilage by moulds is increased.

With regard to pathogenic organisms, some strains of salmonellae will grow at 10°C (50°F), but not at 5°C (41°F). *Staph. aureus* will not grow below about 10°C (50°F); between 15 and 20°C (59 and 68°F) there is growth and toxin production. The sporing anaerobic organism *C. perfringens* will not grow at temperatures much below 15–20°C (59–68°F) and no growth was observed in 6 days at 6.5°C (43.7°F). Most species of *C. botulinum* will grow very slowly at 10°C (50°F) and in some instances toxins may be formed at this temperature; in general, there is no growth or toxin formation at 5°C (41°F), although some strains of types B (non-proteolytic), E and F can grow and produce toxin at 3.5°C (38.3°F).

Unlike most organisms both *Yersinia enterocolitica* and *Listeria* spp can grow slowly at refrigerator temperatures; this fact must be borne in mind when it is necessary to store foods likely to be contaminated with these organisms for prolonged periods at chill temperatures.

Many other bacteria are able to increase slowly at chill temperatures, and under prolonged domestic refrigeration at 4–5°C (39.2–41.0°F) they will gradually spoil foods. Milk, for example, will develop 'off' flavours and odours from the growth of bacteria better adapted to the cold than those which grow and sour the milk at normal temperatures.

Foods of good bacteriological quality may be kept in a satisfactory condition at 4°C (39.2°F) for 3–4 days.

The education of food handlers in matters of food hygiene should include instruction in the correct use of refrigerators and cold rooms. In particular they must be taught that the cleanliness and safety of a refrigerated foodstuff are dependent on the extent of bacterial contamination before refrigeration as well as on the temperature of refrigeration; also that extreme cold merely delays the growth and multiplication of bacteria, which immediately renew their activity when the food is transferred to a warm room.

Deep freeze storage

The freezing of foodstuffs to approximately −18°C (0°F) kills many organisms, and the rate of death of the remainder will depend partly on the temperature of storage. Of the food-poisoning organisms, those of the salmonella group are said to be killed most rapidly on freezing. They disappear in 1 month, and staphylococci in 5 months, from strawberries kept at −18°C (0°F). Yet salmonellae have been isolated after years of frozen storage in whole egg products and meat. The spores of *C. perfringens* and *C. botulinum* are not affected by freezing and the toxin of *C. botulinum* has considerable resistance to alternate freezing and thawing at a temperature as low as −50°C (−58°F). Staphylococcal enterotoxin has been shown to withstand a temperature of −18°C (0°F) for several months.

Moulds and yeasts endure freezing conditions better than bacteria, thus refrigerators and freezers should be kept thoroughly cleaned and free from fungal and yeast growth.

When highly contaminated foodstuffs are kept frozen it is believed that changes may occur in the food owing to the slow activity of surviving organisms over a long period of time. Thus there may be slow spoilage during storage in the frozen state, although far less, of course, than that which would occur in the unfrozen food. Chemical changes in enzyme structures may also take place. Freezing will not restore the freshness of a food already highly contaminated or spoiled by bacterial action. When a frozen food is thawed those bacteria which have survived will recommence growth and decomposition, so that the keeping time of the food is limited and it must not be left at room temperature too long before being eaten, nor should it be refrozen. Manufacturers take great care in the preparation of frozen foods and most of them print instructions on the packet for their correct use. A temporary period of thawing due to power cuts or failure of the freezer cabinet mechanism or even during shopping does not necessarily mean that the partially thawed food should be discarded. Discretion must be used depending on the length of time, rise in temperature, and general condition of the food. When the central core is still frozen the outside will be cold enough to stop most bacteria from growing. Frozen food should be eaten as the manufacturer intends, freshly thawed from the original frozen state. Subsequent thawing and freezing will lead to quality deterioration as well as bacterial hazard.

If a thawed food cannot be used immediately it should be thoroughly cooked and refrozen as a cooked dish.

Cold storage accommodation

There should be ample cold storage space conveniently available in every kitchen. Where the size of the canteen justifies the extra expense, there should be a walk-in cold room with metal shelves and, in addition, one or more household refrigerators, so that foods with strong odours, such as fish, may be kept separate from other foods. Between the cold room and the kitchen there should be a well-defined air space to prevent the hot air from the kitchen reaching the cold room. The temperatures of all domestic refrigerators and cold rooms should be checked regularly by thermometers placed in positions where they can be read easily. Cold rooms and domestic refrigerators ideally should be maintained at temperatures from 1–4°C (33.8–39.2°F). A survey of 559 refrigeration units in 30 hotels in London showed that 110 (19.7%) were set at temperatures above those recommended. The Food Hygiene Regulations have been strengthened with respect to temperature at which cerain foods should be maintained. A two-tier system has been introduced whereby certain foods must be kept at temperatures not greater than 8°C and 5°C for foods in the highest risk group (see Chapter 17).

The life of food in the refrigerator is limited and the longer it remains there, the shorter its life on removal. The results of experiments on the storage life at 0°C (32°F) of meat with different initial numbers of bacteria indicated that meat of good bacteriological quality before refrigeration would keep for many days whereas poor quality meat had a short storage life. When the initial count of bacteria was low, the appearance of slime on refrigerated meat was delayed for 18 days; when counts were high before refrigeration, slime appeared in 8 days. Conditions for the temporary storage of various foodstuffs are recommended by the Ministry of Agriculture, Fisheries and Food; similar recommendations are given in the USA. (See p. 68.)

Refrigerators should be defrosted regularly, automatically or according to the manufacturer's instructions. They should not be overcrowded, so that air circulation may be good; it is advisable to check the contents at the time of defrosting and aged foods should be thrown away. Space in the refrigerator ought to be available for cooked and uncooked foods which are perishable (see below) because they are susceptible to bacterial attack. Aluminium foil, greaseproof paper, or polythene wrapped around foods will prevent loss of moisture and the spread of flavours to other foods. Many foodstuffs commonly stored in a refrigerator do not encourage bacterial growth and they can with safety be kept for a few days in a cool room on a slate or stone slab (see below). Milk can be stored in the cool room during the winter, but in the summer it should be refrigerated whether in bottles, cartons or in churns. Receptacles containing milk should be covered, and the outside cleaned before they are placed in the refrigerator, cold room or larder.

Most unopened canned goods do not require refrigerated storage (see below). Large cans of ham and similar meats which have received a pasteurization process only should be kept under cold storage until required.

At the time of defrosting, walls and shelves of refrigerators should be washed with soap or detergent and warm water and carefully dried before the food is replaced. If refrigerators are switched off at holiday times, they must be emptied and carefully dried to prevent mould growth, and the doors left slightly open.

Suggested conditions for the temporary storage of foodstuffs

Perishable foods

Certain perishable foods should be kept in a refrigerator maintained at a temperature of 1–4°C (33.8–39.2°F). These foods include meat, rabbit, game, poultry, fish, shell eggs, milk and cream. Fish, however, should be kept separately if possible as it readily taints other foods.

Fats, butter, lard, margarine, hard cheese and cured bacon may be kept in a cool room, but if the temperature rises they should be transferred to a refrigerator.

Frosted foods and frozen liquid egg should be kept frozen until required; a temperature of −20.6°C (−5°F) is suitable.

Fruit and vegetables should be kept in a cool and well-ventilated place protected from frost. Sacks of root vegetables should stand on duck boards to allow the circulation of air. Vegetables should be prepared in a separate room or section of the kitchen so that dust and earth from potatoes, carrots, turnips, and other root vegetables may be diverted from the main area of food preparation. (Fig. 8.1, p. 141, gives temperatures for the storage of various foods.)

Milk packs

Refrigerated milk packs are useful in restaurants, canteens, hospitals and other large-scale catering establishments. The machines will accommodate three 5-gallon packs of milk, and they can be used in conjunction with automatic vending of milk.

The use of milk packs overcomes the risk of contamination of milk after arrival on catering premises and resolves the nuisance of empty milk bottles.

Other food storage

In addition to refrigerated and cool storage, there should be dry, light and airy cupboards or rooms for canned and other packaged goods stacked on shelves and marked for rotation. Powdered and granular foods such as flour, sugar, dried milk, tea, oatmeal, sago, rice, egg powder and coconut should be stored in metal bins or metal or glass jars with close-fitting lids. Even these containers should be raised at least 450 mm (18 inches) above the floor to allow space for cleaning. Goods packed in cardboard or wooden crates or cases should be well clear of the floor and preferably on higher shelves. The rooms should be designed to discourage vermin, flies and dust, and they should be easy to clean. Dried fruit, fish or meat should be kept cool and dry. It is important to avoid conditions which could lead to condensation of moisture on the surface.

Storage of cleaning materials and kitchenware and utensils

Cleaning materials require separate accommodation, which should be large enough to allow sufficient space for sinks for washing, boilers for the disinfection of cloths and mop heads, tubular steel racks for drying cloths, mops, brushes and other equipment, and ample light to observe the state of cleanliness of materials. Utility rooms should be carefully designed and used.

Personal clothing and other belongings should not be left in the kitchen, but stored in proper accommodation nearby.

Equipment for cooking, receptacles, utensils, crockery and cutlery should be stacked and protected from dust and stored away from raw foods.

The hygiene of food storage and preparation depends on a knowledge of the habits of bacteria, the food they contaminate most frequently, and the temperatures they do and do not like for multiplication. When these facts are known, a common-sense view will enable the more susceptible foods to be protected against the potential danger of contamination with food-poisoning and spoilage bacteria and their growth.

12

Food hygiene in food manufacture

Malcolm Kane

Introduction

The principles of good hygiene practice in food factories are basically the same as in any other food-handling environment. Nevertheless, the scale of food manufacture and the unique processes developed to produce shelf-stable preserved or sterilized foods for storage, distribution and retail sale provide greater opportunities for food-poisoning agents to become hazardous.

The most important factor for consideration is that attention to the cleanliness of plant surfaces alone is not enough; proper design and control of the complete process are essential to ensure the absence of food-poisoning agents.

The chapter is intended to convey the scope of the considerations necessary for effective hygiene in food manufacture.

Factory hygiene

Although it is generally recognized that good hygiene is essential in food factories as in all food environments, the severity of some food processes can apparently 'overcome' poor initial hygienic practice. Even today it may be considered unnecessary to correct bad practices, because it is assumed that nothing can survive the subsequent baking or other means of cooking and sterilization. Such attitudes must be challenged because they are unethical and potentially unsafe.

Buildings and facilities

Food factories should be designed and built for particular purposes with materials capable of withstanding various physical conditions without deterioration. The principal factors to consider are heat, cold, humidity and vibration.

Interior surfaces should be smooth, non-porous, easily cleaned and not vulnerable to chemical attack by modern detergents and disinfectants or able to sustain biological/microbiological growth.

Good natural light and screened electric light must be provided; plastics are preferable to glass unless the glass is screened to protect from breakage. Electric bulbs and fluorescent tubes should be coated or ensheathed in plastic.

Paints and other surface coats should be non-toxic and not flake; those that contain mould inhibitors must not come into contact with foods.

Overhead structural items, such as beams, conduits and pipes need to be readily accessible for cleaning and screened from open food processes below, preferably with a secure suspended ceiling. The space above such ceilings must be protected from dirt and vermin. Windows require screens against flying insects and birds.

Doors need to fit tightly with air curtains for use when open. A positive air pressure maintained inside the main area of food processing will discourage the entry of insects.

Pipework, drainage ducts, conduits for power supply and other channels should be tightly sealed where they pass through walls, floors and ceiling to prevent entry of vermin and insects.

Water and sewerage

A sound and plentiful supply of potable water is required for all food processes. Generally, chlorinated municipal mains or water privately treated to the same standards of purity should be used. There must be strict control of recirculated water; the amount of chemical treatment it receives will depend on the degree of use and reuse.

Records of the chemical treatment, which is usually by chlorination, and the results of microbiological tests must be kept for all incoming process water. The condition of spent process water will reflect the nature of the process. Free residual chlorine will be present in the retort water after cans have been cooled, but when open foods have been washed, residual chlorine will not always be present as it could create a problem of chlorine taint in the foodstuff. Water supplies to the factory must be adequate to match the maximum capacity of manufacture.

Borehole water is frequently used for industrial processes and can be a satisfactory source of potable water suitable for food manufacture, sometimes without chemical treatment. Where borehole water is used for food processes it must be monitored regularly for chemical and microbiological content. Deep boreholes of 25 m depth or greater can provide pure water, but they should be steel-lined near to ground level to prevent entry of and contamination by surface water. They should be sealed with sanitary fittings of conventional type, and water drawn constantly to maintain freshness and purity. Water from shallow boreholes and open wells should not be used for food manufacture, unless treated with chlorine or other chemicals. All water held in storage tanks should be chemically treated before use for food manufacture, regardless of source.

Sewerage systems must be of sufficient capacity to cope with the maximum demand of the particular manufacturing process. Fat traps and other mechanisms for filtering solid waste materials should be simple, accessible and easy to clean. Back-flush of sewage into the factory must be prevented by non-return flow valves. Human sewage should not mix with factory effluent within the environment of the food factory.

Plant layout and design

Plant for manufacture should be laid out with clear, preferably 'straight through' lines of product flow. Final or intermediate processes must not be in a position where there could be cross-contamination by raw materials.

The preparation areas for all raw meats and all cooked meats should be physically separated by appropriate walls. The movement of operators between the separate processing areas, particularly in meat plants, should be severely restricted by the design of pedestrian facilities, use of allocated coloured clothing and control by management.

Facilities for washing hands with knee- or foot-operated hot water taps should be positioned at all pedestrian entrances in food factories as well as in the toilets. Suitable liquid unperfumed soaps, barrier creams and disposable towels or other suitable drying facilities must be provided at all times, with instruction given as to their proper use.

Manufacturing equipment should be designed to prevent the entry of foreign materials, and the development of 'out-of-sight' dead spots, especially within the operation chamber and associated pipework. Bolts and clips should be attached externally so that all internal product contact surfaces are smooth and easily cleaned. For operator safety, all equipment should be made safe by electrical isolation during dismantling and cleaning.

Materials used for construction must be non-toxic, smooth and with unpainted food contact surfaces; polished stainless steel is preferred. Care is needed to ensure that food materials do not attack the metal; for example, acid reacts with mild steel and salt with aluminium. Equally, the construction materials must not attack the foodstuff, for example vegetable oil with copper. Wood should not be used in open food areas.

Appropriate services for cleaning operations, such as hot and cold water, drainage, compressed air and vacuum lines should be adjacent for cleaning operations.

Cleaning systems

Production operations require careful management on a clean-as-you-go basis to maintain good housekeeping, even during production at peak periods. A structured approach is required to the formal methods and frequency of cleaning of all plant and equipment, including instructions for reassembly. The state of hygiene in a factory should be inspected and reviewed by means of a formal management audit. The subsequent hygiene report will be used to revise cleaning methods and frequency.

Wood and bristle utensils should not be used, and all cleaning materials should themselves be thoroughly cleaned and not be a source of further contamination. It is recommended that the basic principles of plant cleaning be followed in the sequential procedure of five stages, namely, rinse/detergent wash/rinse/disinfectant/rinse.

Cutting knives and similar hand utensils should be stored in a special rack connected to steam or chemical disinfectant.

The use of combined detergent disinfectants needs to be carefully controlled and it must be understood that they are not effective in the presence of heavy soiling, which will require prior cleaning. Cleaning systems which are automatically controlled, including 'Cleaning-In-Place' (CIP) should be designed for the purpose and carefully monitored. CIP, in particular, requires careful matching of pump velocity and detergent flow for efficient operation.

A safety lock-out mechanism must prevent leakage of detergent/disinfectant into foodstuffs during the processing of food. Modern industrial detergents and disinfectants are often strongly acid or alkaline, and if consumed in error in sufficient quantity could give rise to chemical poisoning. All CIP systems need to be supplemented periodically by the more conventional strip-down methods of plant cleaning, with all couplings open and gaskets removed. Chemical disinfectants should be changed at intervals to prevent the development of microbial resistance.

All factories should be subjected to inspection and treatment for the control of vermin and infestation. Care must be taken to prevent the contamination of foods with rodenticide and pesticide residues.

The importance of processing control in hygienic food manufacture

Personnel

Food process control starts with the appointment of suitable qualified and experienced staff. Engineers, production managers and quality control personnel are all involved in and responsible for food safety; food factories should not place reliance on untrained staff. At least one qualified food scientist or technologist is required in a position of managerial responsibility and authority. All technicians must be qualified to take responsibility for the accurate measurement and recording, and understanding of the data they collect. They must understand the risks to food safety and quality if there are failures in processes and observations; the appropriate training of all staff is important.

All staff must be provided with protective clothing including cover for the hair; jewellery should not be worn. Staff must be screened for health risks.

Raw materials

The purchase of all raw materials should be subject to formal written specifications of physical, chemical, organoleptic and microbiological standards. Defects should be quantified, and tolerable limits set for incoming raw materials examined against appropriate specifications prior to acceptance.

Production priorities should not be allowed to override prudent rejection of deliveries of 'Out of Specification' raw materials. Suppliers of raw materials should be selected with care and rejected lots identified to ensure that they are not resubmitted.

Particular hazards, for example aflatoxin, pesticide and herbicide residues, should be monitored in susceptible raw materials.

Process design

It is important that the food product and the process be designed by competent food scientists and technologists who understand the individual importance and inter-relationships of the main chemical and physical parameters.

The water activity (a_w) requirement of the various microbiological agents of food spoilage and food poisoning should be understood and, particularly, the limiting a_w for known pathogens; similarly with pH/acidity, when the inhibitory effects of increasing acidity on the microbiological agents of food spoilage and food poisoning, and the limiting pH for pathogens should be known. A compilation of the factors influencing the growth and death of pathogens in food can be found in books 3 and 5 produced by the ICMSF (see p. 151 and 309).

The integration of time and temperature is important in the context of the relevant thermobacteriology. The determination of process values for pasteurization and sterilization should be carried out regularly where necessary. The establishment of a 'Minimum Botulinum Cook' in terms of F_0 value must be confirmed for every sterilized foodstuff packed in any rigid or flexible hermetically sealed container.

Chemical preservatives

The mode of action of chemical curing and preservative agents should be understood and the maximum and minimum limits for safe usage known; long-term shelf-life studies are required to establish the product conformance.

Synergism

The additive and synergistic effects brought about by the inter-reaction of any two or more of the main chemical and physical agents in a given food system require to be known. In particular, the effect upon the stability and safety of subsequent changes in formulation of a food must be determined.

Pack integrity

Standards of seam and seal integrity must be clearly defined and monitored against specifications. Physical standards should be supported by incubation checks of final products and the establishment of target standards for the incidence of leaker spoilage.

Process monitoring

The maintenance of manual and automatic records is essential for processes such as can sterilization. Formal correlation of all data should be confirmed on all batches prior to release of products from the warehouse; i.e. on a 'Positve release' system. 'Negative release' systems must not be used with any process designed to eliminate a known pathogen hazard. (With a 'Negative release' system, product is routinely released for sale unless a QC restriction is exceptionally applied).

Quality Control (QC) schedules cover all nature of checks and frequencies of checking and recording. Contingencies arising from adverse results in QC reports need to be clearly reported with corrective action taken, recorded and authorized. Full records of all checks, reports and actions must be maintained for at least two years. Contingency procedures for corrective action should be prepared for predictable QC adverse reports. Contingency procedures must be drawn up for withdrawal and public recall of accidentally released unsafe product. Management audits of QC schedules, systems and contingency procedures must be formally carried out at least annually.

Final product

Quality evaluation of the final product cannot control the quality. Control must be applied 'upstream' to the raw materials and processes. Adequate control of the raw materials entering a factory and of all production processes, measured against clearly defined specifications, ought to ensure that the final product is of the standard required. Evaluation of this standard against a clearly defined specification for the final product will therefore support all the'upstream' controls. In this context, the most important standard for the final product is organoleptic. The end product is food which needs to be regularly sampled.

The Hazard Analysis Critical Control Point (HACCP) System

An approach that is being increasingly applied by food manufacturers in order to ensure the safety of their products is that of Hazard Analysis Critical Control Point (HACCP). It is a structured approach that will assess the potential hazards of a food operation and decide which areas are critical to the safety of the consumer. Once identified, the critical control points (CCPs) can be monitored and deviations from the safety limits can be corrected. The HACCP approach can be applied to any food operation, from the large manufacturer to the small caterer. Each food product requires a separate HACCP system.

Application of the HACCP system requires a team effort with personnel involved in all aspects of production. A typical team might include the food technologist, microbiologist, engineer, production manager, quality assurance manager and the hygiene manager.

The system can be broken down into a number of relatively simple stages and an organized approach to the analysis is provided by Hazard and Operability Studies (HAZOP). HAZOP is a carefully considered set of questions designed to ensure that all the relevant safety factors are included in the HACCP process. The sequence of instructions is as follows:

(i) Defining the process;
(ii) Identifying the hazards;
(iii) Assessing the hazards and risk;
(iv) Identifying the CCPs;
(v) Specifying monitoring and control procedures;
(vi) Implementing control at CCPs;
(vii) Verification.

(i) *Defining the process*

A flow chart is produced which defines all aspects of the process from raw materials through processing to distribution, retail and customer involvement. The factors which must be included are times and temperatures of cooking, chilling and refrigeration; chemical, physical and microbiological characteristics of ingredients; quality assurance and management systems; details of shelf life and of handling by the customer. The flow chart should then be audited under actual production conditions and amended as necessary.

(ii) *Identification of hazards*

This involves identification of the hazards associated with each stage of the operation. A hazard is any factor that may cause harm to the consumer: it may be microbiological, for example, the presence of food-poisoning organisms or their toxins, chemical, the presence of toxic chemicals used for cleaning or physical, foreign bodies such as glass. A detailed analysis of the product and its intended use is required. For example, the following points should be considered in relation to possible hazards.

(1) Product formulation

Microbiological profile of raw material

pH a_w preservatives	will these individually or in combination prevent growth?

(2) Intended process
Steps leading to contamination by, or survival or destruction, inhibition of or growth of, food-poisoning and spoilage microorganisms.

(3) Intended use and distribution

Distribution temperature
Packaging – integral to stability of product or not
Expected shelf life – during distribution and storage and in user's hands
Preparation for consumption
Potential mishandling
Susceptibility of the consumer.

Based on answers to these points a preliminary assessment of hazards can be made. Other aspects which must also be considered include hygienic design of food handling areas and equipment, cleaning and disinfection and health and hygiene of personnel.

(iii) *Assessment of hazard*
The severity of each hazard must then be graded. The risk associated with each hazard is the likelihood of its occurrence. Will it always be present or only once a day or once a year? Severity relates to the magnitude of the hazard; whether it is life-threatening, how many will be affected, will there be extensive product spoilage? For example, the growth and toxin production by *Clostridium botulinum* in a shelf-stable canned cured meat would be a severe hazard, but the risk is low. Milk that has been inadequately pasteurized so that salmonellae survive is a severe hazard with a high risk. When a processed product is stored at a temperature that would allow thermophiles to grow and spoil the product the hazard is low, but the risk is high.

(iv) *Identification of CCPs*
Once the hazards have been assessed and ranked according to severity, the CCPs or points at which control of the hazards can be applied are identified. In some processes a single operation at a CCP can completely eliminate one or more microbial hazards and is designated a CCP1. Other CCPs may minimize but not completely eliminate a hazard and are designated CCP2. An example of CCP1 is the thermal process for canned foods or the pasteurization/UHT treatment of milk. While a CCP2 would include chilling after heat treatment. Both types of CCPs are important and must be checked.

(v) *Specifying, monitoring and control procedures*
Often a hazard will indicate its own control procedure. Any process relying upon heat to eliminate pathogenic bacteria will require to be monitored for time and temperature of the heat treatment to ensure that the specified values are achieved and remain within acceptable limits. The procedures should also define the action to be taken when the limits are exceeded, the type of action, when, by whom and the limits requiring further action.

(vi) *Implementing control at CCPs*
Unless the monitoring and control procedures are implemented at CCPs the whole HACCP system becomes meaningless. The main types of monitoring are:

Visual observation	– checking labels to ensure ingredients are within use by date given
Sensory evaluation	– detection of 'off' odours
Physical measurement	– temperature, pH, a_w
Chemical testing	– phosphatase in pasteurized milk
Microbiological examination	– presence/levels of pathogenic or spoilage microorganisms

The effectiveness of monitoring is directly related to the speed with which results are obtained, so visual observation is often the most useful. Results of monitoring should be recorded.

(vii) *Verification*
Verification is the use of supplementary information to check whether the HACCP system is working. An example of verification is the use of incubation tests on packs of UHT milk. If contaminated packs are found then checks can be made to determine whether the failure occurred at a CCP (e.g., inadequate heating or contamination at the filling stage) or whether the monitoring system failed to detect an out-of-control situation at that step.

More tests may be applied at a CCP during verification of an established system which would not be suitable for routine monitoring. For example, equipment hygiene may be monitored routinely by visual inspection, but for verification that cleaning and disinfection are effective microbiological tests will be required to determine whether there has been a build-up of microorganisms to unacceptable levels. End product tests are also examples of verification.

The advantages of implementation of an HACCP process are numerous:

(a) control efforts are concentrated on the critical steps of the operation;
(b) control is often achieved by easily-monitored parameters such as time, temperature and visual assessment;
(c) results are available as faults occur so that prompt remedial action can be taken;
(d) control is affected by the process operator;
(e) potential hazards are taken into account;
(f) the system is flexible and can be applied to changes in or the introduction of new processes;
(g) the system involves all levels of staff concerned with product safety.

The system also has limitations which include high specificity for each operation, the need for input from properly trained personnel and considerable demands on time and resources particularly if a single producer is involved with a wide range of products or processes.

Even if HACCP cannot be fully implemented in a food business the basic principles of identification of hazards and control points should still be applied as far as possible.

Figs 12.1 and 12.2 give examples of flow charts relating to the processing of two foods (Fig. 12.1) the manufacture of UHT milk and the catering-scale production of cooked turkey in a cook-freeze or cook-chill system (Fig. 12.2).

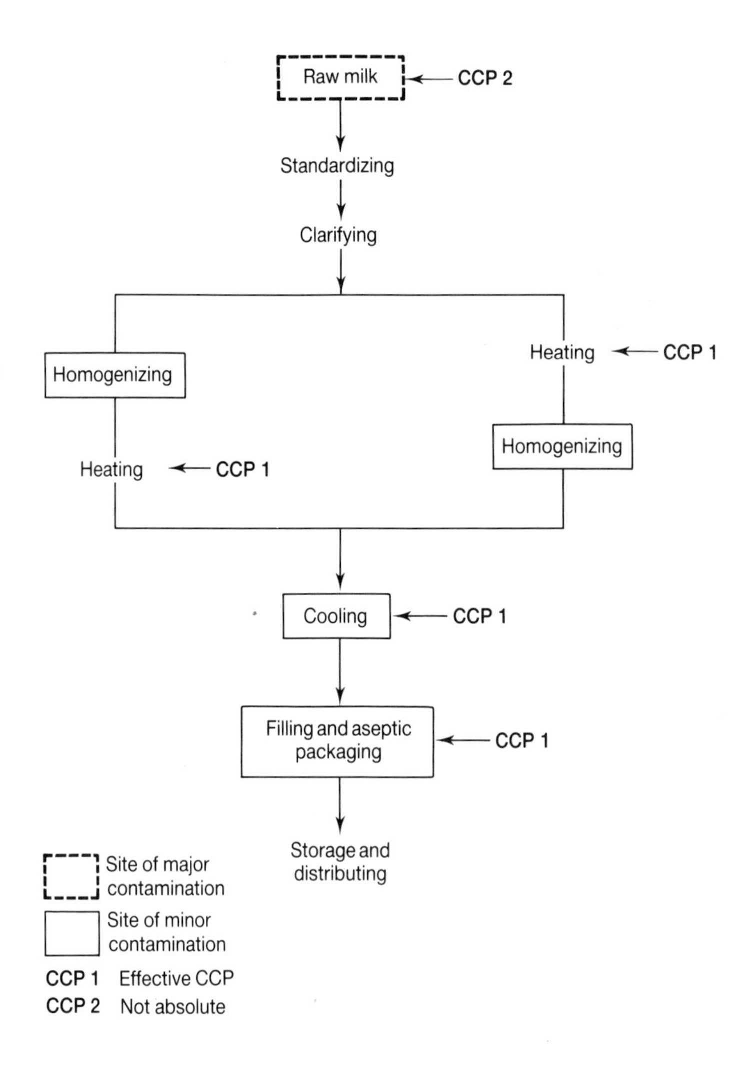

Fig 12.1 Flow diagram for the production of UHT milk (modified from ICMSF, 1988)

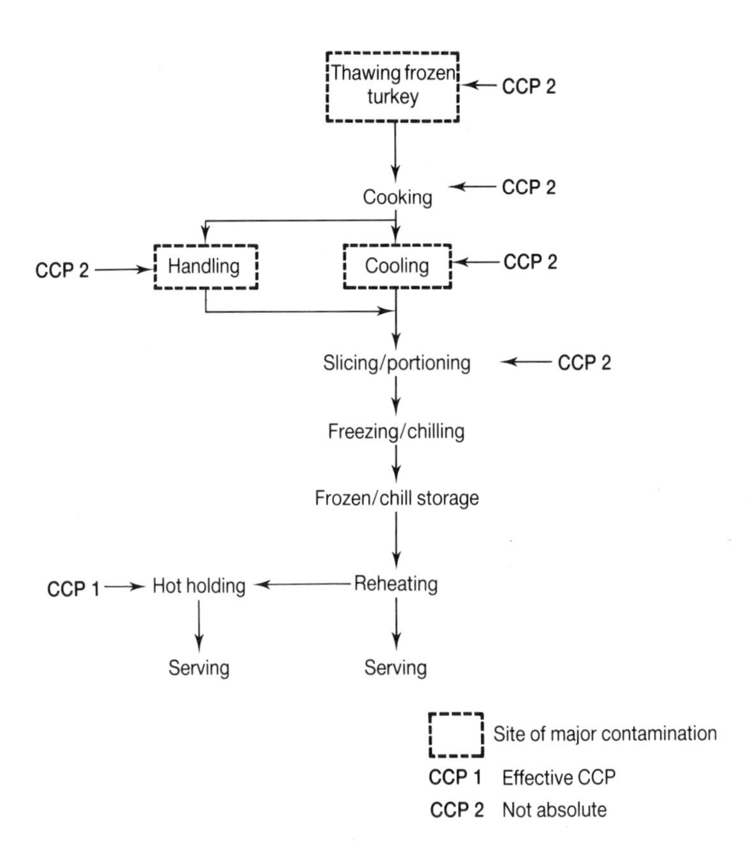

Fig 12.2 Flow diagram for the production of cooked turkey in a cook-freeze/cook-chill catering system (modified from ICMSF, 1988)

13

Food hygiene in the retail trade

Wendy Spence

Introduction

Today, retailing of food covers many different outlets. They vary from the tiny 'corner shop' to the large supermarket, from the vending machine to the farm shop. Each can have its own particular hygiene problem although the overriding factor is to ensure that the food sold to the customer is fit to eat and of good quality. Over the last few years there has been a general increase in reported food-poisoning incidents, but these are rarely traced to retail outlets. One reason may be that sickness and diarrhoea occurring in the family unit is often not reported and secondly, if food is contaminated, the contamination can usually be traced to a stage prior to it entering the retail outlet. It is the retailer's responsibility to ensure that food is kept in a manner and under conditions which will not encourage the growth and spread of any food-poisoning organisms.

The main legislation covering food hygiene in retail outlets is the Food Safety Act, 1990, the Food Hygiene (General) Regulations, 1970 and the various 'amendment' Regulations. (See Chapter 17).

An important aspect in relation to retailing is the definition of 'open food' and 'food' not to be regarded as open food. The Food Hygiene (General) Regulations, 1970 define 'open food' as follows; 'Food that is not in a container of such materials and so closed as to exclude the risk of contamination. . .'. Food not to be regarded as 'open food' can be found in Column (1) of the Schedule attached to the above regulations.

Open food premises

The growth of supermarket companies in the 1960s saw a reduction in the number of traditional retail outlets selling only narrowly defined types of products. Instead, the average supermarket offered a wide range of foods, most of which were pre-wrapped, although some were packed on site but behind the scenes in suitable preparation areas. The 1980/90s have seen a return to the sale of open foods, but in markedly different style. In independent outlets there has been a re-emergence of service counters selling open foods. The standards of hygiene at both types of outlets are similar, being strongly influenced by increasing customer expectations. This trend has necessitated a change of attitude amongst certain traditional trades, most noticeably bakeries, which were not commonly on view to customers. Appropriate standards for individual open-food risks are detailed by the relevant class of food.

Delicatessen and cooked meat sales

The range of delicatessen products (Fig. 13.1) is continually increasing and poses various hygiene risks. Cross-contamination both between raw and cooked foods and between foods of distinctive flavours is rivalled only by inadequate temperature control as the major risk. Foods as diverse as raw salads and roll-mop herring are often sold alongside sausage and cheese, whilst a range of cooked meats and pâtés may be freely mingled in the display.

Food handling

The following points are indicators to be observed during examination of procedures:-

(1) *Separation*. Major groups of foods should be displayed separately and kept apart by display dividers. Individual spoons and other handling equipment must be provided for each individual product, with separate

Fig. 13.1 A delicatessen counter

colour-coded equipment for raw and cooked foods. Scales and slicers should only be used for defined groups of products. It is also preferable to have separate staff for the raw and cooked products.

(2) *Product life*. The types of product handled vary considerably in their shelf lives, some are also quite large initially, for example whole hams, and may take some days to sell. The study of sales and sales projections, becomes necessary to avoid overstocking. Codes are often displayed on outer packaging only and so must be transferred to the individual food items in that package to ensure that the product life is apparent. Pre-sliced foods should not be carried over from day to day. Where 'use by' dates are given it is now an offence to sell the product after that date, stock rotation and checks are, therefore, of the utmost importance. The daily checks should be logged and could become part of the 'due diligence' defence. (See Chapter 17).

(3) *Structural and equipment*. Each counter should be provided with the following facilities:-
- Hand wash basin, liquid soap, nail brush and paper towels. Ideally two hand wash basins should be provided if bacon, sausages or other raw foods are handled
- Double internal sink, particularly if raw and cooked foods are handled
- Adequate chopping boards (not made of wood, and colour coded to denote use)
- Refrigerated counter – screened from customer contamination
- Adequate refrigerated storage
- Access to first-aid equipment
- Refuse receptacles
- Storage for packing materials
- Storage for cleaning materials
- Sufficient spoons, scoops, slices, knives, etc. (not made of wood, and colour coded to denote use)
- Sanitizing cleaning reagent in a bucket or suitable container, with clean wiping material (woven paper types are preferred)

(4) *Temperature control*. In accordance with the new legislation on temperature control (The Food Hygiene (Amendment) Regulations, 1990) (see Chapter 17) chilled counters and storage facilities should be maintained at 5°C. As many display units do not have forced air flow, particular attention needs to be paid to load levels.

Products removed from chill for slicing or packaging must be returned to chill after use and not left on slicers or work tops. The mechanical efficiency and physical condition of all chiller units should be regularly checked. Temperature checks must be carried out with accurate temperature probes and records maintained for inspection.

It will be necessary for all products listed in the new regulations on temperature control of foods to be refrigerated as necessary. The 8°C band is already operational and the 5°C maximum for more vulnerable modules will come into force in April 1993.

Bakeries

Hot bread shops and in-store bakeries (Fig. 13.2) have brought about a dramatic increase in the perceived standards of the bakery trade. This traditionally difficult occupation has had to respond to the eyes of its customers and there have been many advances in hygiene in recent years. The major concerns continue to be the avoidance of 'foreign objects', elimination of infestation, and the control of perishable commodities such as those filled with cream or custard. Ease of cleaning and cleaning practices should also receive particular attention.

Foreign object control

The two golden rules of 'keep it covered' and 'work tidily' should be continually repeated to all staff, whilst equipment design must reflect these principles. More explicit examples of good control of foreign objects include:-

(i) storage of flours, mixes and other ingredients in closed containers;
(ii) exclusion of glass from the unit;
(iii) sieving of white flour (brown flours are unsuitable for sieving);
(iv) filling racks and shelves from the top downwards;
(v) not placing sacks and outer containers on work surfaces;
(vi) not allowing carbon to build up on tins and trays;
(vii) maintaining decoration and equipment in good condition to prevent flaking of paint;
(viii) not allowing the use of loose types of rodent bait;

Fig. 13.2 In-store bakery in a supermarket

(ix) taking care when maintaining machinery, looking for loose nuts and bolts;
(x) elimination of wood;
(xi) carrying out a thorough check of the area where engineers have been working.

Temperature

Cream and custard fillings are obviously perishable (although some custards have quite long lives), care should, therefore, be taken when handling these products, necessary precautions must include:-

(i) monitoring the efficiency of all display cases and the temperature of cream whipping machines;
(ii) ensuring all relevant goods are displayed under chilled conditions;
(iii) avoiding excessive exposure time at ambient temperatures during preparation;
(iv) refrigeration of perishable ingredients.

Cleaning

There should be separate cleaning protocols to cover removal of daily soiling and long-term residual dirt. All machinery, bread trays and bread tins should be included in a separate deep clean scheme, ideally carried out by a specialist firm.

All machinery and equipment should be designed for ease of cleaning, but particular care should be taken to ensure internal access to machinery and the avoidance of crevices and sharp angles. Radius joints are naturally preferred.

Infestation control

The higher standards in the new-style bakeries have helped reduce infestation, through avoidance of residual debris within the machines, and cleaning of inaccessible areas. Commodity storage areas require particular attention to prevent infestation. Flying insects continue to be troublesome, so fly-proof windows and adequate 'ultraviolet' machines are essential.

Silo rooms

These should be cleaned on a 6–12 monthly basis by a specialist company, because apart from hygiene, there are safety implications due to the explosion factor. These areas should be regularly checked to ensure that flour has not escaped and encouraged insect infestation.

Fresh meat sales

In many ways the contemporary butchery department is the most refined of food-production departments, or units, in retailing. The operation has become simpler with the introduction of delivery of primal cuts, rather than whole carcasses from the slaughterhouse. This trade has now the advantage of much well-designed equipment and machinery and new developments taking place in packaging techniques.

Cleaning

Basic cleaning should not be difficult, with day to day soiling of a relatively simple nature. If cleaning is only superficial there will be long-term protein build-up thus the choice of cleaning detergent is important. Since trays are widely used for meats, a tray washer is desirable, the size of which should reflect the throughput of the unit. The dominant food contact surface is the chopping board or block, which should be colour coded, made of a suitable synthetic material and when scored, planed or replaced.

Machinery

With the advent of complex packaging machinery, a different cleaning technique is required, as butchers are only accustomed to readily visible soiling. Many machines have hidden crevices and trays, which require particular in-depth cleaning.

Product control

Many butchers, nowadays, receive vacuum-packed units of meat from wholesalers.

Shelf life control is vital, and also the detection of leaking packs. Controlled atmosphere retail packs have also been introduced, these may be subject to the same problems as 'vac-pacs'. In all cases the control of codes and product life is of importance. 'Foreign objects' are frequently introduced during packaging. Correct storage and strict control of packs prior to filling will help reduce the risks.

Other matters

Personnel hygiene should not be a major problem in these units, but cross-contamination will be a risk if butchery preparation or sales of raw meat are mixed with other goods, in particular cooked foods. All cutting and display of raw meat must be segregated from cooked products both physically and by sound hygiene practices. Layout and structure of cutting rooms is well established, but increasing emphasis towards health and safety must be balanced against ease of cleaning and avoidance of residual meat debris in machinery.

Wet fish sales

The reduction in the overall volume of fish sales nationally has resulted in the polarization of this trade; the small fishmongers have almost disappeared, whilst the quality specialist has remained successful. Various retail chains have also continued to sell wet fish from their superstores (Fig. 13.3). The growth of wet fish sales from non-traditional outlets has been encouraged by advances in packaging and preparation techniques at the wholesaler, most particularly by the introduction of controlled atmosphere packing (CAP).

The hygiene problems relating to fish retailers have remained unchanged over many years, but as in other retail trades, the demands of the customer have resulted in a rise in perceived standards.

Attention to hygiene should include:-

Cleaning

Protein build-up will inevitably occur on all surfaces, and cleaning schedules and methods must take this into consideration. On some metal surfaces aggressive low pH cleaners will be necessary to remove the compound resulting from chemical reaction between metal surfaces and 'drip' fluids from the fish. Attention to detailed cleaning is important due to the unfavourable characteristics of the product; fish scales and protein will build up in all cracks, crevices and inaccessible areas. The corners and sides of display cases will attract similar soiling.

Fig. 13.3 Fish department in a supermarket

Cleaning schedules should therefore detail the thorough scrubbing, with appropriate materials, of all equipment including scales, scale pans, spoons, tongs and price label and bag dispensers.

Cross-contamination

All fish should be screened from the customer and this has now become normal practice. Raw fish, shellfish, and cooked fish should be physically segregated from one another. Care is required to prevent contamination from the external surfaces of delivery pack materials.

Other matters

Ice-making machines are frequently available in modern units, but these machines must be emptied and cleaned nightly. Gutting areas should be provided with running water for sluicing the chopping boards, a suitable sanitizing solution should be available for use afterwards. Careful attention should be given to the disposal of refuse, particularly fish-soaked cartons.

Produce sales

The sale of fruit and vegetables is another trade that has changed over the last 5 years, with the introduction of prepared salads and many exotic fruits and vegetables. Food safety is important and there are several problem areas. Pest control is, surprisingly, neglected in some instances, whilst customer aesthetics, foreign-object control and cut foods are all of concern. Developments have also taken place in prepacking (with the inherent problems of shelf life and coding), and the use of refrigerated displays. The sales of loose, dried products such as figs, dates and apricots also brings the greengrocer closer to the high-risk food hygiene areas.

Cleaning

Many of the difficulties of outlets for produce sales revolve around both poor standards of finish and equipment, and the trade's inherent lack of desire to clean adequately. Attention should be directed towards the design and purchase of display and storage equipment that can be easily cleaned. Cleaning methods should entail the removal of soil and debris, build-up of which is likely to harbour pests, both native and related to the product.

Food handling

Many produce outlets carry out various minor preparation tasks, such as the cutting and wrapping of melons and cucumbers. These operations can only be regarded as open food operations and should receive care and attention especially with regard to knives and cutting surfaces.

Refuse has always been difficult for the produce retailer, but the primary rule remains unchanged; food debris should be kept separately, and well sealed, seepage is particularly attractive to pests. Packaging and cardboard should be separate, but tidy, accumulations provide ideal harbourage for pests.

Staff restaurants

Premises that serve food for consumption by food handlers deserve special attention from a food safety view point. If retail staff handling open food succumb to food poisoning following the consumption of food in their staff restaurant and pass it on to the public the consequences can be widespread and may be far more serious than a similar incident in a public restaurant. The precautions and controls are similar to those for any restaurant, but should be enforced with even greater stringency. Controls should be directed toward prevention of infectious disease among staff, sanitizing systems, and avoidance of cross-contamination. Temperature control is significant, but often there is only one main meal each day, which helps minimize the opportunity for bacterial growth. Food sales displays should be covered, with adequate temperature control. Where practicable food samples should be retained for 48 hours as required in normal catering practice. The use of inferior, waste and out-of-code food from the retail departments should be discouraged at all times.

Non-open food premises

Pre-packed grocery sales

There are a number of retail outlets where only packaged goods are sold. Here, only the basic principles of hygiene apply as it is unlikely that food poisoning will be caused by these products unless the pack becomes damaged by infestation, dampness or bad handling.

All products, whether they are canned or dry goods, must be kept in a dry, ambient condition, with little fluctuation in temperature. If not, cans could become rusty and deteriorate, allowing air to enter. With regard to dry goods, damp packaging could cause mould growth, especially with commodities such as tea and flour.

Good stock rotation is essential, and when shelves are refilled, old stock should be brought to the front. More goods are now coded and therefore stock rotation is easier, but it must be maintained.

Pre-packed perishable sales

These products are particularly vulnerable products and great care must be taken to ensure that they are stocked and retailed in suitably refrigerated cabinets. The display cabinets should be maintained at temperatures of 3–5°C, and not allowed to fall below 0°C or rise above 10°C. When perishable goods are delivered they should be transferred to refrigerated storage with the minimum of delay.

Stock rotation is also of importance, and checks on 'use by' dates must be carried out daily to comply with the new legislation. During code checks, vacuum-packed products which have become 'leakers' should be removed from sale along with other products that have deteriorated or become mouldy.

Correct maintenance of cabinets is important to prevent breakdown and subsequent loss of food. Cabinet interiors require regular cleaning and sanitizing to prevent build-up of debris and fat residue from dairy products.

If milk is sold the retail outlet has to be licenced under 'The Milk (Special Designation) Regulations, 1989', as amended, this can be obtained from the Local Authority.

Frozen food sales

This is an area of retailing that has increased considerably in the last few years. Frozen food outlets often sell other food, including grocery lines, beers, wines and spirits, but rarely open food. However, basic hygiene principles should still apply, especially in relation to good housekeeping and cleaning. A suitable area should be set aside to deal with broken packs, which could be considered 'open food'. Since customers tend to shop infrequently and in bulk, stock rotation is important and great care should be taken by management to ensure that correct stock rotation is implemented.

One of the major concerns is the defrosting of products. This could be the result of bad handling, overloading or breakdown of equipment. Products that have defrosted, should not be refrozen. Maintenance of the cabinets must be carried out on a regular basis to prevent breakdowns. Defrosting of cabinets is an essential part of a maintenance programme and should be done on a 3–6 week rota. Products should be kept as near to −18°C as possible and ice-cream should be held at −23°C.

Health food shops

There has been an increase in the number of health food shops over the last few years, in line with the growing interest by the general public in 'healthy living'. These retail outlets can sell a wide range of products, covering both open food, packaged goods, perishables and others. Generally, the shops are fairly small and, therefore, there could be ample opportunity for cross-contamination. One of the main areas of concern could be infestation, because of the type of dried product sold. Many of the previous comments in this chapter will apply to health food shops, therefore, they will not be reiterated.

Farm shops

Farm shops can vary from the farmer's barn, to a stall at the side of the road, to properly constructed premises with electricity, water and drainage. The products sold will vary from produce to meat, dairy products to home-made confectionery and cakes. Due to the variety of goods and the siting of the shop, prevention of cross-contamination will become a major factor. All food will need to be protected, and should be adequately wrapped or covered in such a way as to prevent contamination by other products, animals and pests. Cleaning and sanitizing of all surfaces and equipment will also be an important consideration. Adequate sanitary facilities must be available.

Again, because of the situation of such shops, infestation could be serious and all necessary precautions should be taken, including expert advice.

Construction and layout of premises

All interior surfaces, fixtures and fittings must be of a suitable impervious material. Today, there are adequate finishes available, so all retail outlets should be able to find something suitable both from acceptability and price.

Floors should be of hard impervious materials, ideally non-slip but to give a cleanable surface. Joints between floors and walls should be coved to prevent dirt traps and facilitate cleaning.

Walls will vary depending on the type of retail outlet. Where open food is sold walls should have a hard smooth surface, usually tiled, although other surfaces are also acceptable. Certain types of sheeting can be used as a wall covering, however, great care must be taken to ensure that it is fitted correctly to avoid insect invasion between the main wall and the cladding. In other areas, hard surfaces, painted with a good washable paint would be acceptable.

Ceilings are essential in open food areas although in warehouses/storage areas they are not so vital. Suspended ceilings should be avoided wherever possible, as they can harbour insects or rodents. If necessary there should be access panels in order to implement pest control. Ideally, the ceiling should be plastered, with a suitable washable paint that will prevent condensation. If acoustic tiles are used, they should be painted with a suitable product.

Doors and windows should, preferably be constructed from metal, or heavy duty synthetic materials which can be easily cleaned. If wood has to be used, it should be hard and gloss painted.

Layout. Inadequate space and badly planned work areas often lead to poor hygiene. It is essential to plan where the food is to be received, where it is to be stored, and in what part of the shop it is to be sold. In premises selling open food care must be taken to ensure that raw and cooked products do not come into direct contact with each other.

Vending of food and beverages

Vending of food and beverages is one of the areas of retailing which has substantially increased over the last few years. Vending machines are to be found everywhere from small offices, which are not big enough to provide their own catering facilities, to large airports. The machines in use range from the simple drink dispensing machines to sophisticated machines providing whole meals. Meals are usually reheated by conventional means or by microwave.

It is important that the machines are sited away from direct sunlight and in an area which can be cleaned adequately including behind, at the side and underneath the machine. The machines must be constructed so that all parts coming into contact with the food are removable to ensure adequate cleaning and sanitizing. All areas within the machine should be accessible so that the machine can be cleaned when it is replenished, a minimum of once a day is advisable. Where perishable foods are sold the machine should be refrigerated. The present 'Code of Practice for Vending Machines' gives temperature recommendations for storage, but does not give the storage-life of the product. Keeping the machine clean and free from debris will also help to prevent infestation, particularly by insects.

Screening and foreign-object control

The active pursuit of strict foreign-object control is thought by many retailers to be beyond their area of responsibility, such matters being regarded as the realm of the producer, not the retailer. Of course this is not the case, food is 'open' for production, packing and display in many units. The principles of the avoidance of foreign objects remain unchanged throughout the food industry although particular risks do seem noticeably prevalent within the retail trades.

The overriding rules must still be to 'keep food covered' and to 'work tidily'. These principles should be instilled into all workers, at all levels, but management and supervisors should be prepared to enforce such principles always.

The common areas of concern are as follows:-

(i) glass should be banned from production areas, and light fittings protected by plastic sheaths;
(ii) wooden equipment should be reduced to a minimum, and worn items discarded;
(iii) flaking decoration should be made good;
(iv) packaging materials should be stored in such a way as to minimize the risk of contamination;
(v) personnel hygiene codes should be introduced with particular reference to jewellery, nail varnish and similar risks;
(vi) maintenance staff should be vigilant against lost nuts, bolts, washers and other articles;
(vii) management should design against foreign objects by reducing cupboards and shelving that could harbour bric-a-brac and by not providing places where personal possessions could be allowed to accumulate, (cigarettes, sweets, or keys, for example).

These principles should be enforced, but the real responsibility lies with management to design work patterns and methods that consider and reduce the likelihood of foreign objects reaching food.

Storage and infestation control

The van salesman and direct deliveries from manufacturers have decreased over the last few years. They have been replaced by independent retailers collecting from 'cash and carry' wholesalers, or for multiple outlets, deliveries from depots of mixed loads covering all requirements. Holding stock costs money and all retailers now reduce their holdings of stock. Such changes have affected both the use and design of storage facilities, and have reduced the risk of infestation by increased throughput, and improved stock rotation.

Storage facilities

The matters of importance in the management of hygiene are:-

(1) Racking to be easily cleaned, particularly at the base of legs.
(2) Adequate facilities for handling pallet boards.
(3) Floor surfaces capable of withstanding any mechanical wear.
(4) Walls capable of being brush cleaned, and washed occasionally.
(5) Lighting not shielded by tall racks, suitable ventilation, not excessive so there is exaggerated chilling of foods in the winter months.
(6) Other areas of note include the construction of cold stores suitable for cleaning (particularly access), finish and construction of shelves, facilities for damaged goods and arrangements for refuse disposal.

Infestation control

Retailers are frequently of the opinion that infestation does not affect them and that they have no such responsibilities. Often this is far from the truth, crawling insects, flying insects and larger animals, for example cats, can be a nuisance, even if rodents are under control. Pest control contractors should be employed at all premises and primary advice given by such specialists. They should offer comments on:-

(1) The type of baiting and bait locations.
(2) Adequacy of proofing arrangements.
(3) Likely sources of infestation.
(4) Necessity of flying insect control.
(5) Management response to infestation or risk of infestation.

Personnel hygiene

Personal hygiene is of the utmost importance in the retail trade, especially where open food is handled.

Hands

Regular hand washing, with a bactericidal liquid hand soap, when beginning work, when changing over from one job to another, and after visiting the toilet, is important when handling or serving open food. A barrier cream, manufactured for the food industry, can be used after hand washing, as this helps protect hands from chaps and cracks, which could harbour bacteria. Disposable paper towels are suitable for drying hands.

It is not advised that rubber gloves be used for handling open food as bacteria can be harboured inside the gloves. In some circumstances, disposable gloves may be worn, but thrown away immediately after use.

Clothing

Clean protective clothing must be provided for all food handlers. Those handling open food should wear head covering that will totally enclose the hair. Clips must not be used to hold the hat in place as they could fall into the food.

First aid

All cuts, boils and other abrasions should be kept covered with a suitable waterproof dressing. It is recommended that blue or green plasters be used, as they are easily identifiable if lost.

Bad habits

Staff should **not**:-

(i) cough or sneeze directly over open food;
(ii) lick fingers in order to open paper bags;
(iii) have long, dirty, or false finger nails;
(iv) wear false eyelashes;
(v) wear rings containing stones, pendants, or earrings other than sleepers;
(vi) touch the nose, ear, mouth;
(vii) eat sweets or food whilst working;
(viii) smoke – this is a legal offence by the individual.

Infectious disease

Staff must report stomach upsets or septic conditions to a nominated person, who in turn should report to the community physician via the environmental health officer illnesses likely to cause food poisoning or a septic condition. Staff should be sent immediately to their GP if this occurs. It is important to interview staff returning from abroad to ensure that they have not been ill whilst away or are still suffering symptoms.

Management

Probably the most fundamental change in recent years has been the growth in professionalism within retail food management. Profit margins are small in the retail food industry and all outlets must be managed skillfully for maximum profit. The new professional retail manager should be expected to include sound hygiene as one of his prime objectives. Packaging and production changes have reduced the need for skilled product knowledge. Frequently, only correct temperature control and obedience to shelf-life principles are required for quality maintenance. It is essential that a manager has an intimate knowledge of the flow of goods through the premises. The passage of goods through the store is the major concern, relating to stock holding, wastage, throughput and many other control parameters. As part of the performance of his store there should be daily and weekly audits of hygiene, which will inevitably produce action lists; the skilled manager will grade the urgency of these tasks.

Skilled tradespersons (butcher, bakers) should receive a higher level of training in relation to food handling, and should be actively encouraged to attend courses such as the Institution of Environmental Health Officers hygiene training programmes.

Until the new legislation with regard to the training of food handlers is passed, it is essential that all such staff receive adequate food safety training.

All staff should receive basic training (Fig. 13.4) in the following areas: personal hygiene; temperature control; cleaning procedures; foreign-object control; infectious disease notification; first-aid procedures.

Fig. 13.4 Training session for food-handling staff

14

Cleaning and disinfection

Peter Hoffman

Definitions

Cleaning	Any process of physical removal of 'soil', i.e. any matter present that should not be part of an item. This matter can contain microbes that are responsible for food poisoning or spoilage.
Disinfection	The elimination of microbes that cause disease or their reduction to safe levels. This can be done by removal (see 'cleaning' above) or by killing them with heat or chemicals.
Disinfectant	Chemical that has a lethal action on microbes.
Sanitizer	Term for disinfectant often used in the food industry.
Sterilization	The total elimination of all microbes (whether harmful or not).

The purpose of cleaning, disinfection and sterilization in food hygiene is to prevent both food poisoning and spoilage. Each of these methods has a role to play in controlling the existence and spread of microbes. Whilst it is not intended that food-handling premises be turned into true sterile zones, an essential element of food preparation should be awareness of the biological nature and behaviour of microbes (see Chapters 3 and 4) and, from this knowledge the ways in which food poisoning or spoilage can be prevented. Areas where harmful microbes are likely to occur should be identified, contained by correct procedures and, if necessary, corrected by cleaning, disinfection or sterilization or disposed of so that there is no transfer of contamination to other food items. Such information is essential to the practice of food hygiene. A random approach of sporadic or haphazard cleaning, disinfection or sterilization will prove unproductive. Hygienic methods of food production and storage will yield a two-fold benefit: the food is more likely to be safe for consumption and will have a longer shelf life. Both of these factors are important in catering. Beware of placing too great an emphasis on cleanliness for its own sake to the possible exclusion of other factors, such as good quality raw materials, procedural and handling aspects, thorough cooking, cooling or reheating and proper storage. A clean kitchen does not necessarily guarantee safe food.

Cleaning

In the context of food hygiene, cleaning is that process which removes those microbes from contact with foodstuff that can cause poisoning and spoilage, or the dirt that protects those microbes from removal and supports their proliferation. Cleanliness is often regarded as the fundamental process of food hygiene. This is, in epidemiological terms, perhaps not the absolute concept that it is often accepted to be. To understand this, it is necessary to explore the relationship between the *Hygienic* and the *Aesthetic*. *Hygienic* refers to practices as they relate to the maintenance of health; in the case of food hygiene this centres around the exclusion and elimination of harmful microbes or their products from the diet. The *Aesthetic* refers to those perceptual factors of the senses; in this case those that relate to the outward appearance of food or its surroundings and acceptability. These two do not always equate. A kitchen can be scrupulously clean and yet contain food items which will give rise to food poisoning when consumed; (when did a sparklingly clean oven prevent a chicken from being undercooked?). Yet there is often a behavioural association between good food hygiene and maintenance of an aesthetically-acceptable kitchen. Failure to observe some of the precautions can indicate lack of knowledge or a lack in motivation to put that knowledge into practice. Cleanliness has a definite role to play, nevertheless one cannot carry out the function of hygiene solely by 'elbow grease'; knowledge, rational thought and practice all have to be coordinated to this end.

The physical removal of potentially pathogenic microbes from any item can be related in importance to hygiene by the proximity of the cleaned item to the consumed product – the closer to the ready-to-consume food, the more vital the cleanliness. For example, consider the importance of the state of cleanliness of a kitchen floor. Whether the floor be either scrupulously clean or exceptionally dirty, food dropped on it would (in an ideal world) no longer be considered fit for consumption. The opposite is the case for surfaces and implements which are intended for food contact, such as chopping boards; the same surface may be used for potentially hazardous foods (e.g., uncooked meat) and foods that require no further preparation before being eaten. The use of separate boards for the two products is a safety procedure or if necessary the same board can be cleaned between processes. The quality of cleaning of surfaces is vital to the hygiene of preparation.

The *QUALITY CONTROL* of a decontamination process is essential to food safety. To examine where this may be so, a basic understanding of the principles of 'Hazard Analysis Critical Control Point' (HACCP) is useful. (See Chapter 12 for a more in-depth appreciation). Briefly, certain processes in the series of steps that lead to a safe food product are more relevant than others. If the processes can be controlled at these points, the safety of the end product may be assured, for example, the safe production of pasteurized milk; contributory factors to the hygiene of milk are: the health of the cows; the hygiene of the milking process; the storage and transport of the raw milk; the efficacy of pasteurization; prevention of recontamination; post-pasteurization temperature control and the hygiene of packaging. The most critical event in this sequence is probably pasteurization to rid the milk of pathogens such as *Mycobacterium tuberculosis, Brucella abortus*, various salmonellae or campylobacters; after this the introduction of such organisms would have to result from a set of unusual circumstances. Thus, the process of pasteurization is probably that stage in milk production in which safety is the most highly controlled and in which the most intensive monitoring occurs, the controls are precise and the 'fail safe' precautions are highly developed, and also the post-process checks; in other words a Critical Control Point (CCP) when applying the HACCP procedure (see Chapter 12). Similarly with cleaning; if it makes a definite contribution to food hygiene, that is where it has to be of the highest assured quality. To go back to the example of the chopping boards; if there are separate and segregated chopping boards for uncooked meats and ready-to-eat food, the cleaning of the respective boards is not such a critical process. Where the same board is to be used for both tasks, the quality control of that process has to be ensured. In practice, the best ways are either to use a good quality dish-washing machine or hand washing by competent, well-instructed staff.

Specific cleaning areas

Floors

As contact between floors and food should not occur, floor cleaning is not closely related to food hygiene. Surprisingly, little transfer of contamination takes place between floors and work surfaces in a kitchen unless by a vector, anything from insects and vermin to food vessels lifted from the floor onto a work surface. Floor cleaning serves a two-fold purpose: safety of the workers – a build-up of grease can cause a slippery surface; vermin control – food residues on a floor provide food for the complete range of kitchen pests. Floor cleaning must be done with cleaning materials (cloths and mops) segregated from those used on surfaces or utensils intimately associated with food. If such segregation does not occur, cleaning can be the cause of rather than a solution to food poisoning.

Drains and gullies

The reasoning here is similar to that of floor cleaning. Dirt or smells from drains may be aesthetically offensive but they are not related to food poisoning except insofar as the dirt can act as food for pests. Acceptability of drains and gullies is a combination of cleanliness and good plumbing. Should their appearance or smell be offensive, the first resort is to clean them using hot water with a degreasing agent (washing soda or proprietary alkaline detergents). If this fails, an inspection for blockages or faulty plumbing is warranted. Use of disinfectants in this situation will not be helpful.

Kitchen utensils

These include knives, forks and spoons, all manner of slicers and mixers, bowls and of chopping boards. Wherever possible, if they are to be used on raw foods, which may harbour food-poisoning and spoilage microbes, they should be separated from those used for ready-to-eat food. Where this is not possible or practicable, they must be thoroughly washed, then quickly and efficiently dried before reuse.

Eating and drinking utensils

Here aesthetic appreciations are readily made by consumers and, without visible cleanliness, conclusions will be drawn as to other aspects of hygiene in an establishment. Lack of cleanliness of these utensils is only a minor food-poisoning hazard, microbial inocula are likely to be small and therefore the contamination in ready-to-eat food is also likely to be small, but there may be person-to-person risk. Should the previous user have an acute infection of the mucous membranes of the mouth, for example a rhinovirus (common cold) infection, contact with the next user's oral mucous membranes via eating utensils may well be sufficient to transfer the infection. (**Note**. This is true only of microbes that can effect this type of salivary transmission; it is not true of all viruses. Other viruses, such as the human immunodeficiency virus, the virus that can give rise to AIDS, will **not** be transferred in this way).

Cleaning utensils, cloths, mops and brushes

Items used for kitchen cleaning are potentially mobile and they are in contact with many different surfaces, utensils and equipment over a comparatively short period of time. This process can itself act as a transfer mechanism for the microbes it is trying to curb. Thus segregation and decontamination of cleaning articles is essential. The aesthetic and hygienic coincide and materials used for floors are not interchangeable with those used for work surfaces. The potential consequences of using a cloth that looks clean, but which has been used to mop up a minor spill of thaw liquid from raw poultry in a refrigerator or to wipe the blade of a knife to be used to cut a cream cake may be far more disastrous with regard to food poisoning than transfer of contamination from the floor. Cloths for cleaning surfaces can be colour coded by their category of use: soiled environment (floors); potentially contaminated foodstuff and their utensils; ready-to-eat foods and their utensils. If the cloths are to be re-used they must be hot washed, rinsed and thoroughly dried at regular intervals (at least daily) (Fig. 14.1). Cloths for different categories of purpose must not be mixed, even in the washing process. In many ways it is both easier and safer to use single-use wipes that are thrown away after a single task, so ensuring that they do not transfer contamination.

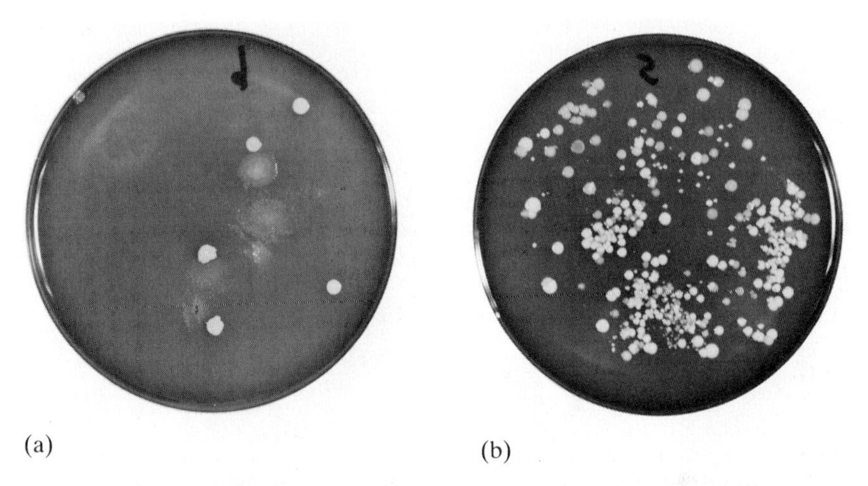

(a) (b)

Fig. 14.1 Bacterial cultures from (a) a clean and (b) a dirty drying cloth

Counter tops and work surfaces

Like floors, these must be cleaned *in situ*; unlike floors, counter tops will make contact with food prior to consumption. Should this contact be sequentially with raw and then cooked or ready-to-eat food, contamination from one to the other can take place. Any cleaning process must be compatible with the surface material and not cause damage, leaving cracks and crevices that can harbour bacteria difficult to remove or kill. Manual cleaning requires a degree of diligence and application; a quick, haphazard wipe will not contribute to food hygiene. Cleaning should take place between preparation of items and not only at the end of a session.

Food storage containers

As these can be used for prolonged contact with foods that will support growth of bacteria responsible for both food poisoning and spoilage, their initial cleanliness is important. Where mechanical dishwashing is feasible for containers (Fig. 14.2, p. 224), it is preferable. When manual washing is used, it must be thorough followed by efficient drying prior to storage or reuse.

Cooking vessels

Although heat disinfected during the cooking processes, cooking vessels should be cleaned thoroughly and stored dry.

Hand hygiene

All skin surfaces carry bacteria; those bacteria that grow and replicate with

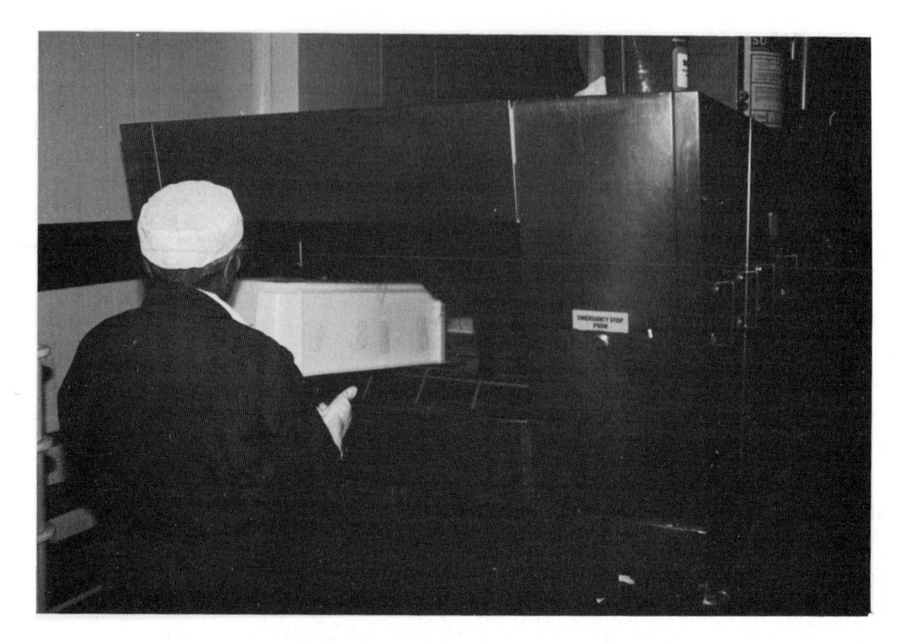

Fig. 14.2 Tray tunnel washer in food-production premises

the skin as their permanent habitat are, in general, harmless. They are present both on the surface of the skin and in the various pores and follicles deep into the skin, and so can never be fully eliminated (i.e. skin cannot be sterilized). These bacteria are referred to as 'residents' of the skin (see p. 163). The other category of microorganism found on skin are known as 'transients', their home is not the skin but they have been acquired by touch, for example by handling contaminated foodstuffs, and they are present only on the surface layers of the skin. This superficial location means that they stay on the skin for a short time and they are easily removed. They can be removed either by touch, and as such are easily transferred to objects subsequently touched, or by washing. To be lost by touch means that they can be transferred by hands from contaminated raw foods to ready-to-eat foods. Washing in-between handling these foods would prevent this transfer (Fig. 14.3, p. 225). It must be realized that hands are a vehicle for transfer of harmful organisms, just as important as chopping boards and other pieces of equipment. By the same logic, it is pointless to attempt to be hygienic by whole-body showering before starting work in catering or on a food-production line. This may remove (or re-arrange) a small proportion of the resident microbes (i.e., those not relevant to food safety), but obviously subsequent hand-transfer of potentially pathogenic transient organisms during food preparation will not be affected. (As a comparison, surgeons pay scrupulous attention to hand hygiene but do not shower before operations as an infection control measure). The only microbe that can sometimes be found on skin that can cause food poisoning is *Staphylococcus aureus* (see p. 60). It is uncommon to find this organism

Fig. 14.3 Hand-wash facilities in retail food premises

as a resident inhabitant of the skin on hands unless there are skin lesions, but it can be a permanent resident of nostrils. Thus when preparing food, the nostrils should not be touched; the hands should be washed after using a handkerchief.

The practice of cleaning: materials

Cleaning is getting rid of dirt (or 'soil' in cleaning jargon). Cleaning has a flexible definition. The degree to which soil should be removed varies, as does opinion as to what 'clean' constitutes. Most people would have higher criteria for cleanliness as it applies to cutlery compared to cleanliness as

Table 14.1 Hard surface cleaning

Soil type	Removal options
Water-soluble	Water (+/− detergent)
Fat-soluble	Water (+ detergent)
	Alkali (forms soap *in situ*)
	Organic solvent (rarely applicable)
Insoluble	Mechanical abrasion (+/− detergent)
	Acid (if limescale)
	Enzymes (if denatured proteins)

applied to floors, for example. Opinions would certainly vary, according to personal fastidiousness, within any given category. One person's idea of cleanliness, as we have all experienced, is not necessarily another's. 'Soil' can be a wide variety of substances or mixtures of substances. They can be broadly classified into groups. Those that are water soluble, those that are fat soluble (and water insoluble) and those that are insoluble in both. This last category comprises those substances that are innately insoluble (such as charred deposits from burnt food) and those (such as proteins in blood) that are initially soluble but can be 'denatured' by heat and thus become insoluble and fixed onto surfaces (see Table 14.1).

There are a variety of ways in which soil can be removed, according to the nature of that soil. The common aim of food-linked cleaning processes is to dissolve and disperse the soil in water and to dilute it to such an extent that it can be regarded as 'removed'. Dilution in water will remove water-soluble substances. Other forms of soil require more complex forms of removal. There are two ways to remove fat and fat-soluble soils. One is to use liquids that are themselves fatty in nature and in which the fat will dissolve. This is the essence of the process known as 'dry cleaning', as used for fabrics. The solvents used are toxic, volatile and expensive, which make it unsuitable for general use in catering. For catering, detergents are used to suspend microscopic fat particles in water, in effect, creating a 'pseudo-solution' with water in the presence of detergent which acts as a fat solvent.

Detergent molecules have two different characteristics within them. One end of the molecule is, because of charge polarization, better able to dissolve in water and other charged or polarized molecules: termed 'hydrophilic', literally *water-loving*. The other end is a long tail without significant polarization of charge which, in turn, has affinity for other non-polarized molecules such as fats and oils (see Fig. 14.4): termed 'hydrophobic' – *water-hating* or sometimes (interchangeably) 'lipophilic' – *fat-loving*. When a solution of detergent molecules in water encounter fat, that portion of each molecule which is hydrophobic will dissolve in the fat, leaving the hydrophilic portion remaining in the water. Detergent molecules will eventually surround the fat, lifting it off the surface as a microscopic ball of fat contained within a coating of detergent molecules half buried within it (Fig. 14.4). These detergent molecules will have their hydrophobic portions buried in the fat and their hydrophilic portions above the fat surface. This will effectively render the exterior surface of the fat hydrophilic and it will behave as if it had dissolved in the water/detergent solution (Fig. 14.4). It will also prevent the dissolved fat droplets coalescing with each other and leaving smears on washed items. So a water/detergent solution is capable of dissolving both water- and fat-soluble soils.

Another way in which fat-based soils can be removed is to use a strongly alkaline preparation. When a fat reacts with a strong alkali such as sodium hydroxide (caustic soda), it forms a soap. This performs a dual function. It renders the reacted fat soluble in water as well as making it capable of acting as a detergent to help solubilize and remove unreacted fat. This method of *in situ* detergent formation is most useful for surfaces with fairly solid baked-on grease, within which there may be insoluble carbon particles from burnt food, as may be found on oven interiors. (Care is needed when working with such caustic preparations).

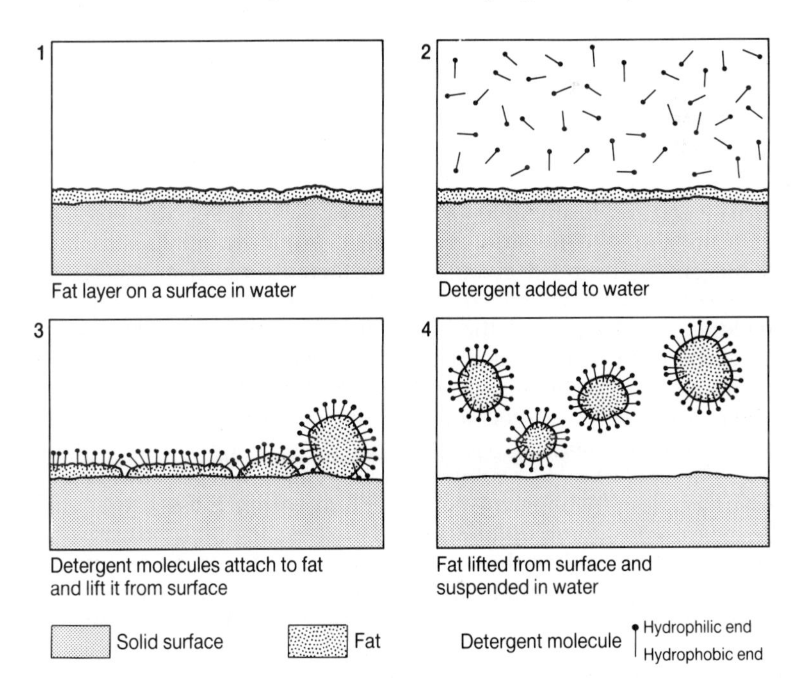

Fig. 14.4 Mechanism of detergent action

This leaves the two insoluble soils to consider. Hard water deposits are insoluble in water. They arise through the reaction of dissolved atmospheric carbon dioxide in rainwater as it filters through chalk in the ground. This dissolves small amounts of insoluble chalk (mostly calcium carbonate) to form calcium bicarbonate, which is carried in solution in water but when heated decomposes back to the insoluble carbonate. This gives rise to hard water deposits in kettles, on glassware and so on. The most practical way to remove these mineral deposits is by a chemical reaction that turns them back to soluble chemicals; such as by reaction with an acid. This turns the insoluble carbonate to a soluble salt. (Care is needed when working with strong acids).

The other common form of insoluble soil is denatured protein. In their natural state, there are both soluble proteins, such as those found in blood and milk, and insoluble proteins, such as those found in muscle (i.e. meat). When soluble proteins are heated, their molecules change configuration and become insoluble. As they do this, they adhere firmly to any surface they are near and in effect become 'baked on'. They can be removed mechanically, i.e., by use of abrasive cleaners on a solid surface, or chemically by use of enzyme-based cleaners if on a delicate surface such as fabrics. 'Proteolytic' (i.e., protein-splitting) enzymes break down the large insoluble protein molecules into smaller, soluble units. This inconvenient, sometimes lengthy process can sometimes itself be caused by an inappropriate cleaning regime. If articles with soluble protein on their surfaces are put straight into hot detergent solutions, rather than remove the protein, the hot water will cause it to coagulate firmly onto the article. A preliminary cool or warm wash stage will remove soluble proteins before they can be denatured by the hot wash.

Formulating a cleaner

Cleaning products are formulated or 'built' with fairly specific tasks in mind. Decisions about formulation will centre around a number of factors. A detergent is the term given to a surfactant, which is useful for cleaning purposes, so which one does one choose? The four main categories of detergent depend on the charge of the molecule. The largest group, the anionic detergents, are negatively charged. There are some good cleaning agents within this group, but they usually foam. Cationic detergents are positively charged and mostly have lower cleaning ability, some have microbicidal activity. Non-ionic (non-charged) and amphoteric (charge depends on pH of solution) detergents can be good cleaners and tend to have low-foaming properties. Even within these broad categories there are considerable variations. It is best to think in terms of *appropriate* detergents rather than as good or bad. For example, when hand dishwashing, a high-foaming detergent is acceptable, whereas for a mechanical dishwasher this property is undesirable. The reverse would be true of efficient detergents used in dishwashers that are corrosive to human skin.

If detergents with incompatible charges are mixed together, they can react with each other and, in effect, cancel each other's activity. The formulation of a built detergent product should be a carefully thought-out process so that its components act synergistically. Cleaning products should never be mixed unless the components are compatible.

Soluble calcium and magnesium salts present in water are responsible for so-called 'hardness'. These metal ions can form insoluble complexes with detergents, which both deplete the amount of active detergent in solution and create deposits that may adhere to washed articles. Such salts can be 'sequestered', that is removed from effective solution, by a variety of additives to detergents. These chemicals, termed water-softeners, are normally sodium carbonate (washing soda), sodium metasilicate, or a variety of sodium phosphate salts. Their need is diminished if water is naturally soft (a reflection of the geology of the substrata in those areas), or is softened artificially in an ion-exchange resin, where the calcium and magnesium ions are exchanged for sodium ions (sodium carbonates and bicarbonates are readily soluble).

Other agents present in formulated-detergent preparations can be acid or alkali according to the function of the preparation and the compatibilities of the surfactant used. Another important aspect of cleaners for many catering items is the finish; they must leave a smooth water-film on a drying object rather than allowing separate droplets to form, which will dry to give a speckled dirty appearance on what should have been a gleaming surface.

The practice of cleaning: mechanisms

In the ideal situation there are four stages that comprise cleaning. Gross soil removal, fine soil removal, detergent removal, drying. This is only a general theoretical outline; in practice cleaning can vary from ultrafine multistage methods used in microelectronics or surgery to a quick rinse under the tap. Sometimes the stages in the process may not be totally distinct from each other but merge into a graded procedure.

Gross soil removal: In the 'materials' section, detergent molecule action was outlined (see Fig. 14.4). Detergent molecules will surround hydrophobic particles, lift them off surfaces, break them up into microscopic particles, each in turn surrounded by detergent molecules and thus keep them suspended so that a rinse will remove them without redeposition onto any surface. This process depends on having an excess of detergent molecules above the equivalent amount of soil present, both to remove soil and in order to keep it in suspension. If this fails due to insufficient detergent, unremoved and redeposited soil will be present on a surface as it goes from the wash to the rinse stage. So unless objects have little initial associated soil or if very high concentrations of detergent are used, washing processes should commence with the removal of gross soil. This will vary according to the washing process that follows. At its simplest, as in hand-dishwashing, it could be pre-rinsing the remains of food off plates before they go into the bowl of detergent and water or batch dishwasher. It could be the first stage in a conveyer belt-type continuous dishwasher (Fig. 14.5). The principle remains the same: where possible there should

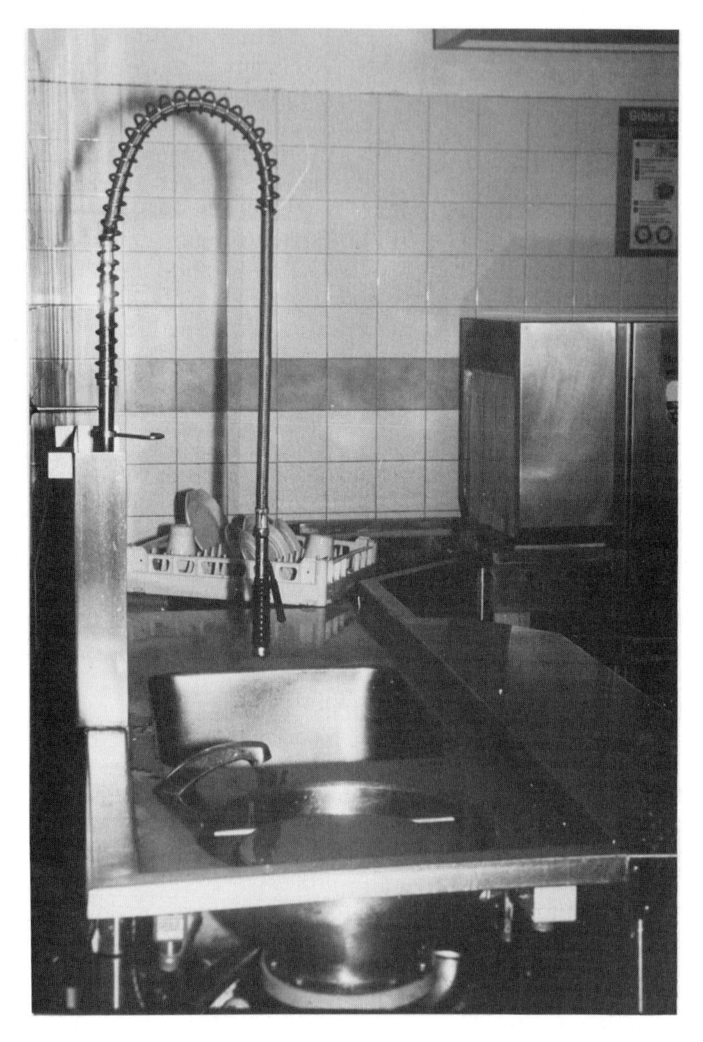

Fig. 14.5 Spray head and waste disposal unit for removal of gross soil before tunnel dish washer

be an initial removal of gross soil by water and mechanical action with or without detergent.

Fine soil removal: This is the stage at which all remaining soil is removed (Fig 14.6, see p. 231)(Fig. 14.7 see p. 232). It involves water, detergent and mechanical action. It may comprise one or more substages. A cool or warm initial wash is desirable to remove protein-based soil, which may be denatured and rendered inextricably adherent by heat. After this, the hotter the wash, the more effective it is. Mechanical action, anything from a washing-up brush to water jets, helps break up remaining lumps of adherent soil so that they can more easily be lifted off a surface by the detergent.

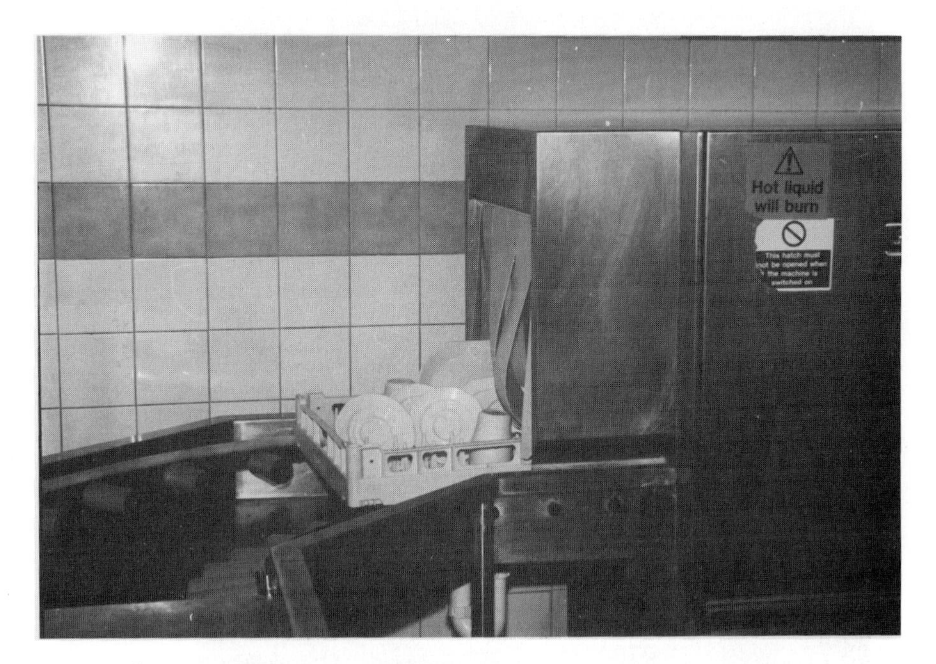

Fig. 14.6 Tunnel dish washer used in commercial premises

Detergent removal: This is the function of the rinsing stage; it aims to remove any toxicity and taint the detergent may impart. It will also remove remaining soil suspended by detergent so that it will not be redeposited onto a surface when drying. Simultaneous functions may be to add an anti-smear 'rinsing aid' to prevent the formation of droplets that would dry to leave a speckled or smeary finish and may include, in the final rinse, a heating stage (Fig. 14.8 see p. 233). This stage will have a dual function: to leave items warm so that they dry quickly and thermally to disinfect the items if the heating is for sufficient time at an appropriate temperature (see later).

Disinfection

Disinfection has a flexible definition (see the beginning of this chapter). In essence, it means the elimination of sufficient microbes with disease-producing capabilities so that safety is ensured. Thus, the definition of successful disinfection will vary according to the situation, along the lines of the relevance of cleaning outlined above. If an object does not present a risk of infection, then disinfection of that object is irrelevant. Most people think of specific use of *disinfectants* as the only means to achieve *disinfection*, yet this is neither the most common nor the most efficient way

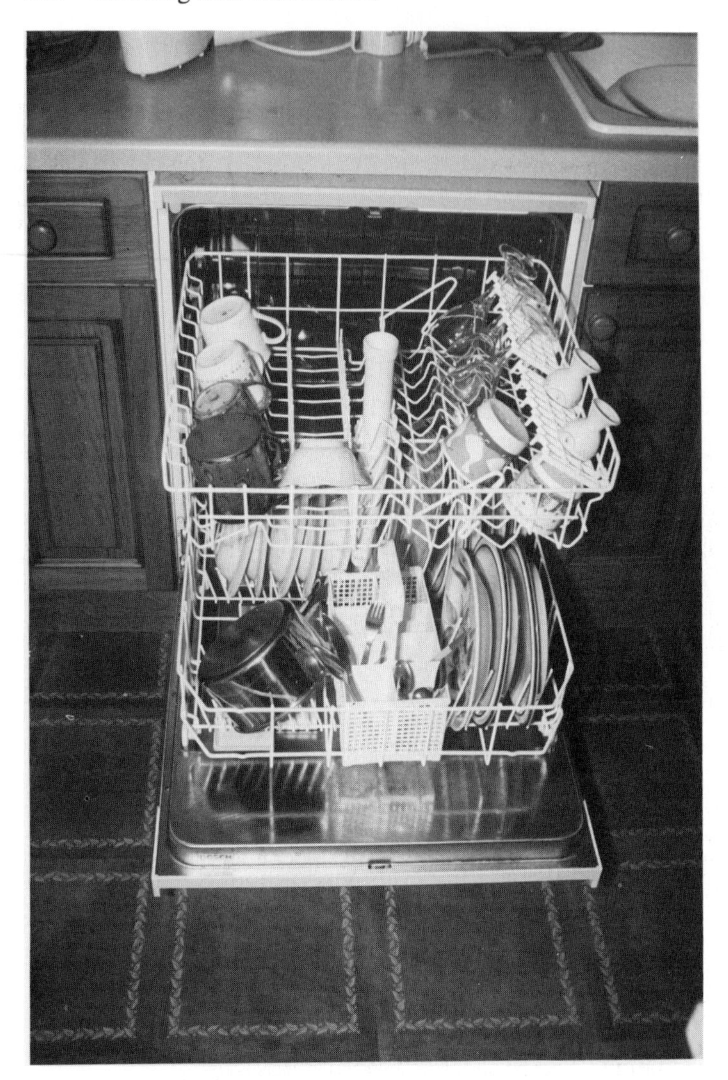

Fig. 14.7 Domestic dishwasher

to do this. A good working definition of disinfection is elimination of all bacteria, viruses and fungi other than bacterial spores.* There are three methods of disinfection; they are *cleaning, heat* and *chemicals*:

Cleaning eliminates microorganisms by their physical removal. This has been discussed above in detail in all its varied forms and there can be no

* Bacterial spores are hardy, heat-resistant life-forms produced by two groups of bacteria only, *Clostridium* and *Bacillus*. Whilst both have definite roles in food poisoning, this is more related to short- and long-term storage of foodstuffs rather than immediate catering food hygiene shortcomings (see Chapter 3 for details of the roles of these organisms in food poisoning).

further information added here. Yet it must be considered along with other, perhaps more obvious, methods of disinfection when making decisions in food hygiene. It can exist as a method used on its own or be combined with one or both of the other methods (heat or chemicals).

Heat disinfection occupies a fundamental position in food hygiene. Cooking is used to make foods simultaneously palatable and safe. Many raw foods, primarily meats and meat products, have an unacceptable probability of containing pathogenic microorganisms. When cooked or thoroughly heated through, all pathogens except for bacterial spores are

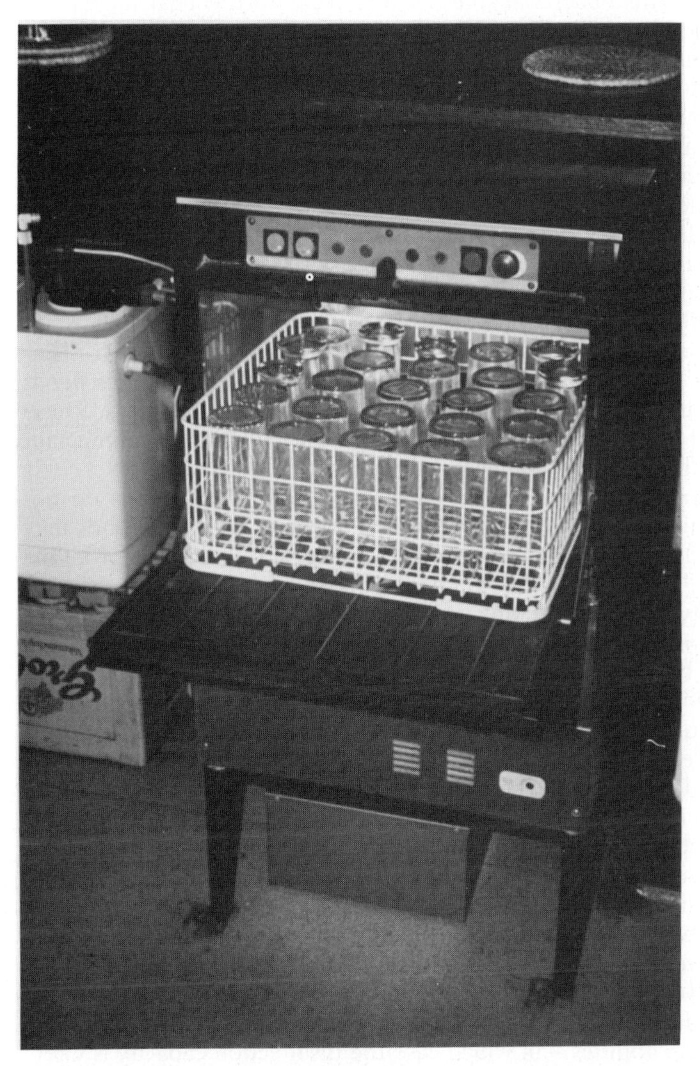

Fig. 14.8 Public house glass washer (fast cycle – hot rinse)

killed. Heat as a more calculated disinfection method is one of the oldest processes in microbiology:- *pasteurization*, (introduced by and named after Louis Pasteur, the nineteenth century microbiologist) was and still is used to eliminate pathogenic microorganisms from milk and other foods such as liquid egg products. In pasteurization, the milk is heated to a variety of temperatures for a corresponding variety of times (the higher the temperature, the shorter the time), for example 74°C for at least 15 seconds, which kills the bacteria responsible for tuberculosis and other milk-associated pathogens. Likewise, heat disinfection can be used for utensils, crockery, cutlery, chopping boards and so on. If one or more parts of a dishwasher cycle involves heat of around 80°C, all the contents will be free from microbes (other than bacterial spores). It is usual for this heat stage to be in the final rinse in mechanical dishwashing as it facilitates drying. Water at these high temperatures causes serious burns on skin, it would be hazardous therefore to attempt heat disinfection in hand-dishwashing. Here, hygiene is achieved by physical removal of contamination rather than by killing it. It is thus a process more dependant than mechanical dishwashing on training and diligence and, as such, is perhaps more fallible. Dishwashing sinks and hand-wash basins should be separate rather than attempting to use the same facility for both. It is a disincentive to wash hands if someone else is occupying the sink for dishwashing purposes.

Chemical disinfection in food hygiene occupies a limited role and should be thoughtfully considered before use. Disinfection by chemicals is, in general, far less reliable than by heat. It requires aptly chosen disinfectants, carefully controlled application for a sufficient time; it often needs pre-cleaning and sometimes needs rinsing and removal afterwards. These are some factors to consider when choosing an appropriate disinfectant (indeed, whether to use a chemical disinfectant at all):

(i) the *microbicidal range* of the disinfectant. All disinfectants do not kill all microorganisms. It is necessary to know what microbes may be present and whether a particular disinfectant is capable of killing them;

(ii) the ability of a disinfectant to *withstand inactivation by a variety of organic matter*. Most disinfectants will be inactivated, to varying extents, by organic matter such as dirt and food residues. Either a disinfectant that can withstand inactivation should be chosen, a higher concentration can be used to compensate for inactivation, or pre-cleaning considered to remove excess organic matter, or a combination of these factors. Detergents and disinfectants can inactivate each other. Never add disinfectants together or add detergents to disinfectants unless their compatability is known. Other causes of disinfectant inactivation can be hard water, contact with a variety of plastics, rubber and cellulose and non-cellulose fibres;

(iii) Disinfectants must be able to reach and make *contact with their target microorganisms*. This will not happen if target microbes are protected within layers of dirt, another reason for cleaning prior to disinfection. Some disinfectants claim to have detergent properties in addition to disinfection abilities – in which case the disinfection capacity is likely to be exhausted by penetration of organic matter layers.

Disinfectants work best at or above normal room *temperature*. If trying to disinfect refrigerated equipment, either it should be allowed to equilibrate to room temperature, a higher concentration of disinfectant used, the disinfectant left in contact for a longer time or an alternate method of disinfection used.

All disinfectants need *time* to kill microbes; this process is not instantaneous. A contact of a few seconds will kill little. Although highly dependant on many of the factors listed above, efficient disinfection needs minutes or hours to take place. Disinfection will only take place in solution, so when a disinfectant applied to a surface dries or evaporates, killing of microbes ceases.

Many disinfectants *decay* once diluted to use strength. When decayed, they are more likely both to fail to kill their target and to become colonized by bacteria. Disinfectants should be made up freshly and regularly. Some disinfectants decay on bulk storage; these should be freshly purchased and stored according to instructions.

In addition to microbicidal questions, factors such as *toxicity, taint* and *corrosion* must be considered. Disinfectants not specifically intended for use in food-handling areas must be screened for toxicity and taint-imparting qualities. If there is any suspicion of a disinfectant on these grounds, it must be excluded. Corrosion, especially of mild and carbon steels, can be initiated and accelerated by certain disinfectants. This can be an expensive error.

The disinfectants

Although there are a large number of disinfectant preparations on the market, they are formulated from a limited range of types of active ingredients. When confronted with an unfamiliar disinfectant, efforts should be made to find out the active ingredient, so that advantages and drawbacks can be ascertained and its suitability for a particular context determined.

Chlorine-based disinfectants are amongst the most useful of disinfectants for food hygiene. They are also termed chlorine-bleaches and hypochlorites. The most common presentations of this group are liquid bleaches (solutions of sodium hypochlorite) and powders and tablets made from sodium dichloroisocyanurate (abbreviated to NaDCC), which give hypochlorite on dissolution in water. Strengths of hypochlorite solutions are often expressed in terms of 'parts per million available chlorine (ppm av Cl)'.** Strengths of hypochlorite solutions generally used are given in Table 14.2.

** It is best to express concentrations as ppm av Cl rather than as percentages as confusion can thus arise; if a bleach contains 10% sodium hypochlorite, it has 100 000 ppm av Cl. If someone refers to a 10% hypochlorite solution, it is unclear whether they mean a 10% dilution of the product (giving 10 000 ppm av Cl), or a solution containing 10% sodium hypochlorite (giving 100 000 ppm av Cl).

Table 14.2 Suggested strengths and dilutions of hypochlorite solutions

Uses	Dilution of stock solution	Available chlorine %	ppm
Disinfection of soiled surfaces	1/100	0.1	1000
Infant and other feeding utensils	1/800	0.0125	125
Uncooked fruit and vegetables (if suspect)	1/1600	0.0065	60–90

Note: These dilutions are from solutions containing 100 000 ppm available chlorine (e.g. Chloros, Domestos, Sterite). Others may vary e.g. Milton (10 000 ppm available chlorine). Adapted from Ayliffe, G.A.J., Coates, D. and Hoffman, P.N. (1984). *Chemical disinfection in hospitals.* London: Public Health Laboratory Service.

Characteristics of hypochlorites as disinfectants are as follows:

Advantages — Wide microbicidal range, rapid action, very low toxicity, negligible taint, cheap, dry powders and tablet forms stable on storage.

Disadvantages — Corrosive to some metals, liquid preparations and use-dilutions unstable.†

Quaternary-ammonium compounds (sometimes abbreviated to QACs) are a family of surface active (and therefore detergent) molecules all based around a similar structure. They are easy to use and can combine the functions of cleaning and chemical disinfection. Their main disadvantage is also related to their surfactant property; if confronted with substances that attract surfactants (such as fabrics and many types of organic matter in solution or suspension) the QAC molecules will combine with this matter, leaving little remaining in true solution, and this will in effect, inactivate the QACs as disinfectants. It follows that, despite being cleaning agents, they work best as disinfectants in already clean environments.

Advantages — Easy to use, stable, little corrosion, all detergent, non-toxic.

Disadvantages — Easily inactivated, smaller microbicidal range than hypochlorites

Iodine-based disinfectants are sometimes used in connection with food. As iodine itself is unsuitable for general use, the disinfectants contain iodine held in a complex with other molecules (most commonly a polymer called polyvinylpyrrolidone, abbreviated to PVP) or held in a complex with detergent molecules. Both these processes allow only a small amount of

† **Warning: Never mix hypochlorites with strong acids, chlorine gas will be released. This is highly toxic.**

iodine to be free in solution, thus diminishing undesirable effects of the iodine. Once the small amount of free iodine has been used up, more will be released from its complex to take part in disinfection.

Advantages	Wide microbicidal range, most have detergent properties, some can be corrosive to certain metals, stable on storage.
Disadvantages	Can be relatively expensive, concentrated solution can be viscous.

Hand disinfection

The bacteria that normally inhabit skin ('resident' microbes) are, for the most part, harmless (see the section on hand hygiene earlier in this chapter for further information). Pathogenic microbes that are picked up by touch (such as handling raw chicken) are easily removed by washing (i.e., disinfection by cleaning). Chemical skin disinfectants have no routine role to play in food hygiene.

Sterilization

This is the process leading to the *complete* elimination of *all* microbes. The most common way to achieve sterilization is by steam under pressure. (Many bacterial spores can survive hours of boiling at atmospheric pressure – 100°C and steam at above this temperature needs to exist in a pressurized vessel). This process is only relevant to the control of food poisoning and spoilage in the process of canning. After the can has been sealed, it is heated by steam in a large pressure vessel to sterilize the contents. This is necessary to kill the extremely heat-resistant spores of the bacterium *Clostridium botulinum* which, if left in a high protein, low-acid food such as meat or fish, can produce a toxin lethal if ingested ('botulism'). (See Chapter 3 for more details on botulism.) Sometimes chemical disinfectants are referred to by those selling them as 'sterilants'. This is misleading. Although in well-thought-out laboratory tests they can be shown to achieve sterility, in normal practice they can be relied on, at best, to give disinfection.

Irradiation of food (see Chapter 8) is primarily intended to extend the keeping qualities of foods by inactivation of enzymes that, for example, break down ripe fruit and reduce bacteria and fungi that cause spoilage. Somewhat higher doses can reduce or eliminate pathogens in products, such as salmonella in raw poultry and animal feedstuffs (this is disinfection rather than sterilization). Although irradiation can be used to produce sterility, the doses needed are normally too high for use on food without unacceptabe changes in taste and other characteristics. (See Chapter 8.)

15

Food premises and equipment

In the construction of buildings for catering purposes a wide range of structural methods and materials is employed, but the basic requirements for all establishments used for the preparation and service of food are the same. They must comply with the ordinary standards of stability, durability and protection against the weather, which are applicable to all buildings. There are traditional forms of construction material, such as brick, concrete, slates and tiles and also lighter forms of construction supported by steel or concrete frames. Insulation and fire protection must meet the required standards laid down in the current building regulations.

There are many types of catering business, and this large and diverse industry serves many millions of meals daily. The catering industry is described as one of the largest employers in the UK and is the most labour intensive sector of the food industry. In 1987 there were about 2 million catering employees compared with about 0.5 million in 1969. Over 50% of the staff work part-time.

Interior

A kitchen should be on the same storey as, and adjoin, the dining room. The delivery of stores and removal of refuse are easier from ground floors, but the natural lighting, ventilation and outlook are usually better on upper floors.

Kitchen staff should work in congenial surroundings. Fatigue and strain from cramped conditions, inadequate equipment, poor light, overwork and noise will predispose to carelessness, whereas good ventilation and lighting, readily accessible and easily cleaned surfaces make efforts to attain hygienic conditions less arduous. The design, construction and equipment of commercial kitchens should satisfy certain basic criteria appropriate for the proposed menu and scale of operation. There should be a logical sequence of safe operations that avoid cross-contamination and be suitable for practicable methods of cleaning.

Layout

The minimum size of kitchens serving the public, and of certain staff canteens, is, in common with those of several other kinds of workplace, subject to requirements laid down by the Offices, Shops and Railway Premises Act, 1963. The rooms must not be overcrowded with the risk of injury to the health of workers. The minimum overall floor area is 3.7m^2 (40 square feet) and 11.3 m^3 (400 cubic feet) the overall air space per person. These statutory figures refer to basic room capacity rather than to the space available after allowing for equipment; since kitchen equipment usually occupies a large amount of space compared with the number of persons using it, it seems unlikely that there will be many kitchens 2.4–3 m (8–10 feet) high where the standards are not met.

The working space required will vary according to the menu, the extent that pre-prepared or convenience foods are used and the type of equipment installed; every plan must be subject to the basic legal requirements. There are heavy-duty, large-scale installations which provide three full meals each day, and also the lighter catering kitchens. The early data on the areas of school kitchens originated from the Ministry of Works which influenced the Ministry of Education. The relevance of the advice is questioned owing to changes in diet, catering and equipment, ergonomics (efficiency of persons in their environment), costs in relation to initial outlay and maintenance and energy consumption.

DHSS and Welsh Office Health Building Note 10, 1986, refers to 'critical dimensions', those which are critical to the efficient functioning of an activity, and component-user data sheets are provided. The Sports Council has produced data sheets 'Designing for Safety in Sports Halls' and Part 7 covers 'Safety in Kitchens and Plant Rooms' (January 1991). It states that the size of the kitchen relates to the form of catering planned, whether snacks or full meals; also whether the food is prepared on the premises or brought in ready prepared, only requiring to be reheated and served, and the number of customers expected. Suggested kitchen and associated store sizes are given in Table 15.1 Space standards are also given in *Planning and Design Data*, Patricia Tuff, David Adler (Eds), Butterworth Architecture, 1990. For the food and beverage department, they are as follows:

Food service areas		
Dining rooms (luxury)	1.7–1.9m^2	per seat
	1.3m^2	per seat
Lounge and bar	1.1–1.4m^2	per seat
Banquet	0.9–1.3m^2	per seat
Staff canteens	0.7–0.9m^2	per seat

Table 15.1 Suggested kitchen and associated store sizes

Number of persons to be served/ Meal type		Cooking area (m^2)	Dry storage (m^2)	Vegetables storage (m^2)	Chilled storage 0–3°C (m^2)	Frozen storage −18°C (m^2)
20		7–8 minimum	Kitchen cupboards		Domestic refrigerator	
50	A	15–20	Kitchen cupboards		1	1
	B					1.5
100	A	20–30	6	4	1.5	1.5
	B			optional		3
150	A	30–35	8	6	2.5	2.5–3
	B			optional		4.5–6

A conventional cooking; B pre-cooked meal. (Sports Council, 1991. Part 7 *Safety in kitchens and plant rooms*).

Service facilities

Facility	Allowance
Kitchen for dining room and coffee shop (exclusive of stores)	60% of dining room and coffee shop or 0.9–1.0m² per seat
Kitchen for coffee shop only	45% of coffee shop or 0.6m² per seat
Food and liquor and china storage	50% of kitchen or 0.5m² per seat in dining room and coffee shop or 0.3m² per seat where coffee shop only
Kitchen or pantry to banquet rooms	20% of banquet facility or 0.24m² per seat.
Banquet storage	8% of banquet area or 0.05m² per seat

The figures are for use in preliminary sketches and estimates. School kitchens are designed and erected to the standards of the Department of the Environment and the Department of Education and Science. Their working space is reasonable: 1.2–1.8 m (4–6 feet) of working space and passage distance is allowed between island cookers and wall sinks or benches. Table 15.2, based on post-war school meals service premises where midday meals only are prepared, may be of interest.

The usual trend in kitchens is to install preparation equipment at the sides, where waste can be conveniently drained away, and to erect island cooking apparatus in the centre where ventilation can be localized. There is need for division into further compartments for the preparation of different kinds of food prior to cooking, and for washing-up. Work and production should flow progressively from delivery of goods to storage, preparation

Table 15.2 Approximate floor areas (in m²/square feet) of school kitchens

No. of meals	Kitchen	Vegetable store	Dry store	Larder	China store	Boiler house	Staff rooms	Refrigerator capacity (cubic metres/cubic feet)
750 in two sittings	176.5/1900 including two services and wash-up	9.3/100	7.4/80	8.8/96	—*	—†	17.1/184	4.6/50
600 in two sittings	131.9/1420 including servery and wash-up	9.3/100	7.4/80	6.7/72	—*	—†	13.9/150	2.3/25
500 in two sittings	111.5/1200 including servery and wash-up	9.3/100	6.7/72	7.4/80	—*	9.3/100	14.9/160	2.3/25
350	69.7/750	8.4/90	8.4/90	6.2/67	4.4/48	6.7/72	6.2/67	1.7/18
250	55.7/600	6.5/70	6.5/70	3.3/36	2.8/30	4.6/50	3.3/36	1.7/18
150	39.0/420	3.3/36	←6.5/70→ (Dry store and Larder)		2.2/24	3.3/36	3.3/36	1.7/18
100	27.9/300	3.3/36	←3.8/42→ (Dry store to China store)			3.3/36	2.8/30	‡
75	19.5/210	1.5/16	←2.6/28→ (Dry store to China store)					‡
40	15.8/170	1.4/15	←2.2/24→ (Dry store to China store)					‡

* Provision made elsewhere than in the kitchen.
† Hot water supply from main building.
‡ Domestic refrigerators included.

and service without return- or cross-traffic. Vegetable storage and preparation should be sited near the point of delivery in an area separated from the other parts of the kitchen to prevent soil from root crops reaching other food. The section for raw meat and raw fish should be well separated from those dealing with cooked and prepared products, including pastry work. Contamination from raw to cooked food should be prevented by any means possible. Stores for each department should be adequate. Most dirt generated in the kitchen comes from organic food material, which attracts vermin. Thus the preparation of food should involve processes that will cause least spoilage consistent with the menu. Frying in deep fat is expensive and generates greasy dirt and therefore it should be reduced to a minimum. For bulk cooking it may be better to use vegetables pre-prepared and packed by the supplier or pre-packed frozen vegetables, so that production of waste associated with used vegetables is reduced.

In general, the size, design and layout of a kitchen depend on the menu, type of service and method of cleaning (by hand, machine, high-pressure lances). The high-pressure lances should be restricted to periodic deep-cleaning to avoid spray and excess water in food-handling areas and where it is known that socket outlets and other electrical equipment is suitably waterproofed. The surface finishes of floors, walls and ceilings are considered first.

Floors

Floor surfaces should be durable, non-absorbent, anti-slip and without joints and crevices in which dirt, bacteria and insects can lodge. They should not be adversely affected by grease, salt, vegetable or fruit acids or other materials used in the preparation of food, and should be capable of being effectively cleaned and light in colour. Angles at floor level should be avoided and the junction between floor and walls coved; the top of the coving sometimes forms a narrow ledge that collects dirt; this should be avoided.

Tessellated quarry tiles are recommended for floors of heavy duty kitchens. With light catering and in dining rooms, vinyl sheet tiles of at least 3.2 mm thickness provide a good floor surface. Sheet vinyls impregnated with graphite particles are available, but may be difficult to clean. Epoxy-resin and granolithic (concrete incorporating granite chippings) coverings are more hardy; but epoxy-resin is costly and the granolithic coverings may crack. The life span of floor finishes and cleaning cost in terms of annual maintenance should be considered. Patent non-slip floors should be inspected under operating conditions before purchase. Non-slip footwear may be a better investment than supposedly non-slip floors that retain dirt and become slippery from deposits of grease and water. Table 15.3 outlines the advantages and disadvantages of commonly-used floor surfaces. Timber floors, whether soft or hardwood, are unacceptable because they are absorbent, wear quickly and the joints harbour moisture and dirt. Rubber and cork tiles are unsatisfactory because rubber is slippery when wet and cork is not durable with heavy traffic. The main objectives for floors are that they should be impervious, easily cleaned and durable. A particular

Table 15.3 Comparison of floor surfaces

Material	Advantages	Disadvantages
Asphalt	Dust-free and waterproof jointless surface which can easily be coved to form skirting. Dull colours only	Will not tolerate excessive heat and concentrated weights. Not resistant to some acids and fats
Quarry tiles	Very hard wearing in all conditions. Non-slip if faced	Numerous joints which can retain dirt and moisture unless well laid
Welded sheet vinyl	Smooth sheet material, impervious provided the joints are welded. Vinyl cove skirting can be incorporated. Wide range of patterns	Subfloor must be smooth and level. Will wear in dense traffic areas. Can be slippery when wet
Granolithic concrete	Low initial cost. Non-slip surface	Can be subject to dust and cracks. Difficult to colour
Epoxy-resin composition	Virtually indestructible. Non-slip finish to degree required. Joint-free and can incorporate cove skirting. Wide range of colours	High initial cost

floor surface may not be suitable for all food rooms, a floor surface in one room may be totally unsuitable for another. If the floor is graded to a floor gully, care must be taken to avoid the formation of pools.

Walls

Smooth, impervious, light in colour and durable wall surfaces from floor to ceiling are required to allow cleaning without deterioration. The basic materials for wall structure are brick or concrete blocks. Faced stud partition walls of plasterboard should not be used as they tend to harbour vermin which can reach the cavity by gnawing through the plasterboard, skirting or duct work; furthermore, fittings or equipment cannot be attached because of the low strength. A plastered finish will provide a smooth and non-porous surface, especially if the final skin coat is 'hard' plaster. Sheet materials with a vitreous face are acceptable, such as the standard glazed tiles. Other wall surfaces include resin-bonded fibreglass, ceramic-faced blocks and rubberized paint on hard plaster or sealed brickwork. Some paints incorporate a fungicidal additive, but absorbent emulsion paint should not be used on walls. The better wall surfaces, such as solid-bedded ceramic tiles, have no hollow places; any spaces should be filled with lightweight concrete. Proprietary wall-cladding materials, if properly fixed, can be an hygienic alternative to the traditional wall surfaces.

It might be necessary to provide localized protection against damage. Glazed tiles will protect walls behind working surfaces and should extend at least 450 mm above the surface. External corners subject to contact with trolleys or other moveable equipment can be protected with suitable polyvinyl chloride (PVC) extrusion, which can be fixed to the wall with adhesive. Areas such as trolley parks should be protected by a PVC or timber rail at handle level or at any other height where there are projections liable to damage the wall surface. Stainless steel provides an impervious surface almost as indestructible as tiles or sheets for splash-backs and angles. Galvanized or stainless steel crash rails can be used also. The resin-bonded fibreglass provides a finish that withstands severe impact blows without damage; it remains impervious and withstands heat, provided that fire-retardant resins are used. There should be special provision for wall surfaces near sources of heat. Where separating walls are provided in food rooms, the tops should be rounded to assist cleaning and to prevent their use as shelves.

Ceilings

A ceiling is necessary in food preparation areas to prevent dust falling from the roof or upper structure. Ceilings should be smooth, fire-resistant, light-coloured, coved at wall joints and easy to clean. They may be solid or suspended. Plasterboard finished with a skim coat of plaster to eliminate joints may be combined with a hot-air blower suspended from a corner of the ceiling to eliminate condensation. Patent suspended ceilings in absorbent 600 mm square acoustic tiles (fibre-board) may be used. The tiles laid loose on aluminium bearers are cheap and easy to replace. Disadvantages include cleaning difficulties, deterioration when saturated, attack by vermin and corrosion of the alloy suspension grids. With suspended ceilings there should be access for pest-control purposes and service such as extraction ducting and water tanks. Services to equipment are usually accommodated in floor zones for easy access. Services to lighting run in the ceiling void; they are usually free from trouble. Insulation to the roof should be placed between the plasterboard or tiles and the structural roof.

Surface finishes for walls and ceilings

Decorative finishes for walls and ceilings protect the wall finish from damage caused by cleaning and provide a visually acceptable and pleasing environment. Oil-bound eggshell paints are the most widely used with a gloss finish on fittings. Other paints may help to reduce condensation, but they may not seal surfaces as well as oil-based paint. White ceilings are generally best as they reflect the light from artificial sources. All-white walls may be hard. Pastel colours with bolder tones on fittings are better. Paints should be used with caution with the danger of flaking and the possible contamination of food in mind. Proprietary ceiling systems are available for food-handling areas and some wall materials may be suitable for ceilings.

Lighting

Good lighting in kitchens improves concentration and safety; it also deters insects and vermin. The Offices, Shops and Railway Premises Act, 1963, requires that effective provision be made for sufficient and suitable lighting in every part of the premises. The Health and Safety Executive (HSE) publication *Lighting at Work* (HSG38) deals with lighting and how it affects the safety, health and welfare of people at work. The Chartered Institution of Building Services Engineers (CIBSE) Code for Interior Lighting gives recommended values of 500 lux for food preparation and cooking areas, 300 lux for serveries, vegetable preparation and washing-up areas, and 150 lux for food storage and cellars, with a limiting glare index of 22 lux. Artificial illumination is necessary even though the suggested standard could be met during normal working hours by natural lighting when there is high window-to-wall ratio. Good natural lighting is desirable, but without glare from direct sunlight causing reflection from polished surfaces. North light is preferable, otherwise overhanging eaves, tinted glass or solar film should be fitted to reduce direct light; Venetian and roller blinds are difficult to clean.

The running costs of fluorescent lighting are lower than those of tungsten filament and the light is more evenly distributed. The tubes should be fitted with diffusers to prevent glare and also to protect food, if there is breakage. Twin tube 'Kolor-rite' or 'Trucolor 38' lamps are recommended for food preparation areas as they give equal emphasis to all colours. The tubes lose 20% efficiency within a year or earlier and should be replaced when flicker and black ends are observed. The Chartered Institution of Building Services Engineers, 1984, revised Code, 1992 for interior lighting gives guidelines on design to improve texture, contrast and direction. Light fittings (luminaires) should be vapour proof and consideration should be given to the sites of equipment and preparation areas including sinks, stoves and tables. Spotlights can be placed directly over the servery counter to complement the overall lighting. Additional lamps may be required to aid cleaning of less accessible dark areas. In kitchens, proofed 'luminaires' are necessary to withstand both the heat and the possibility of explosions if tubes are defective.

Windows and doors

Window boards and ledges situated behind kitchen equipment are not easily accessible for cleaning. Sills should be higher than equipment and constructed with an angled ledge to prevent their use as shelves. Louvered windows in kitchens are not recommended as they may be difficult to clean and maintain. Fly-screens where fitted over window or other openings should be removable for cleaning purposes. The windows in a kitchen should not face due south unless precautions are taken to eliminate glare and the effect of solar heat.

Doors should be tight-fitting and self-closing; where necessary they should be proofed against insects and vermin. Doorways must be large enough to allow movement of mobile equipment. Swing doors should have sight panels and also hand and kick plates.

Ventilation

Efficient ventilation and extraction systems are essential. The operations of preparation, cooking, serving and washing-up generate large amounts of water vapour, which if not extracted will condense to create moisture dripping from ceilings or running down walls. In addition to high humidity there may also be volatile fat. To control humidity and cooking odours, localized systems of extraction may be used. Extractor fans in ducting should be chosen with care at the design stage and they must be maintained otherwise they can give rise to noise nuisance within the kitchen or to neighbours. Internally, if the noise from fans irritates food handlers, they may be switched off. Outlets to extraction systems need to be carefully sited to avoid malodours, which may arise even with proper filters if badly maintained.

The commercial fish fryer for fish and chips includes ventilation matched to the equipment; the system may still contravene the law by releasing objectionable smells. A ventilation system cannot operate without sufficient air for the extraction fans. With gas appliances, adequate ventilation is essential so that the open flames burn steadily.

The grouping of cooking appliances in islands is convenient for the

Fig. 15.1 Efficient and hygienic grouping of equipment

Fig. 15.2 Canteen servery

extraction of steam and odours. The equipment shown in Fig. 15.1 is modular and assembled like unit furniture with all the services located in a central spine. The arrangement saves space and lends itself to a localized extraction system, but the items cannot be moved for cleaning, and it may be difficult to reach the services. Cantilevered or mobile equipment with approved flexible connections, say, to gas appliances, will assist cleaning. Fig. 15.2 shows the design of a modern canteen servery used to serve a wide variety of lunches and snacks. Fans should be sited outside food rooms and be capable of extracting through grease filters. The filters should be so placed that they can be readily removed for cleaning or replacement, and a spare set should be available. The extraction should be up to 20 m^3/minute per m^2 of hood area. The ventilation should flow from a clean to a dirty area. Ducts should be as short as possible, airtight with properly sealed joints. Simple baffle boards are sometimes placed below extraction fans. They are usually made from block-board covered with plastic laminate. They are about 1 metre square and are suspended below the ceiling on adjustable straps. The distance between the board and the ceiling should be fixed so that the desired extraction velocity can be achieved. Such boards can be removed and cleaned and replaced within minutes. Fig. 15.3 shows simple access to ceiling extract ducts. In small kitchens wall extraction fans can be placed in walls or windows. Their position should be carefully chosen to prevent air circulating between window and fan. When the fan is running, windows should be closed. The use of pressure vessels, microwave ovens and increased insulation on ovens will reduce heat production. Air conditioning may be installed.

If the vegetable store is not situated on an outside wall, ducted mechanical ventilation must be introduced. The dry store will also need ventilation although to a lesser degree. Where mechanical extraction units are incorporated in food storage areas to form 'rapid cooling larders', it is important

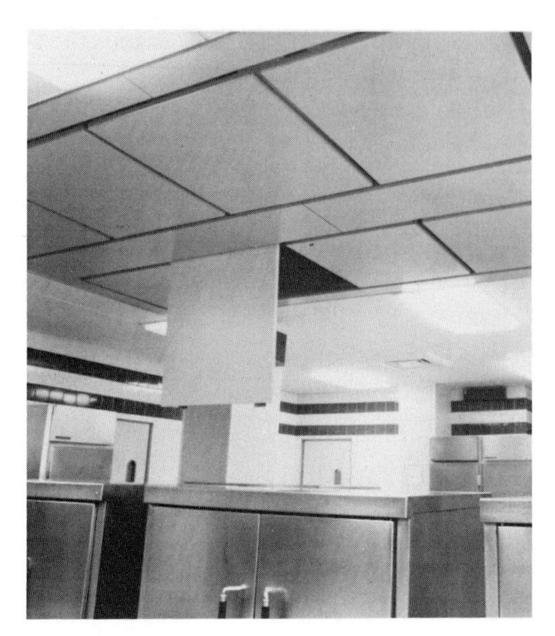

Fig. 15.3 Ceiling extract duct

to ensure that air inlets draw in fresh air and not air from the general kitchen area.

Heating

The law requires a minimum air temperature of 16°C (60.8°F) in work rooms, to be reached within the first hour of starting work. Usually, heating appliances are necessary only in staff rooms and offices. The law conflicts with the Department of Health (DH) publication *Chilled and Frozen – Guidelines on Cook-Chill and Cook-Freeze Catering Systems* and some food industry Codes of Practice, where a maximum ambient temperature of 10°C is recommended for rooms where food is handled after cooking, for example, where food is portioned. There are discussions in progress between the DH and the HSE (Health and Safety Executive) aimed to reach some agreement.

Services

Services for kitchens include water, drainage, electricity, gas and ventilation; numerous points are required so that services may be tapped readily. Ducts can provide routes for vermin to enter, leave and infest buildings. Holes for services must be sealed and proofed. Floor ducts liable to be flooded by wash water or to harbour vermin should be filled with lightweight concrete or similar inert material. Services should be chased into walls, where possible, or fitted clear of walls to allow proper cleaning. The use of wooden ducts around services encourages insects. With suspended ceilings, electricity cables may be sited above them. Appliances should not be permanently connected so that proper cleaning and maintenance can be carried out.

Gas

Gas is popular with chefs, but pressure in installations must be carefully balanced using a water gauge. Automatic ignition to burners is essential to reduce waste, and devices to respond to flame failure will help to prevent accidental build-up of gas. The possibility of explosions can be eliminated by the installation of an alarm system for gas detection.

Flexible supply pipes with bayonet fitting connectors for each appliance have in-built mobility that allows effective cleaning and maintenance. Gas appliances should have governors fitted to prevent variations in gas pressure affecting the operation of the burners. Supply pipes should be mounted clear of the floor, and of other pipes, for cleaning purposes. The Gas Company or an approved gas installation technician should ensure that the pressure is satisfactory.

Electricity

Electricity may be expensive, but it is the most suitable source of energy for many catering processes. As a fire precaution, one master switchboard, clearly identified, is necessary indicating the supply to each piece of equipment; it should be situated close to the kitchen. Single-phase electricity may be satisfactory for a few light catering kitchens, but usually a 415 volt supply consisting of three phases and a neutral wire is essential. Electrical wiring should be either insulated copper-covered cable or water-resistant heavy gauge conduit with outlets about 1.5 m above the floor. Electrical wiring should be protected by waterproof conduits and switches fitted flush to walls. Main switches should be placed outside food-preparation rooms. Where switch boxes serve individual appliances, they should be splash-proof, readily accessible, but sited away from areas of excessive soiling. There should be no low-level sockets in danger from wet floors (cleaning) and sockets should not be sited above sinks or hand basins.

Water supply

To ensure constant and adequate supplies of hot and cold water, pipes should be installed as a ring main, whereby water circulates continuously; stop taps should isolate every fitting. A supply direct from the mains is required in kitchens except at wash basins, showers, baths and dish-washing machines. A cold water tap for culinary use can be conveniently sited adjacent to cooking appliances. Softened water is of benefit in hard water areas to reduce the amount of detergent used. Boiling of water for tea and coffee rapidly produces hardness scales, hence boilers should be supplied with soft water.

An ample supply of hot water, 60°C (140°F) at least, is essential, but the actual quantity of hot and cold water needed depends on the type of menu and the scale of equipment.

The temperature of the water used for a disinfecting rinse may be raised by means of a thermostatically controlled gas burner or electrical unit, or by steam. This compensates for loss of heat from the rinse water while work is in progress. Alternatively an independent boiler may provide hot water directly to the disinfection sink from which there may be a constant overflow to the washing sink as the water cools. Where a piped supply is not practicable from the main hot-water system, for example, for hand-washing, small gas geysers or electrical appliances are useful.

An external water supply should be available for flushing refuse areas and loading bays.

Sanitary and washing facilities

In all premises where persons are employed, including buildings used for the preparation of food, toilet and washing facilities must be incorporated to satisfy the Sanitary Convenience Regulations, 1964, and the Offices, Shops and Railway Premises Act, 1963. Local Authorities are empowered to require that sufficient specified appliances (water closets, urinals and wash basins) be provided and maintained in a clean condition at places for public entertainment, or where food and drink is sold to members of the public to take away or for consumption on the premises. The suitability and adequacy of sanitary accommodation is considered before a licence is granted by the Brewster Sessions. In general terms the provision of one lavatory for every multiple of 15 males or females will meet the regulations, but under special conditions more may be required. Low-level WC cisterns are recommended as they are more readily adapted to a pedal-operated flush.

The siting of toilet accommodation in relation to food-preparation areas and other rooms in danger from pollution is subject to statutory control. All toilets must be entered from an intervening and separately ventilated lobby or corridor. Toilet areas should be adequately lighted and ventilated either by natural or by artificial means. Artificial ventilation should provide six air changes per hour, and supply air for fans. Two fans should operate continuously and an emergency fan be available. For small places, individual fan units, operated by the light switch, continue to function 20 minutes after the light is switched off. Mechanical ventilation for sanitary facilities must not communicate with other ventilation systems within a building. The Washing Facilities Regulations, 1964, require basins or equivalent facilities in troughs or fountains to be provided on the same scale as toilet accommodation.

Hand-wash basins must be placed in or adjacent to toilet cubicles as well as in food-preparation areas where hands will be soiled from contact with raw food and other materials; foot-operated control is recommended (Fig. 15.4).

When food handlers are working on more than one floor of a building, conveniences and adjacent wash basins are desirable on each storey; they should at least be situated so that each serves not more than two floors. For

Fig. 15.4 Foot operated hand-wash basin

troughs, a length of 0.6 m is regarded as equivalent to a wash basin, and for fountains 0.6 m in circumference. The circular bowls should measure at least 0.9 m in diameter.

For hand-washing there must be supplies of running hot and cold water, or water blended to a suitable temperature. Where a piped supply of hot water is not available thermal storage units and small instantaneous water heaters are convenient.

Spray taps that discharge water at a temperature of approximately 50°C (122°F) are economical in the volume of water delivered and consequently in cost. Lever-arm taps, operated by the elbow or forearm, reduce chances of cross-contamination; alternatively, pedal-operated taps or sprinklers should be used (Fig. 15.5). The automatic tap, whereby water flows at a pre-determined temperature when hands are placed in front of the tap and ceases when the hands are removed, is an alternative to the foot or lever-arm-operated tap. All basins and troughs should be made of stainless steel and connected to the drains by trapped waste pipes.

Washing facilities include the provision of soap and nailbrushes and some means to dry the hands. Liquid or dry soap dispensed from fitted containers

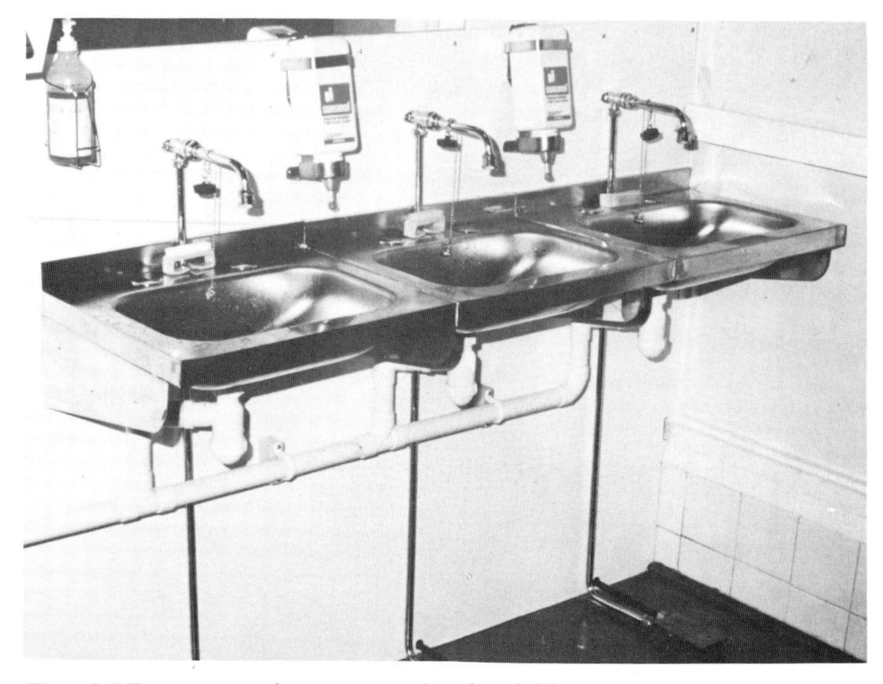

Fig. 15.5 Foot operated water control and sprinkler

is preferable to tablets of soap in wet dishes. Liquid soap should be provided in disposable containers where it is unlikely that dispensers will be washed and disinfected before re-use. Disposable plastic nailbrushes are available also. A paper towel cassette system, similar to the continuous roller towel, is thought to be better than the electric hot-air blower, which may not effectively dry hands without patience. Some plastic nail brushes with nylon bristles may be disinfected by heat, although many plastics do not withstand heat. They should be stored dry, up-ended and not in disinfectant solutions. They may be suspended from a wall attachment by means of a magnet. Satisfactory means for drying hands include paper towels, the patent continuous roller towel providing a clean portion for each person, and electric hot-air blowers.

Sanitary and washing facilities must be separate for each sex, unless the number of persons employed does not exceed five at any one time. The facilities provided for customers may be used for employees, but the number must be increased by one extra closet and wash basin when more than ten persons are employed. A notice drawing attention to the necessity to wash hands must be displayed in all sanitary facilities for staff, and brought to the attention of customers also.

Sparge pipes, particularly in hard-water areas, should be thoroughly cleaned periodically to remove build-up of scale at the water outlets and subsequently inadequate swilling of urinals leading to malodours.

BS6465, Sanitary Installations, Part 1, Code of Practice for the Scale of

Fig. 15.6 Double sink with small section for hot disinfection

Provision, Selection and Installation of Sanitary Appliances, 1984, suggests the number of hand basins, urinals, and water closets that should be provided in restaurants and canteens for customers. The Building Regulations, 1985, as amended, require that in new buildings, where customer toilets are provided, at least one should be available for the disabled.

Sink and washing-up units

Various dish-washing machines and sink units are available for crockery, cutlery, glassware, pots, pans and utensils. A separate sink for vegetables and salad preparation is essential. Guards of 150–300 mm integrated with the back of sink units and preferably made of stainless steel protect adjacent walls. Sinks should be installed under wall-mounted taps with stopcocks fitted at a low level on the water pipes. When the taps and plumbing are not fixed directly on to a sink, the waste trap can be unscrewed and the unit removed to allow concealed wall areas to be cleaned.

Articles can be disinfected in a dish-washing machine or by means of a stainless steel, double-sink unit with the water temperature in the rinse sink maintained at 77–82°C (170–180°F). An ample supply of deep wire baskets is required. A sink unit with a small section for hot disinfection of knives, choppers and other articles is useful (see Fig. 15.6). Usually clean, dry crockery, cutlery and glassware is acceptable without thermal or chemical disinfection. Where disinfection is required thermal methods are preferable to the use of chemicals; alternatively, disposable items may be used. Where dish-washing machines are not provided, for example, in institutions such as HM Prisons, there is a continuing need for heated sinks to remove grease and to disinfect the eating utensils. Sinks and hand-wash basins may also be mobile. Automatic lettuce- and vegetable-washing machines with integral dirt filters are available.

Drainage

The design and construction of drainage systems for all classes of buildings have to meet the requirements of the Building Regulations, 1985, as amended. The design drawings must be inspected and approved by the local authority before construction takes place. Grease discharged into kitchen drains comes mostly from the washing-up area; it is molten as it enters the pipes, but solidifies unless intercepted. Grease traps are tanks in the ground that hold sufficient water to cool the inflow of washing-up water to a temperature below the melting point of the grease. The inlets and outlets are submerged so that the solidified grease forms layers that can be skimmed off at regular intervals. Such traps are not entirely satisfactory; with correct procedures, accumulation of grease can be avoided. It is important for drains to be rodded from the fitting to the entry into a manhole so that blockage can be cleared easily. Large cast-iron and copper waste pipes should be used, 50–75 mm in diameter. Plastics able to withstand high temperatures are now available. One large pipe can be used in common to drain several fittings; it should discharge to a back inlet gully. Wherever possible drainage channels should be avoided and only used where necessary; floors could be graded and drained to a floor gully. Where tilting bratt pans or similar equipment are in use, a small section of stainless steel floor channel, with widened lip and properly sealed at the edges should be sufficient. The grating should be easily removable and not screwed down. Gullies serving hand-wash basins and sinks should be sited outside the building and there should be no access to drainage systems in food rooms, or soil pipes passing through them from above. Drainage channels are required around cooking equipment; they should be covered with galvanized gratings which can be easily removed. Sinks and basins should discharge via trapped gullies. Items of equipment such as potato peelers, dish-washing machines, or waste-disposal units that are connected directly into the drainage system should be trapped to prevent waste pipes acting as vents.

Waste disposal

Unlike the waste from many industrial and commercial processes, kitchen waste is organic and requires particular care pending final disposal.

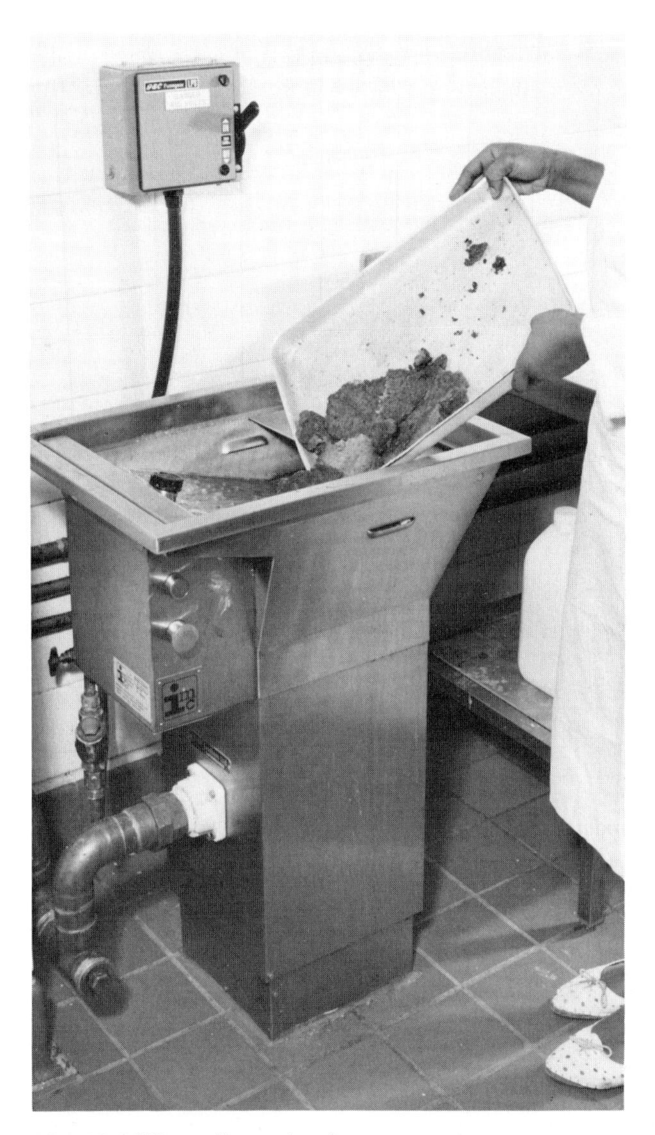

Fig. 15.7 Waste disposal unit

Immediate disposal by disintegration and flushing to the drainage system is the most convenient method. Such disposal units may be fixed under metal sinks or stand in a frame with their own receiving hoppers. An electrically driven macerating unit breaks down the waste food to a fine suspension which is washed away through a trapped waste pipe. One type of machine is automatically switched on only when water is flowing through the grinding chamber (Fig. 15.7). Units for the domestic market have motors of one-third horse power, but heavier duty machines

Fig. 15.8 A refuse compactor

are needed for commercial premises. Where waste disposal units are to be fitted it may be necessary to obtain local authority approval, to ensure that the drainage system is adequate. Commercial waste disposal should also have a fail-safe device so that if, for example, a fork were to fall into the mechanism, it could not be retrieved until a baffle, held in place by a clamp knob was removed. The removal of the clamp knob switches off the motor. Refuse compaction is economical in large food businesses where skips are provided; machines are available to reduce bulky refuse to a smaller size (Fig. 15.8).

Paper or plastic sacks filled with waste are light and clean to handle, they can be sealed easily with a bag-twist tool and 254 mm (10 inch) wire twists, both available from agricultural merchants. They should be placed inside bins to prevent attack from vermin, domestic animals and birds.

The British Standard Specification for dustbins states that they should be made of stout and impervious material; 22 gauge galvanized iron is recommended. Lids should fit closely with an overhanging lip; lids of hard rubber reduce noise. Mobile paladin bins are often used for the collection of waste from business premises, although the heavy lids may be left off and the bins can be difficult to clean. Alternatively, wheeled 'Eurobins' with hinged lids are easier to clean and they are protected from birds.

Staff should be aware of the necessity to remove refuse daily and to take care with storage. A properly constructed storage area for refuse is essential, with a concrete floor and a hose tap, suitable for water to flush the yard; the drainage gully should be fitted with a straining basket. The concreted storage area for refuse needs to be graded to the yard-floor gully and if it consists solely of a concreted platform, there should be a bund wall to prevent spillage/washings accumulating on the surrounding earth and giving rise to the breeding of flies. The wall surfaces should be smooth, impervious and able to withstand water from high-pressure hoses. Refuse bins need to be washed each time they are emptied and stored inverted to drain dry on metal racks above the yard surface. A daily collection is necessary for large quantities of kitchen waste; if frequent collection is impossible, a secure refrigerated refuse store, away from food, should be provided.

Equipment

The European Commission has issued a Council Directive (89/392/EEC) on laws relating to machinery and essentially deals with machine safety. Also there is a section on 'Agri-foodstuffs machinery' which contains rules of hygiene. Working groups representative of member states are formulating hygiene standards for different types of food equipment/machinery.

Equipment must be designed and sited so that all surfaces are accessible for cleaning. Every item should be reached and removed easily. Large equipment should stand free from walls and floor. Mobile equipment with wheels which can be locked facilitates cleaning, and also helps to reduce the movement of foods (Fig. 15.9). It is recommended that there should be at least 300 mm (12 inches) space under appliances, otherwise they should be sealed to a solid base such as a concrete plinth approximately 150 mm (6 inches) high, topped with quarry tiles and with a coved tile skirting. Dirt may accumulate in the framework and outer casing, which should be as smooth as practicable, and spaces in the supporting structure totally enclosed and sealed or adequately open for cleaning purposes. For example, legs of standing equipment formed of sealed tubes are better than angle or channel section. The basic material for food equipment must be non-absorbent and not subject to injury by cleaning or by disinfectants. There are a number of trolleys available with temperature control, which are widely used in institutions. Some are able to regenerate cook-chill foods in accordance with the time/temperature combination laid down in the DH Guidelines. Wooden duckboards and pallets should not be used.

Surfaces

The surfaces of equipment should be smooth, continuous, sealed and easily cleaned and if necessary disinfected. Internal angles should be easily cleaned and without voids and dead spaces. Cleaning chemicals or acid foods should not lead to deterioration. To service and clean,

Fig. 15.9 Food mixer mounted on mobile table

equipment must be accessible or mobile. Tables and worktops should be constructed for ease of cleaning, strength and durability. Stainless steel tables with tubular legs and braked castors are recommended. Working surfaces/tables in 18/10 gauge stainless steel may be cantilevered, centralized or back-to-back. Wood is unsatisfactory because it is porous and cannot be effectively cleaned and disinfected. Teak sinks still harbour numerous intestinal bacteria even after washing and scrubbing with soda solution. A stainless steel upstanding back, 150–300 mm (6–12 inches) at the rear of worktops, as for sinks, will keep adjacent wall surfaces clean and undamaged. Movable worktops will facilitate cleaning, or if fixed at least 300 mm (12 inches, broom width) space should be left behind them.

Open shelves of stainless steel or cantilevered plastic laminate are recommended for the storage of small items of equipment and foodstuffs in preference to cupboards or drawers. Fixed shelves sited over cooking equipment should be made of stainless steel and not wood. Hollow plinths under shelves gather dirt and may attract mice for nesting. Free-standing wire shelves allow air circulation and dust to fall through, but they are unsuitable for open food. Detachable shelves are convenient for cleaning and redecorating.

Cooking equipment

There are many kinds of cooking equipment available either gas or electrical where the appliances may be arranged in combination. Some equipment may be wall-mounted or back-to-back with fitted splashbacks. Induction cookers rely on electromagnetic waves to heat cooking vessels directly; thus there is a greater degree of heat control than from the traditional electrical elements. A vitreo-ceramic surface on which the vessels stand allows simple cleaning and does not become heated in areas where the vessels are not placed.

Polished stainless steel, for example type 302 and thickness 14/16 gauge, is the best construction material for catering equipment with rounded corners and deburred edges. The design should allow heavily soiled parts to be removed for cleaning. Each piece should fit easily into a sink, or better still, into a dish-washing machine directly before emptying the tanks. Several washes may be necessary for heavily soiled trivets, trays and shelves; if the washing process is carried out daily, the finish can be maintained. The traditional method of cooking, using pots and pans on an open ring, is wasteful of energy. As much as half the heat produced is lost by ventilation and so wasted. A large amount of food in one container above several open burners requires much water and constant stirring. It is hazardous for containers full of hot food to be carried by the staff. Also, excessive heat in the kitchen causes unpleasant and debilitating working conditions and the warmth and humidity encourage microbial growth. Cooking large amounts of food too far ahead of requirements without adequate cooling and refrigeration facilities may lead to food poisoning.

Equipment is available to cook food in batches with only small losses of flavour and nutriment, for example, pressure steamers, bratt pans that can be tipped to empty, and forced air-convection ovens. Such equipment can be raised above the floor surface and cleaned underneath. Thus the safety and efficiency of cooking equipment are improved by separating the boiling top from the oven, which should be situated at waist height on movable stainless steel tubular frames. Aluminium or light alloys should not be used for burners and handles because they are difficult to clean and are damaged by caustic agents. Convection ovens, microwave ovens and pressure cookers are described in Chapter 11. Pass-through cabinets, either heated or chilled, allow the passage of ready-to-eat food from the kitchen to the self-service area on the other side.

Cooking vessels

Cooking vessels and other utensils are made of a wide range of materials, including various metals and vitreous substances. There have been no reports of illness following the consumption of food prepared in contact with the more common metals such as stainless steel, aluminium and its alloys. Iron and tin cans for food are mostly constructed of tin-plated iron sheet, although aluminium is increasingly used as a can material for a variety of foods as well as soft drinks. Prolonged storage of acid fruits in unlacquered cans occasionally results in absorption of iron and tin and high concentrations of tin can be poisonous. Copper, also, can be attacked by acid foods. Lead is a poisonous metal and the glaze of earthenware vessels, which contains lead oxide, may react with acid products; similarly, zinc may be dissolved from galvanized equipment. Antimony is a component of enamel used for coating food vessels. Cadmium is used for plating utensils and fittings for cookers and refrigerators, and may be a component of earthenware from some countries. Both antimony and cadmium can be poisonous when vessels are used for acid foods; care should be taken with chipped enamel vessels and food should not be laid directly on the shelves of refrigerators. The use of cooking pots and pans which are heavily carbonized wastes energy.

Fryers

Deep-fat fryers hold large quantities of cooking oil, which undergo a physical change at prolonged high temperature when the quality of fried food deteriorates. Fat deposits may build up in kitchens and ventilation systems and even outside buildings. The breakdown of cooking oils is caused by factors such as:

(i) high-temperature frying and oxidation;
(ii) the introduction of excessive moisture, common salt, potato whitener and charring by food particles; charring can be reduced by providing a deep cool zone in frying equipment;
(iii) the catalytic action of copper and brass in old chip fryers or filter valves, which rapidly darkens fat or oil.

Portable fat filters may be used for removing suspended particles from the oil or fat. They may also be used to hold the fat while the fryer is being cleaned. Several small-capacity fryers give a greater flexibility and better results. All frying equipment should be fitted with thermostatic controls and heat applied slowly. Thermostats should be calibrated and checked against a thermometer at least weekly. Enclosed filter systems are useful for clarifying oil; open systems may oxidize the oil. Manufacturers recommend a minimum level of oil during frying and frequent topping-up with new oil. For fish and doughnuts a complete change of cooking oil is required regularly. All deep-fat equipment needs careful operation and supervision.

Hot cabinets

The hot cabinet is satisfactory for heating plates but not for keeping food warm since food dries quickly, losing flavour and nutritional quality. The appliances are not designed to cook food and the system of holding food in this manner can provide the warm conditions necessary for the growth of food-poisoning and spoilage organisms. Nevertheless, provided that heated cabinets or display cases for hot-food storage are thermostatically controlled and hygienically designed they are important for the maintenance of relevant foods at temperatures greater than 63°C (145°F). It is generally necessary to pre-heat these units prior to use, in order to achieve the required temperature. Food should be served directly from cooking appliances, hot *bain maries* and cold cabinets.

Cutting boards

All kitchen equipment traditionally made of wood, for example, mincer plungers, spoons, paddles and sieves, should be manufactured from impervious, easily cleaned materials. Synthetic, polypropylene, non-absorbent cutting pads are easily cleaned and disinfected and can be put in a dishwasher, although they may sustain cuts. Hard synthetic rubber is good although it may be deformed by heat; this can be corrected if it is softened in hot water and allowed to harden on a flat surface. Hardwood, if used, should be constructed without joints.

Separate cutting blocks and boards should be used for different foods to avoid the risk of cross-contamination when raw and cooked foods are prepared in the same area. Cutting boards should be colour-coded as well as the kitchen knives. They should be kept in suitable racks. Hard synthetic rubber chopping blocks or table tops are useful in many food trades. The surface is suitable for boning meat, cutting pastry and filleting fish without damage to knives. Rubber is impervious, unlike wood, where the cellular structure permits the absorption of food particles and juices. Sandpaper may be used to re-surface rubber pads after prolonged periods of heavy use. Rubber table tops of 25 or 50 mm (1 or 2 inches) thickness should be supported on stands made of tubular aluminium, stainless steel, or galvanized steel, and of adjustable height.

Slicing, mixing and mincing machines

Slicing and mixing machines should never be used for both raw and cooked foods without thorough cleaning in-between. The practice of using mincing machines for both raw and cooked meat even with washing in-between should be discouraged because of the difficulty in the thorough cleaning of parts of the mincer. All these machines can be responsible for cross-contamination. Food machinery should be designed for easy cleaning and disinfection, readily dismantled and reassembled; safety is also important in the selection of machines. Hygiene and safety considerations should come first in the choice of equipment and outweigh the initial cost.

The risk of cross-contamination will be increased in machines with interchangeable parts and used for more than one purpose, such as mincing and mixing.

Food storage

Refrigeration

The growth and multiplication of bacteria in food will be reduced by rapid cooling and refrigerated storage. The temperatures at which relevant foods must be kept are laid down in the Food Hygiene (Amendment) Regulations, 1990 and 1991. These temperatures are above 63°C or below 8°C. The lower temperature will be reduced to 5°C for certain foods on the 1st April, 1993. Refrigerators should provide separate storage space at the temperatures recommended for each particular type of food; separate refrigerators may be required. All commercial refrigeration equipment should have in-built thermometers, which should be checked periodically for accuracy. In some situations it is necessary, to have a temperature history of refrigerated foods which may be in various forms – such as electronic recall, as on some vending appliances, to chart recorders. A suitable digital thermometer with appropriate probes should be available together with a supply of disinfectant wipes. This thermometer should be checked for accuracy at regular intervals against a reference thermometer or the sensor checked with a wet ice mixture. Cold-storage facilities should be provided for vegetables and fruit, at a temperature of approximately 10°C (50°F). Cooked and uncooked food should be refrigerated and stored separately, not on the cold-room floor or beneath other foods that may spill or drip. Cold-store evaporators should be defrosted regularly, otherwise temperature control cannot be maintained and the temperature will rise inside the refrigerator. Automatic defrosting is recommended. Refrigeration is a heat exchange process and heat removed by the refrigerant must be dissipated elsewhere. Ample ventilation to the condenser is required and no obstruction of the plant should occur. Wherever possible, large refrigerator condensers should be sited outside the building and always outside the kitchen. Refrigeration equipment, if outside the premises, should be sited in such a position that it does not cause noise nuisance to neighbours. They are a source of heat and good nesting places for mice. A warning device for refrigerator or deep-freeze failure is essential. Two separate monitors are valuable in case one fails. Mobile refrigerated cabinets are recommended for use in large-scale catering (Fig. 15.10) with spacious preparation areas. Blast chillers and blast freezers are required for cook-chill and cook-freeze catering and are available in several sizes; some allow trolleys holding food to be wheeled in directly. They may also be used with conventional catering, provided the caterer is knowledgeable and has carried out a risk analysis. Such units allow greater flexibility with an even distribution of work, thus avoiding 'peak' periods and menus may be wider and properly planned. DH guidelines recommend for cook-chill foods a maximum life of 5 days, including the day of production and the day of consumption, and for cook-freeze foods up to 8 weeks. Blast chillers should be able to reduce the temperature of cooked foods to between 0–3°C (32–37°F) within 1.5 hours and blast-freezers should allow the food to reach −5°C (23°F) within 1.5 hours of entering the freezer, and subsequently to reach the storage temperature of −18°C (0°F).

Deep freeze

Frozen foods should be maintained at a constant temperature in a cabinet operating at −18°C (0°F); the temperature should be checked frequently.

The position of the cabinet should be such that it is not affected by draught, heat from the sun or radiators, and with no obstruction to the compressor, motor and condenser.

Rapid cooling larder

The provision of facilities for rapid cooling is strongly recommended, particularly for foods refrigerated after cooking. They are especially necessary for meat and poultry dishes intended to be served cold. A

Fig. 15.10 Mobile chill cabinet

room should be allotted on the cold side of the kitchen provided with an extractor fan and means to maintain the temperature at 8–10°C (46–50°F). Such facilities are also useful for the storage of dairy products. The mobile cabinets provided with fans and air filters are desirable in any situation where food is cooked in bulk for large numbers of people. It cannot be over-emphasized that rapid cooling from cooking temperatures and before refrigeration would prevent many outbreaks of food poisoning. The provision of this room except for domestic or small catering may no longer be appropriate with blast chillers and consequent stricter temperature control of foods.

Larder for dry goods

Equipment used for the storage of dry goods, such as cereals, should be constructed of inert materials, easily cleaned, mobile and lidded. Shelves should also be mobile, preferably stainless steel and easily cleaned. The lowest shelf should be at least 36 cm (12 inches) above floor level to facilitate cleaning and reduce the risk of infestation by insects and harbourage for vermin. Many foods are labelled 'Store in a cool, dry place' which should be less than 10°C (50°F); although a temperature is rarely specified. Some advice/guidance is required on acceptable temperatures for 'cool places'.

The vermin-proof room or cupboard, depending on the size of the kitchen, for dehydrated and canned goods should have impervious surfaces to the floor, walls and shelves and it should be screened against flies and other insects. Good ventilation can be supplied by 229 mm (9 inch) air bricks in an external wall together with an extractor fan to aid air movement. The inlet and extract vents should be on opposite walls, the inlet at a lower level than the extract to ensure good circulation of air.

Cleaner's equipment room

The various articles used for cleaning, although nominally regarded as clean, are often profusely contaminated with bacteria and they should not be stored in food rooms. Cloths, mops and brushes can recontaminate otherwise clean surfaces, and it is essential to wash and disinfect them after use. Such equipment should be colour-coded to reduce the risk of contamination and the use of the same equipment in toilets and food-handling areas.

Provision should be made for racks to drain buckets after washing and for lines to dry cloths. There is a risk that spillage of chemicals may contaminate food, therefore separate storage areas are essential for cleaning materials. Sinks and wash basins in kitchens should not be used for the disposal of wash water used for cleaning. Fig. 15.11 gives a plan for the cleaner's equipment room.

A separate slop sink or gully should be provided together with supplies of hot and cold water, hand-wash basin and extractor fan. If cloths are used for equipment, cutlery, surfaces and glasses they should be disinfected at least daily in boiling water or freshly prepared bleach solution. After disinfection in the bleach solution, cloths must be thoroughly rinsed. Paper swabs and disposable cloths are recommended. Twin bucket trolleys are available for washing floors manually using a mop. One bucket contains detergent water and the dirty water is wrung out into the second bucket (Fig. 15.12). Powered floor-cleaning machines may be used; hot water is sprayed on the floor with a cleaning agent and a rotating brush loosens the dirt. Some types of machines are able to remove the dirt from grooves in the floor, but it may be necessary to follow the path of a machine with a mop and 'squeegee', particularly if the machine cannot remove dirt from the intersections of the floor and walls or around machinery. Other types of floor-cleaning machines scrub and dry the floor in one operation. A small domestic washing machine would be useful in the cleaner's room, for washing mops and cloths at a disinfecting temperature.

Staff room

Facilities away from food rooms must be provided to store and dry the outdoor clothing of employees. Where lockers are provided they should be heated, constructed of wire mesh and situated close to the staff sanitary and washing area. It may be preferable to hang clothes freely on rods and to supply a source of heat for drying purposes. It is desirable to install showers for the staff and also a receptacle for discarded protective clothing. Facilities may also be required for the

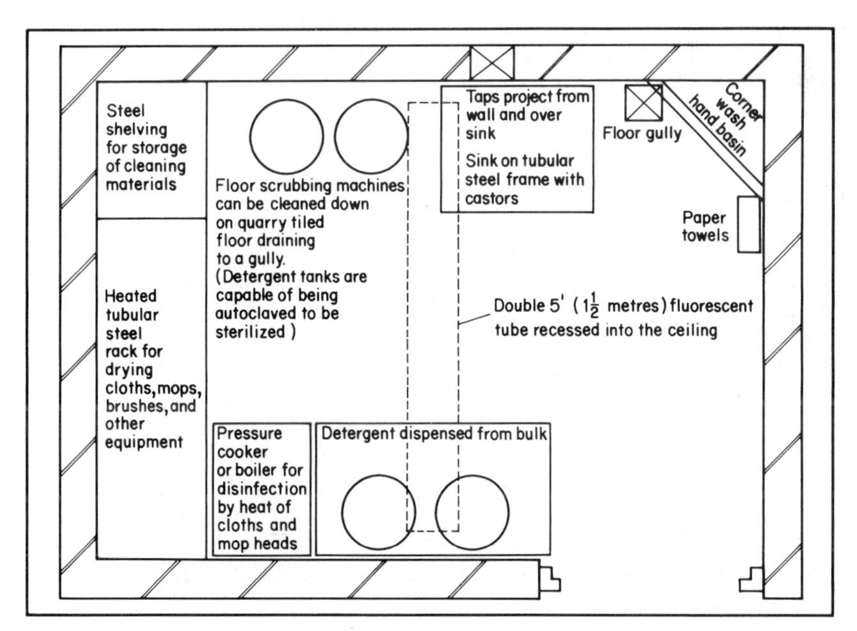

Fig. 15.11 Plan for cleaners' equipment room

additional protective clothing provided for employees who work in cold stores or in refrigerated rooms, and for the cleaning of special footwear, such as Wellington boots.

First aid

The provision of first-aid materials is a legal requirement laid down in the Food Hygiene (General) Regulations, 1970, and in greater detail in the Offices, Shops and Railway Premises First Aid Order, 1964, Number 970. The Health and Safety (First Aid) Regulations, 1981, are applicable. The HSE published in 1990 *First Aid at Work* which is an approved Code

Fig. 15.12 Mobile twin bucket trolley

of Practice giving guidance on these Regulations. Blue rather than skin-coloured dressings will be noticed more easily and are less likely to become extraneous matter in food.

Licensed trade

In the public house or hotel bar there may be a mixture of business and domestic arrangements, but the preparation of food intended for customers should be separated from domestic occupations. As for all households, laundering should be carried out in a separate room away from food preparation. If there is no special accommodation, the bathroom could be used.

Food displayed on bars, or elsewhere, should be kept cold and a rapid turnover assured. The DH has issued for Enforcement Officers, *Guidelines on the Food Hygiene (Amendment) Regulations, 1990* and *Guidelines for the Catering Industry on the Food Hygiene (Amendment) Regulations, 1990 and 1991*, also with MAFF, Welsh and Scottish Offices, the DH have issued under Section 40 of the Food Safety Act, 1990, Code of Practice No. 10, *Enforcement of the Temperature Control Requirements of the Food Hygiene Regulations.*

16

Control of infestation

It is probable that estimates of the material damage caused by rats and mice are often overstated, and the implication of these rodents in the spread of infection in Britain is probably not considerable, although not insignificant, at the present time. Nevertheless, the presence of such animals in and around food stores is repugnant, apart from being a source of material loss, and a potential danger to health.

Most local authorities maintain a staff of pest operators trained in methods of rodent destruction; some rely upon private companies to carry out the work. The occupier of infested premises can call for practical assistance. It should be pointed out that the onus for keeping premises free from infestation rests upon the occupier or owner. Councils are entitled to charge for such services, but rarely do so in domestic premises. Where there is serious damage to food caused by any infestation the occupier is under an obligation to report it to the authorities (Prevention of Damage by Pests Act, 1949, Section 13).

Block control of rats and mice

The Black Death in the fourteenth century was bubonic plague caused by *Yersinia pestis* which was spread from rat to rat and from rat to man by fleas carrying the organisms; 25 million people died. Certain countries still suffer this disease. The rat is clever, breeds prolifically, gnaws voraciously and may carry agents of other diseases such as leptospirosis, and viral and food-borne infections.

A survey conducted by the Institution of Environmental Health Officers in the late 1980s revealed a 20% increase in the number of surface sightings of rats in the urban areas of the UK. Contributory factors included mild winters, a proliferation of fast and take-away food premises giving rise to increased refuse, demolition and building works and the deterioration of the sewerage system, combined with a marked decrease in the number of 'treatments' for rat-infested sewers carried out in recent years.

An infestation is not always confined to one part of a building which may be occupied by different families or firms, nor to a single building; the rat or mouse colony may spread through several adjoining premises. Thus a tenant might kill off the rodents on his own premises, only to find them replaced in a short time by an overflow from his neighbour's property. Concerted action over the whole infested territory or block is necessary before all the rats and mice can be destroyed. Usually this is carried out most conveniently by the local authority and it is known as block control (Fig. 16.1). A contract with a commercial pest-control organization will minimize the risk of reinfestation.

Vertical block control

An infestation may extend below ground to drains and sewers, so that vertical block control schemes must include treatment for the destruction of rats in underground pipes as well as in the buildings above. Black rats may be present in the roof and brown rats at ground level.

Rodent-proofing

Some comments have already been made on structural materials and their properties of resistance to penetration by rodents. Naturally, stony and metallic materials have basic advantages over those of a fibrous nature. Impenetrable materials, however, are of little value if the building is

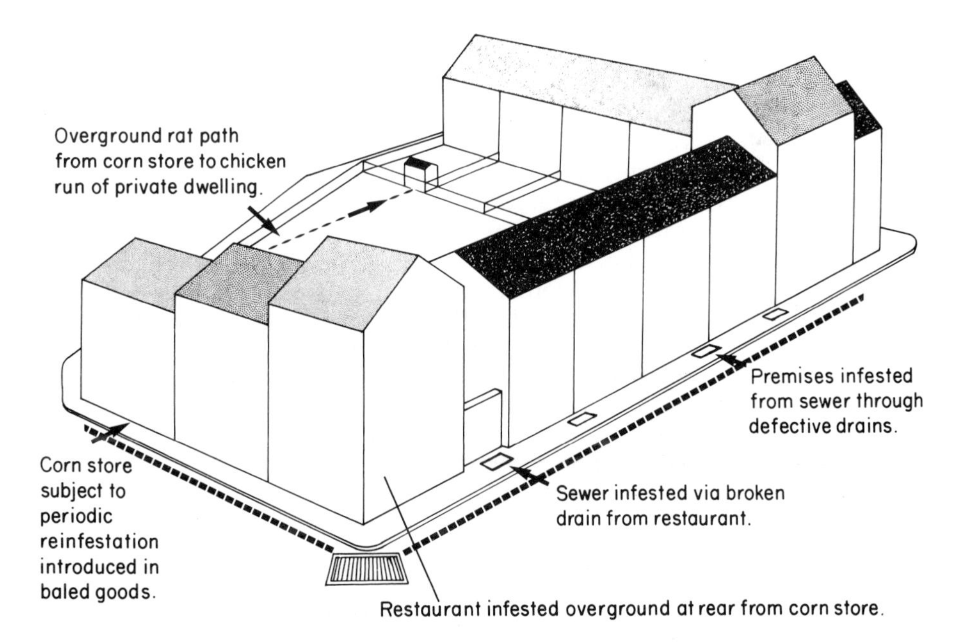

Fig 16.1 Block control: an example of related infestation in separate buildings which require simultaneous treatment.

constructed so that apertures or cavities are left for rodents to use as a means of entry or passage, or as a place to nest. Furthermore, thoughtless methods of food storage, careless stacking and general untidiness may allow a considerable infestation to become established in an otherwise sealed building; a rat or mouse family may be introduced through the store entrance, for example, in straw packing materials.

The major requirements for survival and family life in the rodent world are food, warmth and housing, and to deny any or all of these is a means of discouraging infestation. Therefore, material which is likely to be suitable for rat food, e.g. cereals, starchy vegetables and fatty compounds, including even tallow and soap, should be kept in rodent-proof metal bins or containers, and refuse of the same type in properly covered metal dustbins, whilst awaiting removal. As far as possible, access to water should be denied by attending to dripping taps and keeping grids on gully traps, for example. Articles temporarily unused should not be allowed to accumulate in odd corners nor remain undisturbed for more than a week or two, thus providing refuge for rats; anything which may afford cover to rats should be eliminated from the building and attached yards. Mice obtain sufficient water from their food and their small size enables them to enter buildings and move about easily.

Cartons, boxed goods, or sacks should be neatly stacked close together on wooden pallets at least 0.6 m (2 feet) from a wall. Although rats hesitate to cross open spaces, the space behind should be inspected. Incoming goods must be inspected and segregated from packaging and existing stock.

Drains often provide an entrance to a building and should be maintained in good repair with all inlets, manholes, and rodding eyes properly sealed or covered. There are other vulnerable points such as broken air-gratings in walls, openings for pipe-work sometimes below ground, ill-fitting doors, worn thresholds, the gnawed bottoms and jambs of doors, and corrugation in the wall or roof. Structural harbours may occur below hollow floors, in hollow walls and partitions, in pipe ducts and casings, and less commonly in roof spaces above false ceilings.

Fig. 16.2 illustrates remedies for some of the common defects mentioned above. It shows some of the means by which the passage of rodents from one part of a building to another (for example, from a sub-floor space to the room above) may be prevented. It will be seen that to proof a building against rats, openings must be sealed by cement mortar, sheet metal, or mesh. Similar materials should reinforce vulnerable points such as door edges and junction points of walls and floors. Coarse wire wool should be embedded in the cement used to block holes in walls. Metal plates can be used in the base of ill-fitting doors.

Rat destruction

Rats are creatures of habit and are suspicious by nature; they avoid new objects. They may be trapped, but poisoning, when practised methodically, is generally more effective.

For some years warfarin has been used in suitable bait (e.g. oatmeal).

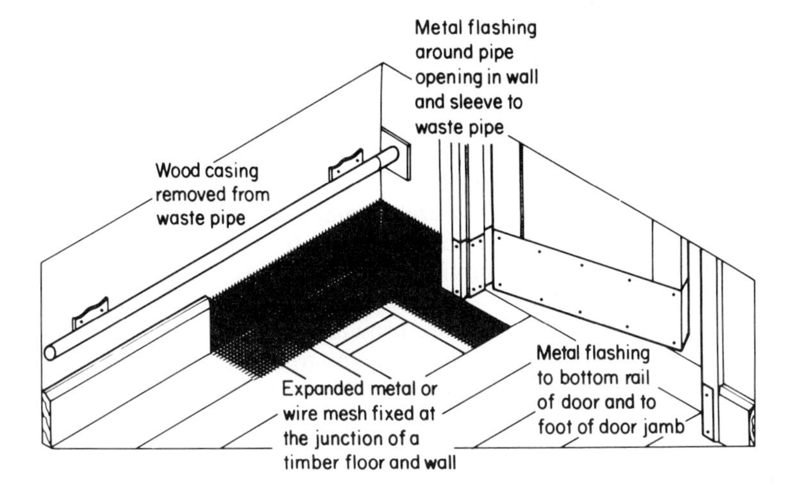

Fig 16.2 Rat-proofing: methods of protecting some vulnerable points

After several doses it prevents the coagulation of blood and produces spontaneous haemorrhage. During the feeding period of 3–8 days before death, the rat appears to experience no pain or significant loss of appetite and no 'bait-shyness' develops. As the poisoned bait is acceptable immediately, pre-baiting is unnecessary, but it must be replenished or renewed for a period up to 10 days or until feeding ceases. The proportion of warfarin used in the bait is small, giving a final concentration of commonly 0.005–0.05%, so that there is a large margin of safety in case of accidental ingestion by man or domestic animals. Also, as the poison is effective only after repeated doses, the danger to creatures other than rodents is remote. It is considered advisable, however, to lay the baits in containers or under cover in positions where the poison will be readily accessible to rodents only; thus the bait, approximately 85 g (3 oz), should be laid in established and currently used rat runs or where smears or droppings indicate rat activity. Liquid warfarin is also available. Before baits are laid it is necessary to carry out careful surveys of the area to determine the extent of the infestation, rat runs and harbourages, and to estimate the number of baiting points required. Each point should be baited with 227–454 g (8–16 oz) of material; frequent inspections are necessary.

Populations of rats in various parts of the UK have become resistant to warfarin; resistance is reported in Denmark and in other parts of the world. The selective destruction of non-resistant strains by the continued use of an anticoagulant reduces the natural competition amongst the rats in a neighbourhood, and can lead to a less restricted reproduction and overflow of resistant rat colonies into wider areas.

When rats become resistant to warfarin it should be assumed that other anticoagulants such as chlorophacinone and coumatetralyl will also prove to be ineffective. In such areas, difenacoum or calciferol should be used;

some resistance to difenacoum has been observed. The two anticoagulants, brodifacoum and bromadiolene are not so freely available as warfarin and at present are subject to restrictions in use. It is anticipated that brodifacoum and similar compounds will replace warfarin for rodent control. Acute poisons such as zinc phosphide are not recommended unless it is essential to obtain a quick kill – for example on agricultural land where swine vesicular disease has occurred. Success with zinc phosphide may be limited, however, by problems of palatability and bait shyness. It should be handled by a trained operator only and care must be taken to ensure that food and water for human consumption are not contaminated. It should not be placed in dwelling houses or in any establishment where food is prepared. Difenacoum and calciferol are proving to be useful alternatives. The vasoconstrictor norbormide has not lived up to its early promise. It is a selective rat toxicant and expensive; it is not toxic to mice. Poisonous contact dusts are a potential hazard as contaminants of foods and food surfaces, when carried or blown about. They are increasingly replaced by rodenticidal gels or devices such as mouse tubes (see below), which function by the same principle of ingestion during grooming. Lindane, a persistant organo-chlorine pesticide is unsuitable and approval for its use is likely to be withdrawn.

Fluoracetamide is a liquid rodenticide used in sewers only, and may be used only by fully trained operators. It should not be used in agriculture, horticulture, the home, or any place where food is handled. The use of fluoroacetamide is increasingly banned by water authorities, for example Thames Water Authority, and sewers are under the control of water authorities. Local authorities may continue to carry out sewer treatments on an 'agency' basis and must conform to their requirements. Sodium monofluoroacetate (1080) may be used in ships only.

All rodenticides, insecticides, ascaricides and other such substances are now subject to the approval process set up under the Control of Pesticides Regulations, 1986. (See p. 279).

Mice

The mouse appears to be less a creature of habit than the rat, less suspicious of unfamiliar objects, and less consistent in travel routes and established feeding points. Thus individual traps may be useful, whilst wholesale extermination by poisoning may be more difficult than for rats. Mice appear to be on the increase, due to the development of resistance to warfarin, which has given rise to the so-called 'super mouse'.

Of all traps, the spring break-back type is most used although it suffers from the disadvantage that once sprung it is of no further use until reset. A substantial number of traps set in a mouse-infested area over a period of several nights usually has a good effect. A treadle or plate is better for holding bait than a prong, because cereal such as oatmeal, dried fruit, nuts or chocolate may be used, and, contrary to belief, they are more attractive to mice than the traditional cheese. It is desirable to place the trap at right angles against the wall so that it will be sprung from either direction, and the treadle should be nearest the wall or vertical surface. A newer device is the mouse tube. The tubes are plastic and attractive to the inquisitive mouse; they contain wicks impregnated with brodifacoum and the mouse brushes against them on its way through the tube.

Baits containing 0.25% of warfarin and in quantities of 28 g (1 oz) are also used with reasonable success against some mice, although resistance is developing. The hypothermic poison, alpha-chloralose, is now in use; it kills quickly and humanely. The metabolic rate is reduced and the mice die from heat loss. Alpha-chloralose is used indoors only where floor temperatures are less than about 18°C (65°F); at higher temperatures mice may recover from the stupefying condition. A new rodenticide in bait form which contains 1% calciferol is proving to be effective.

Food animals, especially pigs, and domestic pets have been known to eat dead rats and mice and themselves to be killed by the poison used to destroy the rodents. Therefore, it is essential that the minimum effective dose of poison be used, so that it gives the minimum effective toxicity. There is a legal requirement that the dosage/concentration must be stipulated on the product label.

The Animals (Cruel Poisons) Act, 1962

Red squill, phosphorus and strychnine have been prohibited in the UK by a regulation under the above-mentioned Act, and certain other poisons may eventually be prohibited in the same way. Thallium compounds are dangerous to human life besides being very inhumane. Permits for their use for rodent control must be obtained from the Ministry of Agriculture, Fisheries and Food; their use as insecticides was prohibited in 1967. Rodenticides may not induce apparent symptoms of pain or suffering.

Flies

House-flies, bluebottles and greenbottles are still prevalent and potentially dangerous, although nowadays less significant as a result of two major changes in our way of life over the past half century. These changes are: firstly, the disappearance of horse-drawn traffic and the consequent diminution in the amount of horse manure, which is the fly's first choice of breeding ground; secondly, the water-carriage system of drainage in use throughout urban areas, which greatly reduces the risk of access by flies to infectious material.

In spite of these improvements there are often far too many flies about in their season, and any uncovered and undisturbed vegetable refuse, fish and meat offal can serve as breeding grounds. The chance that a fly may spread communicable diseases by infecting food is a possibility and there are circumstances where flies may gain access to infected material. For example, this may occur in earth closets or pail closets where they are still used; in crêches and day nurseries where very young children are still at the 'pot' stage; at slaughter-houses and knackers' yards; and in places where there are stools of rats, mice, cats, dogs and other animals which may excrete food-poisoning organisms. Food should be protected from contamination by flies.

Whilst the danger of fly-borne contamination of food must not be overemphasized, and certainly not to the extent of diverting the public mind from the really vital safeguards of personal cleanliness, temperature control, animal husbandry and refrigeration, it would not be safe to ignore completely the possibility of contamination and all steps should be taken to deny flies access to food intended for human consumption.

Four methods of approach may be suggested, but since none of these are likely to be effective independently, a combination of all four should be practised. Control may be exercised (a) by the elimination of breeding places; (b) by measures to destroy the fly at some period of its life cycle; (c) by protecting foodstuffs, or better still by making food premises fly-proof as far as practicable; and (d) by obtaining practical advice from the environmental health officer or contractor responsible for pest control.

The elimination of breeding places is the responsibility of the food trader, but others have a duty in the matter too. Refuse is universal and the public should be aware of the sources of material used by flies for breeding, so that they too may play a part in its control. Likewise the local authorities should be vigilant in detecting and dealing with accumulations of offensive material. With modern methods of on-site disposal, refuse collection, and the availability of suitable refuse containers, there is no reason why waste which is attractive to flies should be left putrefying and uncovered anywhere.

Measures against flies, as with all pests, depend primarily on a knowledge of the life cycle and habits of the insects. Flies pass through four stages, the time spent in each depending upon the species and weather conditions, especially temperature. The stages are: (a) the egg, which is laid by the female in some material which provides food for (b) the larva or maggot which hatches from the egg after 8 hours to 3 days. The larva burrows into its food supply, eating vigorously until fully grown, which takes from 42 hours to as much as 6–8 weeks, when it seeks a dry and cool spot in which to pupate. The pupa (c) or chrysalis remains motionless in its early stages (puparium) for a period of 3 days to 4 weeks, when (d) the adult fly emerges.

The most vulnerable stage is the adult. Although there are many forms of fly trap and many insecticides, the most effective way to destroy adult flies is by sprays containing natural or synthetic pyrethrins, iodofenphos or fenitrothion, bromophos, or dichlorvos. Liberated as a fine mist, a spray of pyrethroids immediately knocks down flies on the wing. The other compounds sprayed on walls, ceilings and hanging fixtures have a residual action, killing flies which come in contact with the surfaces over a period of weeks. Deposits of insecticides on food or on equipment in direct contact with food must be avoided because of possible toxic effects. Aerosol devices which release pyrethrins at regular intervals and resin strips or blocks impregnated with dichlorvos can be used to control flying insects. The manufacturers' instructions with regard to excess ventilation should be observed.

There is a danger that flying insects affected by insecticides may drop into food mixes and escape detection. An electrical device avoids this risk by using ultraviolet fluorescent tubes as a lure to attract flying insects on to an electrified grid; the insects are immediately killed and fall into a collecting tray.

Migrating maggots may be destroyed by boiling water or contact insecticides, and where manure bins are used there should be a larva trap either beneath or around them.

Thorough cleanliness and an absence of uncovered food scraps both inside and outside the kitchen will make the premises less attractive to flies. Whilst it is not always easy to fly-proof premises, at least all external doors should be self-closing, and windows and ventilators covered with fine gauze (Fig. 16.3). Flexible fly screens (Stixscreen) using the Velcro fastening system are suitable and effective for all types of window. It is said that a net, which moves in the wind, may be hung loosely over doorways to exclude flies for even though the mesh be quite large flies fear entanglement; hanging plastic strips are used widely on the Continent for the same purpose. In buildings which are not fully proofed against flies – and many are not – larders should be safeguarded and displayed foods covered; crockery and cutlery should be protected after they have been cleaned.

Cockroaches

Cockroaches most commonly found in Britain are the Oriental cockroach or 'black beetle' and the German cockroach or 'steam-fly', which is dark

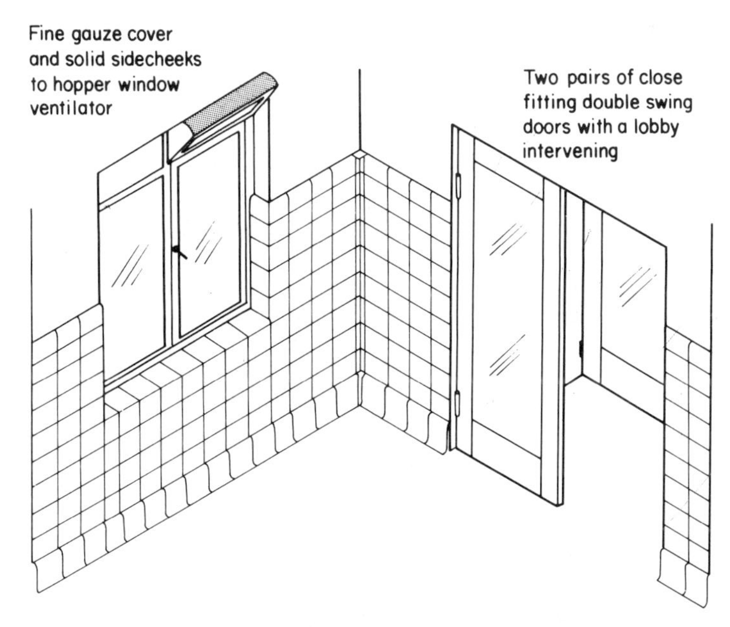

Fig. 16.3 Fly-proofing: protecting common places of entry

yellow-brown in colour. The two species are sometimes present in one building at the same time. They are both nocturnal in habit, occupy crevices in walls and floors, especially in warm places, and they are catholic in their choice of foodstuffs, with a notable taste for beer. Hitherto they have rarely been implicated as the proven vector in outbreaks of disease; nevertheless, salmonella organisms were isolated from cockroaches infesting a children's ward in an hospital where there had been an extensive outbreak of gastroenteritis. The creatures are known to carry other pathogenic organisms, including dysentery and tubercle bacilli, the cholera vibrio, streptococci and staphylococci.

Before commencing an insecticidal treatment it is important to determine the extent of the infestation by means of a thorough inspection carried out after dark when the premises are quiet. Cockroaches are readily repelled by some insecticides and for this reason treatment should start beyond the infested area so that insects attempting to disperse cannot avoid contact with the insecticide.

Cockroaches are susceptible to most classes of insecticide, including organochlorines (dieldrin and chlordane), organophosphates (fenitrothion, diazinon, iodofenphos), carbamates (propoxur, bendiocarb, dioxacarb), and pyrethroids (pyrethrins, resmethrin). Organochlorines have restricted approval and should be avoided, because of the increase in resistance of creatures and persistence in the environment. In choosing an insecticide it is important to remember that most strains of the German cockroach (*Blattella germanica*) possess a high level of resistance to organochlorine insecticides and treatments with these compounds against such strains are doomed to failure. Insecticides are formulated and can be applied in various ways to make optimal use of their properties. For example, dieldrin formulations are extremely persistent, particularly the lacquers, and retain their effectiveness for many months. Chlordecone should not be used as it is considered to be carcinogenic. The organophosphorus and carbamate insecticides are usually applied as sprays. In hot, dry ducts and voids a light dusting of boric acid dust will remain active for long periods, whereas other insecticides may break down and become ineffective in a short time. A residual insecticide dust called 'Drione' is also available. It is a combination of synergized pyrethrum and silica gel. The gel damages the cuticle so that water cannot be retained and there is a fatal loss of water. It is claimed that insects are unable to develop resistance to this substance. The selection of the appropriate insecticide and formulation, and its application so that maximum effect is achieved, require skill and experience, and are often best entrusted to professional pest-control personnel. This is particularly so if the more toxic insecticides are to be used, when great care must be taken to prevent the accidental contamination of food and food-preparation surfaces during application.

Ants

Two kinds of ants occur in buildings – garden ants which nest outside but enter in search of food, and Pharoah's ants, which nest in warm buildings.

Garden ants are not known to be of public health importance, but they may be a nuisance in kitchens and other food premises. Pharoah's ants have been shown to carry organisms of significance in human disease and they constitute a health hazard in hospitals by roaming from open wounds and soiled dressings to sterile equipment and dressings or to food. Ants are frequent pests in West Africa and in India. If spilt food is not cleared away immediately then large numbers of ants, both large and small, will appear on the floor.

Much can be done to avoid attracting ants, by careful observance of cleanliness, removal of waste food, even the smallest crumbs, and the repair of structural cracks and crevices in walls and floors which afford nesting sites to the ants. Garden ants can be destroyed by treating the nest with boiling water, paraffin, malathion, or lindane. Established infestations of Pharoah's ants indoors will involve many nests, each with several queens, over a large area. Control must be carried out systematically. The infested area should be determined by placing small pieces of fresh liver to attract the ants. Treatment should start beyond the infested area and consists of applying bands of insecticide at the wall/floor junction of rooms and corridors, around pipe exits, sinks, air vents, cracks, sills and on the undersides of cupboard shelves in infested buildings.

Lacquers containing dieldrin or sprays of chlordane, dieldrin, propoxur, or fenitrothion are used. Lacquers and sprays cannot eradicate infestations of Pharoah's ants, but may temporarily reduce the number of workers. Methoprene, a juvenile hormone analogue mixed with liver, honey and sponge cake, when fed to worker ants is passed on to the queens and their broods and causes sterility in the queens. It takes at least 20 weeks, the life span of the ants, to wipe out the colony. The main action of methoprene is to disrupt the metamorphosis of ants, thus preventing the broods reaching sexual maturity and producing the next generation.

Wasps

Wasps can be a nuisance in food premises although there appears to be no evidence that they have been implicated in the spread of food infections. From mid-summer onwards the foraging workers are attracted by and feed on fruit juice, sugar, syrup and other sweet substances, but in early summer they will also take insects, fresh and decaying meat or fish to feed the grubs in the nest. Wasps seem to find their way into premises that would normally be regarded as fly-proof, although they cannot penetrate mesh as fine as 3 mm (⅛ inch).

The workers may range a mile or more from their nest and when an infestation occurs a search should be made for nests in the vicinity and steps taken to destroy them. This work is best carried out in the evening when most of the wasps are in the nest and drowsy, and it is a wise precaution for the operator to wear gauntlets and a beekeeper's hood. When the nest is in a suitable position it can be soaked with a rapid 'knock-down' liquid insecticide and burnt or broken up. Equally quick results can be obtained for nests outside by pouring or syringing

1.4–2.8 dl (¼–½ pint) of carbon tetrachloride into the entrance hole of the nest and then plugging the hole, but this must be done with care and not in a confined space. Carbon tetrachloride is regarded as highly dangerous to the pest control operator, and again only approved substances may be used for the work.

A rather slow process is to place derris powder dust into the entrance hole so that the insecticide is carried into the nest by the insects. It is not necessary to remove the disused nest as reinfestation does not occur. Baits containing carbaryl in an attractive material may be placed around the perimeter of vulnerable premises. While lindane dust will kill wasps in a confined space, where there is restricted movement the repellent effect is likely to drive them in an angry mass away from the nest, therefore carbaryl is a better choice. See earlier comments on organochlorines generally, and lindane in particular.

Where it is impossible to find or to deal with the nests, attractive baits set outside the premises will often intercept and divert wasps from the building. The bait may be jam, syrup, molasses, fermenting fruit, or beer, mixed with enough water in a wide-mouthed jar to drown the insects. The addition of detergent (about a teaspoonful per 4.5 litres or 1 gallon) causes the trapped wasps to sink more quickly. The use of chlordane, applied by brush to alighting points, is also effective. The ultraviolet light traps used for flies are also effective against wasps inside buildings.

Other pests

Weevils

Large numbers of insects and mites are of significance in the spoilage of stored food; they are often imported in the raw materials. Although they are severely destructive and may impart an odour to the infested food, they are not a hazard to health. Bread baked from mite-infested flour has an unpleasant musty smell. The grubs of many moth species produce large amounts of webbing which causes granular products to adhere and form sticky masses which may block elevators, augers and screening equipment in flour mills. Most of these spoilage insects can fly, so they can reach neglected areas where there are deposits of waste food, and thus they reach other areas of food plants. Weevils infest cereals in hot climates; one remedy is to freeze the food and sieve out the dead weevils.

Methods of control include careful surveillance of raw materials and even fumigation of incoming goods. The general hygiene and warehousing should be inspected frequently.

Birds

Birds are hazards in food establishments because their droppings may contaminate food with organisms of the salmonella group and other pathogens. Also they may be vectors of pests which spoil stored products. Screens on windows and doors would help to exclude birds, but this may be impracticable. The treatment of windowsills with certain materials would help to deter the birds from alighting on strategic points. The Wild Life and Countryside Act, 1981, provides a general protection for birds (with some listed exceptions). It also restricts the use of certain control methods for the non-protected species. Thus the legality of deterrent materials could be questionable.

Safety precautions

A wide variety of formulations is available to obtain effective control of all insect pests found in Britain. They can be purchased under the various proprietary names as sprays, dusts, lacquers, or aerosols. The mixtures as purchased contain the correct percentage of insecticide and should be safe to use provided the manufacturers' instructions are followed. The use of any rodenticide or insecticide in food premises can present risks to those who apply the materials, to those who work or dwell in the buildings and to those who consume the food that is present when the premises are treated.

Pesticides must always be applied strictly in accordance with the instructions given by the manufacturer. It is now required by law that instructions on labels of pesticide containers must be followed carefully. Legislative and statutory approval systems are now in force together with codes of practice. These codes, like the Highway Code, do not have the force of the law, but transgression of their requirements may be regarded as evidence of an offence.

In the past reputable manufacturers and suppliers have cleared their materials through the Pesticide Safety Precaution Scheme which relied upon their voluntary co-operation. This scheme has now been superseded by a statutory approval system for pesticides and their use as laid down in The Control of Pesticides Regulations, 1986. The Regulations and The Food and Environment Protection Act, 1985, under which they are made, are concerned with safety in an environmental context and operate in conjunction with the Health and Safety at Work etc. Act, 1974, which places an obligation upon employer, employee and the self-employed to ensure they do not jeopardize human health and safety as a result of work activities.

17

Legislation

Michael Jacob

The Minister of Agriculture, Fisheries and Food, the Secretary of State for Health and the Secretaries of State for Scotland and Wales act jointly for all food legislation, but responsibility for individual sections of legislation is taken by the department most concerned. The Ministry of Agriculture, Fisheries and Food (MAFF) main concerns are food production and quality, correct labelling, the absence of adulteration and unapproved ingredients in food. MAFF also holds the major responsibility for research and association with other countries exporting food to the UK. The MAFF Food Safety Directorate formed in 1990 acts separately from commodity divisions and consists of the Food Safety Group, the Food Science Group, the Animal Health Group and the Pesticides Veterinary Medicine and Emergencies Group.

The Department of Health (DH) is responsible for the safety of food with regard to the health of the consumer. Both food and health departments work together where there is known contamination of food. DH Health Aspects of Environment and Food medical and administrative divisions are concerned with microbiological and toxicological aspects of food and water supplies.

The Food Safety Act, 1990, and its various regulations dealing with food poisoning and food hygiene are enforced by local and port health authorities and MAFF, who are responsible for dairy farms under the Milk and Dairies Regulations, 1959. The officers who carry out the main tasks of inspection and enforcement are local and port health authority environmental health officers and MAFF (Agricultural Development and Advisory Service) officers on dairy farms. Medical expertise is provided by consultants in communicable disease control (CCDCs) who are employed by health authorities, but act as local authority officers when they are needed in outbreaks of food-borne disease and to provide medical support. CCDCs are regarded as 'proper officers' under the Public Health (Control of Disease Act), 1984, for those functions carried out by the local authority where medical expertise is necessary. CCDCs are accountable to directors of public health who are responsible for the control of communicable disease (including food poisoning) in health authority districts.

Public health in England

Powers to deal with food poisoning

A committee of enquiry into the development of the public health function was set up by the Secretary of State for Social Services in 1986 (the Acheson Committee). The report recommended a review of the law concerned with infectious disease control; the DH subsequently produced a consultation document in October 1989. Some 27 specific control measures exist to deal with notifiable or infectious disease, including food poisoning, within the Public Health (Control of Disease) Act, 1984, and the Public Health (Infectious Disease) Regulations, 1988.

The provisions most widely used (Sections 19, 20 of the Act) exclude persons from work if they know they have a notifiable disease yet carry on a trade, business or occupation that is likely to spread the disease. The proper officer may request such persons to discontinue work. Likewise Schedule 4 of the 1988 Regulations empowers the local authority to stop someone working with food. This applies to anyone suffering from food poisoning or found to be a carrier of food-poisoning organisms including those able to give rise to typhoid and paratyphoid fevers, salmonella or shigella infections or staphylococcal intoxication; the local authority may take any measures as advised by the proper officer to prevent the spread of infection. The local authority must compensate persons who suffer financial loss.

Other powers within the Act allow application to a Justice of the Peace (JP) for the medical examination of a person or persons believed to be suffering from a notifiable disease or infection. The JP may order the removal of such persons to hospital.

The district health authority and the chief medical officer of the DH must immediately be informed of any serious outbreak of food poisoning or any case of serious food-borne disease.

The consultation document stated that 'having considered the responses . . . the Government will develop proposals with a view to introducing legislation as soon as space can be found in the Parliamentary timetable'.

Powers to deal with contaminated food

Sections 7, 8 and 9 of the Food Safety Act, 1990, provide substantial powers for local authorities to deal with offences caused by foods, which may have been contaminated with food-poisoning organisms.

Section 7 provides that it is an offence for food to be rendered injurious to health. This Section has been revamped from its earlier counterpart in the Food Act, 1984, and now allows food authorities to deal with situations such as occurred in 1989 when an unsuitably processed ingredient (hazelnut purée) was used in yogurts produced by a number of farm dairies.

Section 8 declares that food fails to comply with food safety requirements if:

(i) it has been rendered injurious to health;

(ii) it is unfit for human consumption;
(iii) it is so contaminated (whether by extraneous matter or otherwise), that it would not be reasonable to expect it to be used for human consumption in that state.

Courts, therefore, now have wide discretion under Section 8 to deal with what they may consider to be contamination of an unacceptable nature.

Section 9 empowers an authorized officer of a food authority to inspect any food intended for human consumption and if it appears to him that it fails to comply with food-safety requirements he may seize it or require it not to be moved for up to 21 days pending investigations. If he is satisfied that it does not meet food-safety requirements he may take it before a JP with a view to its being condemned. He may also impose the prohibition upon removal of the food and its use for human consumption without inspecting it if it appears to him that the food is likely to cause food poisoning or other communicable disease. This power might be exercised for example as a preventative measure in response to a food-hazard warning received from the DH that a particular batch of food was implicated in a food-poisoning outbreak.

Compensation may be payable to any owner of the food in such cases where the authorized officer has detained or seized food which subsequently is found to be fit for human consumption.

The reference in Section 9 to 'any disease communicable to human beings' allows action to be taken in respect of parasitic infections such as *Cryptosporidia* or *Giardia*, which do not necessarily fall under the definition of 'food poisoning'.

Section 13 – Emergency Control Orders. Currently the DH uses its food-hazard warning procedure in situations where nationally distributed foods are found to be hazardous. Food-enforcement authorities follow up with the wholesale, retail and catering outlets involved to ensure the removal of the food from distribution points, supermarket shelves and kitchens.

The Emergency Control Order power gives Ministers ordinarily of the DH, MAFF or territorial departments in Scotland and Wales the ability to make orders prohibiting the carrying out of commercial operations with respect to food, food sources or contact materials that may involve imminent risk of injury to health. Such action would apply where the normal voluntary procedures for the withdrawal of food are not appropriate. Where necessary Ministers can direct food-authority staff such as environmental health officers (EHO) or trading standards officers to act to implement the terms of an order.

The microbiological examination of food – food examiners

Prior to the Food Safety Act, 1990, statutory sampling of food was limited to milk, cream, skimmed and semi-skimmed milks, milk-based drinks and pasteurized liquid egg in accordance with the requirements of regulations. Samples of food have been traditionally collected during epidemiological investigations into food-borne disease and food-poisoning outbreaks to assist in the identification of foods causing outbreaks.

The Act creates a new entity – the food examiner – to perform the statutory function of microbiological examination of food. In this task he will correspond to the role of the public analyst who has a long-standing traditional role in carrying out statutory chemical analysis of food samples related to nature, substance and quality, correct labelling and description of food being sold.

Evidence by food examiners (including formal certificates) can be used by authorities to support prosecution cases in legal proceedings following microbiological contamination of food. Such evidence could also be used by the defence.

The Food Safety (Sampling and Qualifications) Regulations, 1990 – qualifications of examiners

There is, at present, no single qualification that uniquely indicates the level of practical and academic competence sufficient to demonstrate to a court that the holder is an expert in food microbiology. These regulations therefore require an examiner to possess a combination of practical and academic credentials.

The Food Safety (Enforcement Authority) (England and Wales) Order, 1990

This Order assigns responsibilities for enforcement of certain provisions of the Food Safety Act, 1990, as respects non-metropolitan districts in England and Wales. Section 12 is the responsibility of district councils and Section 15 the responsibility of counties. A Code of Practice under Section 40 of the Act on Division of Enforcement Responsibility deals with the respective duties. Any case involving contamination by microorganisms or their toxins, such as salmonella, listeria or botulism will be dealt with by district councils. Chemical contamination which poses no immediate risk to health will be dealt with by county councils and they will undertake routine checks and analyses for chemical contamination. However, where there is evidence of gross contamination which potentially poses an immediate hazard to human health, the facts of the case should be passed to the district council (environmental health department) for further action. Where imported food is concerned in a port health district the port health authority or at other ports the district council, should take the appropriate action in all cases.

Food Safety Act, 1990 – enforcement action related to premises

Section 10 – Improvement Notices. This section provides the power for an authorized officer to serve a formal notice where he has reasonable grounds for believing that the proprietor of a food business is failing to comply with any regulations to which the section applies (e.g., the Food Hygiene (General) Regulations, 1970). Proprietors of premises can be required to take measures within specified periods (not less than 14 days) to comply with the notice. This formalization of the notice procedure regarding defects found by EHOs in food premises puts emphasis on the legal nature of the inspection and failure to comply with an Improvement Notice is itself an offence.

Section 11 – Prohibition Notices deal with 'health-risk conditions' in the operation of a food business. A health-risk condition is one that involves the risk of injury to health. A court is able to apply a prohibition on the use of equipment, a process, part of a premises, or the whole premises where they are satisfied that risk of injury to health applies. Additionally, an order can be applied by the court on the proprietor or manager managing any food business to prevent his controlling the premises or any food business for a period of not less than 6 months.

When the health-risk condition is no longer fulfilled, the proprietor must apply for a certificate lifting the prohibition. There is a right of appeal against any decision of the local authority not to lift the prohibition order.

Section 12 – Emergency Prohibition Notices and Orders. Under this section environmental health officers can take necessary and immediate action themselves where conditions pose an imminent risk to health. The service of a notice may impose the appropriate immediate prohibition on equipment, a process or the whole premises, but the notice must be followed up by an application to the court for an Emergency Prohibition Order within 3 days. Any food authority not complying with the procedural requirements in Section 12 will be liable for compensation to the proprietor of the business. This may apply also if the court feels that the health-risk condition was not fulfilled. There will also be a liability to compensate if the court does not declare itself satisfied as to the existence of the health-risk condition at the time of the service of the notice.

The defence of due diligence

Section 21 introduces this new defence to replace the defences available under previous legislation. In future proceedings a defendant can plead a defence that he took all reasonable precaution and exercised all due diligence to avoid the commission of the offence by himself or by a person under his control.

Future case law will probably concentrate on the extent to which the importer or manufacturer thought about and guarded against the possibility of mishaps and their consequences.

The Implementation Advisory Committee

Section 40 of the Act provides for Ministers to issue Codes of Recommended Practice on the execution and enforcement of the Act and regulations and orders made under it. Codes must be laid before Parliament after being issued. Food authorities must have regard to provisions of the codes and shall comply with any direction given by Ministers. Ministers are obliged to consult with organizations which appear to them to be representative of interests likely to be substantially affected before issuing any code. The food industry and consumer organizations are therefore consulted as well as enforcement interests.

An Implementation Advisory Committee was established to advise Ministers on Codes of Practice to be issued under Section 40 of the Act.

The Committee is jointly chaired by officials of the DH and MAFF and consists of nominees from the local authority associations and main professional bodies (the Institution of Environmental Health Officers and the Institute of Trading Standards Administration). The main aim of the Codes of Practice is to ensure consistent enforcement nationwide of the food-safety legislation. By April 1992, 13 Codes of Practice had been published by Ministers covering the following:

(i) legal matters;
(ii) responsibility for enforcement;
(iii) inspection procedures, general;
(iv) prohibition procedures;
(v) improvement notices;
(vi) inspection, detection and seizure of suspect food;
(vii) sampling for analysis or examination;
(viii) food-hygiene inspections;
(ix) food-standards inspections;
(x) enforcement of the temperature control requirements of the Food Hygiene Regulations;
(xi) enforcement of the Food Premises (Registration) Regulations;
(xii) division of Enforcement Responsibilities for the Quick Frozen Foodstuffs Regulations, 1990;
(xiii) enforcement of the Food Safety Act, 1990 in relation to Crown premises.

Other codes proposed cover food-hazard warning systems; inspection of imports and meat products, fish and shellfish (related to EC Directives).

Regulations under the Food Safety Act, 1990

The main regulation-making powers in the Act are in Sections 16–19, but additional powers also exist in other sections.

The Food Premises (Registration) Regulations, 1991

The objective of these regulations is to provide information for food authorities about food businesses in their area so that they can target their enforcement resources more effectively. It is intended that food authorities should use the information provided by registration to help plan their inspection programmes. The obligation to register and to notify changes to the authority falls on the proprietor of the food business (or the owner of premises used by more than one food business). There is no charge for registration and food authorities cannot refuse to register premises. Registration does not need periodic renewal, but changes in the nature of the business or a change of proprietor must be notified.

Any new business has to apply to be registered at least 28 days before it opens. After 1 May 1992 it became an offence to use unregistered food premises.

Premises that are not used for more than a few days or not used regularly, do not need to register. Registration is required if premises are used for the purpose of a food business for 5 or more days (whether consecutive or not) in any 5 consecutive weeks. This rule also applies to any premises used by two or more food businesses.

The regulations list certain specific exemptions. These include premises where crops are harvested, but not processed, those used for egg production, those which are already registered or licensed for certain other food-law purposes, staff and vehicles that operate from premises, which are already registered or are exempt by virtue of them being registered under other schemes. There are also exemptions for some businesses run from domestic premises.

The 'Registration Authorities', i.e. local authorities, must keep a register open to inspection by the public containing the names and addresses of the food premises concerned and the type of business operated there. Records must also be kept on other details supplied by food businesses on the registration form.

Apart from the information that the regulations require to be available to the public, authorities are required by the regulations to keep the remaining information confidential. The information cannot be revealed except to an authorized officer of another enforcement authority or to a police constable.

Details of the application form for registration of food premises and other general information on the regulations is contained in MAFF/DH Code of Practice No. 11 on Enforcement of the Food Premises (Registration) Regulations.

Information acquired by food authorities through registration will be used in statistical returns that are required to be sent to government departments in accordance with the requirements of the EC Official Control of Foodstuffs Directive Article 14. European Community Governments have to send to the Commission annual returns of information relating to food premises, the inspections carried out and priorities for inspection on an annual basis from 1991.

Food-hygiene training

In 1989 the DH issued a consultation paper on the subject of food hygiene training to take views on aspects of necessary training prior to making regulations on the subject under the Food Safety Act, 1990. Approximately 300 responses to the consultation document were received.

The DH issued a Press release on 28 July 1992 stating that it proposed to introduce a basic requirement that food businesses must instruct, train and supervise their staff in food hygiene in a way commensurate with their individual tasks in the preparation and handling of food. Businesses would be required (depending on their size) to–

(i) maintain a training plan;
(ii) train new employees within a month of starting work;
(iii) keep training records.

The Government will be producing a Code of Practice for enforcement officers on food hygiene training.

The Food Hygiene (General) Regulations, 1970

These Regulations aim to secure a standard of construction and equipment in food premises and home-going ships where food is handled and of conduct of food handlers to protect foodstuffs against contamination. Permanently moored ships will likely be brought within the scope of the Regulations by order under Section 1 (3) of the Food Safety Act, 1990.

The Regulations contain comprehensive requirements relating to the premises in which the food is handled, the personnel employed in the premises – their personal hygiene and the way they handle food, and the food itself – the contamination risks and the precautions which need to be taken to ensure absence of food-poisoning risks. A Code of Practice (No.9) setting out details on the conduct and timing of hygiene inspections has now been published.

The Food Hygiene (Markets, Stalls and Delivery Vehicles) Regulations, 1966

These Regulations prescribe food-hygiene requirements for markets, stalls (including vehicles used as stalls) and automatic vending machines and delivery vehicles; they are similar to those required for fixed premises and ships contained in the Food Hygiene (General) Regulations, 1970.

Food Hygiene (Amendment) Regulations, 1990/1991

The General Regulations and the Markets, Stalls and Delivery Vehicles Regulations have been amended by the Food Hygiene (Amendment) Regulations, 1990/1991. These Regulations set new standards for effective temperature control of perishable foods throughout the chill chain from manufacture to service to the consumer in retail and catering outlets. Regulation 27 of the 1970 Regulations and Regulation 12 of the 1966 Regulations are amended by the new Amendment Regulations and deal with temperatures at which certain foods are to be kept.

The Government has produced a Code of Practice (No 10) on the Regulations under Section 40 of the Food Safety Act to aid enforcement aspects.

Chilled convenience foods, including ready-meals, have been a major growth area in food retailing in the UK in recent years. Additionally, many prepared meat products and dairy products being sold as 'fresh' foods may have reduced preservative levels because of consumer demand. Storage temperatures and restriction in shelf life for many foods may therefore be more important controlling factors related to risks of bacterial growth in retail and catering premises.

After considerable consultation, the government made regulations imposing a maximum of 5°C (with an upward tolerance of 2°C for 2 hours, for events such as the defrost cycle) for high-risk foods intended to be consumed without further cooking or reheating. For foods such as prepared salads, or foods to be consumed after cooking or reheating, a maximum 8°C (again with an upward tolerance of 2°C for 2 hours) is prescribed.

The Regulations come into force in three stages:

(i) all specified foods to be stored at 8°C from 1 April 1991;
(ii) application to small delivery vehicles from April 1992;
(iii) 5°C for certain foods from 1 April 1993, and the remainder remaining at 8°C.

Foods to be stored at or under 5°C

The foods to be stored at or below 5°C from 1 April, 1993 include:

(i) ripened soft cheeses, which have been cut or otherwise separated from the whole cheese from which they were removed;
(ii) cooked products ready for consumption containing or comprising meat, fish, eggs, substances used as a substitute for meat, fish or eggs, cheese, cereals, pulses or vegetables whether or not the food also includes other raw or partially cooked ingredients;
(iii) smoked or cured fish;
(iv) smoked or cured meats that have been sliced or cut after smoking or curing;
(v) sandwiches, filled rolls and similar bread products containing any of the foods mentioned in this section unless they are intended to be sold within 24 hours from the time of preparation.

Foods to be stored at or under 8°C

From 1 April, 1991 the foods listed above are required to be stored at or under 8°C. Other foods to which this requirement refers include:

(i) desserts, an ingredient of which is milk or anything used as a substitute for milk and which have a pH value of 4.5 or more;
(ii) prepared vegetable salads including those containing fruit;
(iii) cooked pies and pasties containing meat, fish and any substitute for meat or fish or vegetables encased in pastry except (in each case) those into which nothing has been introduced after cooking and which are intended to be sold on the day of their production or the day after that day;
(iv) cooked sausage rolls, other than those intended to be sold on the day of their production or the day after that day;
(v) uncooked or partly cooked pastry and dough products containing meat, fish or any substance used as a substitute for meat or fish and;
(vi) dairy cream cakes.

Other requirements and exemptions
For foods to be held hot before consumption, the minimum temperature for holding the food is 63°C.

There is an exemption for hot food which is to be sold within 2 hours of the conclusion of its preparation and for food on display in catering premises for a period of not exceeding 4 hours.

For cold food there is an exemption if the food is sold within 4 hours of the conclusion of its preparation.

In addition there are specific exemptions for some traditional products such as fresh baked pies and sausage rolls which are sold on the day of preparation or the day following preparation.

Following representations received, Ministers announced on 21 February, 1991, other changes to the Regulations that would ease restrictions on industry without compromising public health. These changes:

(1) Allow 2 hours after baking of certain products before chilling need begin (this includes pastry products such as custard tarts, Yorkshire curd tarts, quiches, cream cakes, certain cooked pies and pasties and cooked sausage rolls).
(2) Exempt certain uncut baked egg-pastry products such as custard tarts and Yorkshire curd tarts from temperature controls if intended to be sold on the day of their production.
(3) Exempt cheese as a filling encased in pastry from temperature control on the day of production and the next day.
(4) Extend time-limited exemptions for display of food in catering and retail premises to markets, stalls and delivery vehicles.
(5) Give specialized delivery vehicles for airline meals and railway carriages until 1 April, 1992, to comply with temperature controls.
(6) Exempt foods delivered by mail order while further discussions are held on practical means of control.
(7) Bring non-dairy cream cakes within temperature controls.
(8) Will permit products such as cheese scones to be outside temperature control.
(9) 'Warm' foods such as Hollandaise sauce should not be kept outside temperature control after the conclusion of preparation for longer than 2 hours before sale and after that time should not be sold or offered or exposed for sale.

In February 1993 Ministers announced that further revisions would be made to the Regulations with respect to certain short shelf life products and partially stable products.

Delivery vehicles
From 1 April, 1991, delivery vehicles over 7.5 tonnes gross weight must hold and deliver foods covered by the Regulations at or below 8°C or at or above 63°C.

From 1 April, 1992, delivery vehicles of 7.5 tonnes or under gross weight must hold and deliver foods covered by the Regulations at or below 8°C or at or above 63°C.

From 1 April, 1993, delivery vehicles over 7.5 tonnes gross weight must hold and deliver relevant foods at or below 8°C or 5°C or at or above 63°C.

From 1 April, 1993, relevant foods held and delivered locally in small vehicles may be delivered at up to 8°C rather than 5°C if delivered within 12 hours. Otherwise the relevant foods must be held and delivered at or below 5°C.

DH have produced comprehensive guidelines on the Food Hygiene (Amendment) Regulations, 1990. The guidelines provide much detail on scope and content of the Regulations and on methods of measuring and monitoring temperature. It represents an essential document for food manufacturers, distributors, the retail and catering industries and environmental health officers. Guidelines specific to catering have also been published (see Bibliography).

The Food Hygiene (Docks, Carriers etc.) Regulations, 1960

These Regulations require the hygienic handling of food at docks, warehouses, cold stores, certain stores and also premises used by goods carriers. The Regulations were introduced at a time when most of food imports arriving at British ports were of a bulk-break cargo character requiring fairly substantial manual handling at docks and warehouses. The growth of containerized cargo now results in most imported foods being held in containers to their point of destination so that handling of products is minimal.

EC proposal for a Directive on the hygiene of foodstuffs

The European Commission have now published a proposal for a Directive on the hygiene of foodstuffs (COM(91)525 Final) as part of the measures for the completion of the Single Market. The proposal, which is horizontal in nature, covers general principles of the hygienic handling of food. The measures are additional to EC vertical legislation dealing with specific commodities.

Amongst the requirements in the proposed Directive are the following:

(1) Production, processing, manufacturing, packaging, storing, transportation, distribution, handling and sale of foodstuffs must be carried out in a hygienic manner.
(2) Food businesses must comply with rules set out in an Annex which in broad terms define the conditions necessary for hygienic production, distribution and sale of food to the final consumer. There are provisions on hygiene training.
(3) Businesses must identify, control and review any activities which are critical to ensuring food safety. There is reference to HACCP (The Hazard Analysis Critical Control Point procedure), but the proposal stops short of imposing a statutory obligation on all businesses to operate an elaborate HACCP plan. Nor does it entail the approval of a Hazard Control plan for each business as a condition of authorization to operate.

There is a requirement that member states should encourage the production of industry Codes of Practice and these Codes would amplify the general hygiene requirements specified in the Annex and also have regard to the international *Codex Alimentarius* general hygiene commodity standards.

The general approach in the proposal requires detailed application to be set out in industry Codes of Practice. These Codes would be tailored to the practicalities and requirements of particular sectors of the food industry. The proposed Directive envisages a hierarchy of codes, the preference being for European-wide agreement. The mechanism proposed for drawing up codes is an established EC system of committees CEN, which is currently mirrored in the UK by the British Standards Institution.

Implementation of the proposed Directive in England and Wales would affect the Food Hygiene (General) Regulations, 1970, and the other hygiene regulations relating to markets, stalls, delivery vehicles, docks, carriers and the Food Hygiene (Amendment) Regulations, 1990/91, relating to specific temperature controls.

Slaughterhouses and meat legislation

Meat hygiene legislation

The whole field of meat hygiene legislation is currently undergoing change as part of the harmonization of food law in the Single Market. Currently where premises are approved for trade between EC Member States they have to comply with Community standards of hygiene and supervision including export health-certification arrangements on consignments of meat. After 1992 common standards will apply to premises operating for both domestic and intra-Community trade. The period up until the end of 1992, therefore, saw much change in the industry, with some existing premises not capable of meeting the new standards (or not wishing to for commercial reasons) ceasing to trade. UK hygiene and meat inspection arrangements are already close to EC requirements: poultry meat will be already harmonized. The main changes will be the introduction of specific structural rules for increased veterinary supervision over slaughtering and cutting premises and in cold stores, in the red meat sector.

The Single Market rules expected to be adopted (or already adopted) by the Council and to come into effect on 1 January, 1993, will cover red meat (adopted as Directives 91/497/EEC and 91/498/EEC); poultry meat (COM (89) 668); farmed game and rabbit meat (adopted in Directive 90/495); and wild game (COM (89) 496).

In addition, there are proposals covering the production of minced-meat preparations (COM (89) 671) and animal fats (COM (89) 490); and some premises handling fresh meat (e.g. retailers, who will be exempt from the rules described above) will need to comply with more general hygiene rules set out in a separate 'catch all' proposal to cover all products of animal origin (COM (89) 492).

Proposed National Meat Hygiene Service

Following a review of fresh meat hygiene enforcement, the Minister of Agriculture, Fisheries and Food announced on 9 March, 1992, a proposal to create a National Meat Hygiene Service for Great Britain. This will be an agency of MAFF. Local authority responsibility for enforcing other food legislation will not be affected. The transfer of existing meat enforcement functions from local government to the new agency will depend on the availability of parliamentary time for the necessary primary legislation.

Red meat

Licensing requirements for slaughterhouses are contained in the Slaughterhouses Act, 1974. Hygiene and inspection standards and procedures are set out in secondary legislation made under the Food Safety Act (and its predecessors).

The Slaughterhouse Hygiene Regulations, 1977, (and subsequent amendments in 1990 and 1991) and the Meat Inspection Regulations, 1987, apply to all red meat slaughterhouses. The Fresh Meat Export (Hygiene and Inspection) Regulations, 1987, apply supplementary rules to slaughterhouses, cutting premises and cold stores wishing to be EC export approved.

These will be amended and consolidated to implement the new Single Market rules. These require the approval of all premises which slaughter or cut or store red meat (except retail and similar premises). They require ante- and post-mortem examination at slaughterhouses, and set rules on hygiene, operation, structure, supervision and qualifications of inspection personnel in all premises.

Currently, although licensing and control of premises is the responsibility of local authorities, this responsibility is moving to MAFF. Officers of the State Veterinary Service of MAFF monitor standards in export-approved premises on a monthly basis and advise local authorities and plant management on compliance with EC and UK requirements. They also visit other meat slaughterhouses every year. In addition, EC Commission representatives visit a sample of export-approved premises annually to assess compliance with Community requirements.

Poultry meat

The health rules governing the production of, and trade in, fresh poultry meat are currently laid down in Directive 71/118/EEC on Poultry Meat Hygiene, which is implemented in England and Wales by the Poultry Meat (Hygiene) Regulations, 1976, as amended.

The Commission's Single Market proposal to replace Directive 71/118/EEC is aimed at ensuring that all poultry meat, except that supplied in isolated cases direct to the final consumer, is produced to Community standards in licensed premises. The proposals in the main repeat existing detailed provisions and continue to permit the production of uneviscerated poultry. Many small producers, who are currently exempt from the existing requirements, will however need to be licensed and will be subject to the full hygiene and inspection requirements. The proposals extend the scope of the requirements to include pigeons, pheasants, quails and partridges.

Meat products

The production of meat products is currently covered by the Food Hygiene (General) Regulations, 1970, as amended. Earlier EC legislation has been amended by Directive 92/5/EEC which sets out improved standards of hygiene and inspection requirements for meat products, extending to the domestic trade current requirements for plants involved in intra-Community trade.

Although the temperatures at which meat products are cooked and held whilst cooling are highly important in relation to microbiological risk factors, there is no detailed legislation that applies process controls in this area. Following information on a number of food-poisoning outbreaks in 1989 related to meats cooked in small establishments and mostly due to *Salmonella typhimurium*, the DH issued a '10 point plan' for cooked meat products. This recommended that the centre of the meat must reach a core temperature of at least 70°C for 2 minutes or the equivalent. Assurance that cooking equipment achieves this performance consistently should be sought. Cooking processes should be monitored by using a probe thermometer with records of core temperatures being obtained. Products should be effectively cooled through the temperature zone of 50–10°C. Cooked products should be stored at 5°C or less in a designated refrigerator. Other points covered in the '10 point plan' include hygiene principles to reduce cross-contamination risk, cleaning aspects and personal hygiene. Reference should also be made to earlier DH guidance on the processing of large canned hams and the Code of Practice on canned foods (under review).

Animal waste

The Meat (Sterilization and Staining) Regulations, 1982, as amended, control the handling and movement of unfit meat in slaughterhouses. Licences are required to operate knackers yards, and these are issued by local authorities under Section 4 of the Slaughterhouses Act, 1974. Knackers yards will have until 31 December 1995 to comply with EC hygiene standards under the newly-adopted Directive 90/667/EEC on the Disposal and Processing of Animal Waste and the Prevention of Pathogens in Feedstuffs. This new Directive does not cover compound feedstuffs containing animal and vegetable products, for which there will be separate provision at a later date. The Directive designates all animal waste as either 'high-' or 'low-risk'. If rendered the disposal of 'high-risk' material under the Directive's requirements entail heat treatment to a core temperature of at least 133°C for 20 minutes at a pressure of 3 bar. The particle size of the raw material must be reduced to at least 50 mm by means of a pre-breaker or grinder before processing. Other heat treatment systems may be used, but only where approved under the Directive. Annex 2 to the Directive contains microbiological standards for salmonellae (absence in 25g) and enterobacteriaceae, which must be met in final products in high- or low- risk processing plants. In addition, finished products in high-risk processing plants, sampled directly after heat treatment must be free from *Clostridium perfringens* (absent in 1g of the product).

Following advice from the Tyrell Committee and a review of the Animal Feed Industry and its relation to food safety by an expert group chaired by Professor Lamming, the Bovine Spongiform Encephalopathy (BSE) Order, 1991, now requires all movement of protein derived from specified bovine offal to be under licence issued by an officer of the Agricultural Departments of England, Scotland or Wales. In addition, the new Order consolidates existing legislation on BSE and provides for the seizure of carcasses suspected of having BSE and the payment of compensation. The Feedingstuffs Regulations, 1991, implement a number of EC Directives. Full declaration of ingredients for all compound feeds are now required, with declaration of minimum storage life and batch number or date of manufacture.

The Processed Animal Protein Order, 1989

Protein of animal origin must be tested for *Salmonella* and no contaminated material may be used for animal or poultry feed.

Importation of Processed Protein Order, 1981

Licences under this Order have been reviewed and further controls imposed on those countries frequently found to be exporting contaminated products to the UK.

General animal health aspects

The Zoonoses Order, 1989, requires all isolations of *Salmonella* from samples taken from an animal or bird, or from the carcass, products or surroundings of an animal or bird or from any feedingstuff, to be reported to a veterinary officer of MAFF. The Order also allows veterinary inspectors to enter any premises and carry out such enquiries as are considered necessary to determine whether *Salmonella* is present. Powers are also available to declare a premises infected, to prohibit the movement of animals, poultry, carcasses, products and feedingstuffs in or out of the premises except under licence, and to serve notices requiring the cleansing and disinfection of premises where *Salmonella* is known to have been present. The Order also applies certain provisions of the Animal Health Act, 1981, to organisms of the genus *Salmonella*, including powers relating to the compulsory slaughter of poultry.

In March, 1989, the compulsory slaughter with compensation of laying flocks infected with *Salm. enteritidis* and *Salm. typhimurium* was introduced. This was further extended to broiler breeder and layer breeder flocks in November 1989.

Early in 1991 the automatic slaughter policy for laying flocks infected with *Salm. typhimurium* was discontinued.

The Poultry Laying Flocks (Testing and Registration etc.) Order, 1989, requires owners of flocks comprising not less than 25 birds which are kept for the production of eggs for human consumption (including birds being reared for this purpose), or flocks comprising less than 25 birds the eggs from which are sold for human consumption, to take samples from their flocks and have them tested for *Salmonella* at an authorized laboratory in accordance with the schedule laid down in the Order. Owners of flocks comprising 100 or more laying hens must also register with the agriculture departments.

The Poultry Breeding Flocks and Hatcheries (Registered and Testing) Order, 1989, requires the owners of all layer and broiler breeding flocks comprising 25 or more birds, and hatcheries with an incubator capacity of 1000 eggs or more, to register with agriculture departments and to comply with specified operational standards. Owners of breeding flocks and/or hatcheries must also take samples from birds or their progeny and have them tested for *Salmonella* in an authorized laboratory in accordance with the schedules laid down in the Order.

MAFF Codes of Practice

Five voluntary Codes of Practice supplementing the above legislation and designed to control *Salmonella* in feedingstuffs and raw materials were introduced in 1989. They cover the storage, handling and transportation of raw materials intended for incorporation into animal feedingstuffs, the rendering industry, the fishmeal industry and the production of final feed for livestock.

EC proposals

There is an EC proposal on Controls for the Prevention of Specific Zoonoses COM(91)310 requiring the testing for *Salmonella* of poultry flocks, hatcheries and, at the point of production, compound feedingstuffs intended for poultry. It has been drafted largely in response to pressure from the UK for salmonella controls to be introduced on a Community-wide basis. The proposal is wide ranging and covers:

(i) collection of information on the incidence of certain zoonotic agents in humans and animals;
(ii) sampling plans for detection of central zoonotic agents in animals, feeding stuffs and products of animal origin. (These include both *Salmonella* and *Listeria*);
(iii) measures to control specific zoonotic agents.

Eggs

Egg products

The Liquid Egg (Pasteurization) Regulations, 1963, prohibit the use as an ingredient in the preparation of food of liquid egg which has not been pasteurized. New Regulations will be introduced to implement the provisions of Directive 89/437/EEC on hygiene and health problems affecting the production and placing on the market of egg products. In addition, the Regulations will set conditions for the pasteurization of egg albumen for the first time. Currently in England and Wales separated whole egg and separated yolk are pasteurized at 64.4°C for not less than 2.5 minutes and then immediately cooled to a temperature below 4°C. The Liquid Egg (Pasteurization) Regulations, 1963, only apply to liquid whole egg. Different combinations of time and temperature may be used in other Member States.

The Ungraded Eggs (Hygiene) Regulations, 1990

Because of the greater hygiene risk cracked eggs represent, these Regulations now make it an offence for producers to sell or to offer for sale ungraded eggs that have visible cracks. Generally speaking, the sale of cracked eggs to consumers is prohibited by the EC Egg Marketing Regulations but these do not cover the sale of eggs by producers to consumers at the farm gate, at local markets and through door to door selling. These Regulations now deal with this gap in the Community legislation. Visibly cracked eggs are those which the producer should be able to identify and remove from sale. It includes leaking eggs and eggs in which the shell when viewed in ordinary light by the naked eye is visibly cracked (whether or not it is leaking). Environmental health officers are responsible for enforcement of these Regulations.

Milk and milk products

The Milk and Dairies (General) Regulations, 1959

These provide for the registration of milk producers, distributors and their premises. There is provision for the veterinary inspection of cattle and the procedure to be used where milk is contaminated or liable to be contaminated with disease organisms communicable to man. When there is evidence that milk has been responsible for disease or food poisoning, a notice can be given requiring its treatment before consumption so that it is safe.

Buildings used during the handling and processing and for storage of milk must be constructed and maintained in such condition that there is no risk of contamination; premises must be provided with a satisfactory water supply. Care must be taken throughout milking and subsequent handling, processing and storage to prevent contamination of the milk.

Other hygienic requirements relate to personal hygiene of milkers, cleanliness of cows during milking and handling temperatures for milk. Effective cleaning and sterilization of dairy equipment is also required.

The Milk (Special Designation) Regulations, 1989

These Regulations deal with the requirements for the production and sale of pasteurized, sterilized and UHT milk and untreated milk. Milk producers and dealers are allowed special designations for milk treated in different ways. 'Untreated' milk must come from a herd accredited as free from brucellosis; it has undergone no heat treatment. For 'pasteurized' milk the raw milk must be either heated to a temperature of between 62.8 and 65.6°C and held at this temperature for a period of at least 30 minutes or heated to a temperature of not less than 71.7°C for at least 15 seconds. After heat treatment, the milk must be immediately cooled to a temperature of not more than 10°C and put into bottles or other appropriate containers.

'Sterilized' milk must be filtered or clarified, homogenized and heated to a temperature of not less than 100°C for a time which will satisfy the turbidity test. Bottles must be aseptically filled. 'Ultra-heat-treated' milk must be heated to a temperature of not less than 135°C for not less than 1 second. Sterile containers must be filled aseptically and sealed so that they are airtight.

The Milk and Dairies (Semi-Skimmed and Skimmed Milk) (Heat Treatment and Labelling) Regulations, 1988, the Milk and Dairies (Heat Treatment of Cream) Regulations, 1983, and the Milk Based Drinks (Hygiene and Heat Treatment) Regulations, 1983, specify process requirements for these products.

The Milk (Special Designation) (Amendment) Regulations, 1990, and the Milk and Dairies (Semi-Skimmed and Skimmed Milk) (Heat Treatment and Labelling) (Amendment) Regulations, 1990, amend the above legislation by deleting provisions permitting a 'sell-by' date and introducing provisions requiring a 'use-by' date in labelling conditions for milk. New offences are introduced for selling milk bearing an expired 'use by' date and for anyone other than the person originally responsible for applying the date mark to change it.

The Milk (Special Designation) Regulations, 1989, end the use of the methylene blue test for untreated and pasteurized milk. Both types of milk require a plate count and a coliform count. Pasteurized milk now also requires a plate count on a sample preincubated at 6°C for 5 days. The phosphatase test remains. UHT/sterilized milk is pre-incubated at 30°C for 15 days for a plate count designed to check sterility.

A particularly important change regarding untreated milk came into effect from 1 September, 1990, whereby all containers of untreated milk must carry the statement 'this milk has not been heat treated and may therefore contain organisms harmful to health.' In England and Wales, sales of untreated milk now represent less than 1% of the liquid-milk market.

To aid enforcement sampling and testing of pasteurized milk for the purpose of the Special Designation Regulations, the Institution of Environmental Health Officers, The Association of Public Analysts, the PHLS, the Milk Marketing Board, the National Farmers Union and the Dairy Trade

Federation jointly produced guidelines on sampling and testing. These guidelines have no statutory authority, but are aimed at sensible application of the Regulations.

The Special Designation Regulations were amended as a result of Directive 85/397/EEC which laid down standards for intra-Community trade in heat-treated drinking milk. By applying the higher Step 2 EC standards to domestically produced milk, the UK has been able to require imported milk to meet the same high standards. The EC has now issued a Directive (92/46/EEC) which will extend the principles of Directive 85/397 to all Community production of heat-treated milks.

The definition of milk treatment establishment in this proposal will include all farm producers of pasteurized milk currently controlled by licensing and inspection by EHOs.

There are approximately 500 farm-pasteurization plants and these may have variable standards requiring a higher level of monitoring. The British Standards Institution issued a draft British Standards Code of Practice for the pasteurization of milk on farms and in small dairies in 1990. This Code of Practice will provide comprehensive guidance for farmers operating pasteurization equipment and EHO's enforcing the regulations.

Milk products

A further part of completing the European internal market involves Directive 92/46/EEC which lays down the health rules for the production and placing on the market of raw milk, heat-treated milk and milk-based products. This Directive will come into effect on the 1 January 1994. The health rules include microbiological standards for the production, processing and manufacture of other milk and milk products. It covers milk from cows, sheep, goats and buffalos and deals with raw milk for manufacture and milk-based products. One element of this proposal will require the labelling of products containing untreated milk in their manufacture to indicate that fact on the label.

Currently in England and Wales, other than for heat-treated cream, the manufacture of all milk products fall under the broad requirements of the Food Hygiene (General) Regulations, 1970. In future, however, cheese, butter, dairy desserts and yoghurt will be covered by the new Directive.

The Food Safety Act, 1990, provides the power to make regulations on sheep's and goats' milk. These milks are increasingly being used in the manufacture of dairy products, particularly farmhouse cheeses.

The Ice-cream (Heat Treatment etc.) Regulations, 1959 and 1963

Ingredients used in the manufacture of ice-cream are required to be pasteurized by one or other of three specified methods or sterilized and thereafter kept at a temperature below 7.2°C until the freezing process has begun. In the pasteurization process, after the ingredients have been mixed, the mixture shall not be kept for more than 1 hour above 7.2°C before being raised to and kept at a temperature of pasteurization (normally 71.1°C for 10 minutes or 79.4°C for at least 15 seconds). It must then, within 1.5 hours, be reduced to not more than 7.2°C and kept there until freezing has begun.

A sterilized product, where the mixture must be raised and kept to a temperature of not less than 148.9°C for at least 2 seconds, need not be cooled similarly to pasteurized products providing it is immediately canned under sterile conditions. Ice-cream which has an elevated temperature above −2.2°C must be subjected to a reheating process prior to sale. Some ice-cream products may fall within the scope of the EC proposal on milk products, but the position is not yet clear.

Fish and shellfish

The hygiene of handling wet fish, the use of fish and shellfish in catering and the processing of fish are covered by the Food Hygiene (General) Regulations, 1970. The Public Health (Shellfish) Regulations, 1934, cover the sale for human consumption of molluscan shellfish (both live and heat treated). From the beginning of 1993, however, new European standards will be introduced to cover fish and fishery products and shellfish.

Directive 91/493/EEC lays down health conditions for the production and placing on the market of fishery products. It sets out detailed hygiene requirements for fish and fish products from landing up to, but excluding, retail sale. Factory vessels are also covered and separate measures dealing with hygiene on board fishing vessels are to be adopted before the end of 1992. Directive 91/492/EEC lays down the health conditions for the production and placing on the market of live bivalve molluscs and deals with the condition of harvesting, relaying, purification, processing and distribution of molluscan shellfish such as oysters, clams, mussels and cockles. Regulations on fish and shellfish will implement the EC Directives and there will also be Codes of Practice for enforcement authorities under Section 40 of the Food Safety Act, 1990.

Directive 91/67/EEC concerns the animal health conditions for aquaculture animals and products and deals with controls over fish farming to minimize risks of transfer and development of fish diseases.

Other regulations

Food (Control of Irradiation) Regulations, 1990

These Regulations revoke previous regulations that prohibited the sale of irradiated food and permit the sale of specified types of irradiated foods where the irradiation has taken place under licence. Licences are issued jointly by MAFF, DH and the Welsh Office in respect of plants in England and Wales and by the Scottish Office for facilities in Scotland. Licence conditions are enforced by specialist staff from these departments. The Regulations are enforced jointly by the agriculture and health departments, EHOs and trading standards officers. The Regulations set out the terms, procedure for grant variation and suspension of licences, and also provide for charges to be payable following an application for grant and agreement for variation of licences and in relation to the carrying out of inspections.

Where irradiation takes place in another Member State, food can only be imported if the irradiation has taken place in a plant subject to formal recognition by agriculture and health ministers.

The specified types of food which may be irradiated or imported after irradiation are fruit, vegetables, cereals, bulbs and tubers, spices and condiments, fish and shellfish and poultry. The Regulations specify the limits of overall average dose of ionizing radiation which apply. Licences will include requirements to segregate irradiated food from other food in the premises to which the licence applies and a prohibition on re-irradiation.

Food Labelling (Amendment) (Irradiated Foods) Regulations, 1990

These Regulations require indications of irradiation in the labelling of pre-packed food sold to the ultimate consumer. Irradiated foods used in catering establishments and other food sold without pre-packing must also be appropriately labelled.

Food Labelling (Amendment) Regulations, 1990

These Regulations implement provisions of Directives 79/112/EEC and 89/395/EEC. They further amend the Food Labelling Regulations, 1984, and delete national provisions permitting the use of a 'sell-by' date and introduce provisions requiring a 'use-by' date for highly perishable foodstuffs. New offences are introduced for selling foods bearing an expired use-by date and for anyone other than the person originally responsible for applying the date mark to change it. The Regulations remove the exemption from date marking for deep-frozen foods, certain cheeses and long-life foods with effect from 20 June, 1992, and extend the current limited labelling requirements for small packages to include indelibly marked glass bottles intended for re-use.

Quick-Frozen Foodstuffs Regulations, 1990

These Regulations implement Directive 89/108/EEC in Great Britain. They define and list conditions applicable to quick-frozen foods and equipment used to store and retail them.

The Imported Food Regulations, 1984

These Regulations contain measures for the protection of public health in relation to imported food. They are enforced by port health or local authorities at the point of entry, but examination may be deferred until the food reaches its final destination, in which case the port health or local authority must inform the receiving authority which then assumes responsibility for inspection.

It is an offence to import food that has been rendered injurious to health, is unfit for human consumption or is unsound or unwholesome. Most consignments of food imported into the UK originate from the EC. EC legislation resulted in the disappearance of the Imported Food Regulations, 1984, in their present form before the end of 1992. New domestic legislation will be required to implement two Directives concerned with the organization of 'veterinary checks' (defined as checks for both animal and public health purposes) on products of animal origin. Directive 89/662/EEC deals with veterinary checks on intra-Community trade and Directive 90/675/EEC with products originating from third countries (countries other than EC Member States).

This latter Directive is of particular significance to port health authorities. It requires that from July 1992 all foods of animal origin imported from third world countries shall be subject to comprehensive checks, usually at the place at which they first enter the EC, under the direction of a veterinarian. The veterinarian may be assisted by other officers.

The Importation of Milk Regulations, 1988

Imports of milk, cream and milk-based drinks are subject to the above Regulations, which are enforced by port health authorities. These require permitted imports to be accompanied by an appropriate health certificate and prescribe testing and enforcement procedures. These regulations will need to be revised to take account of Directives 89/662 and 90/675.

General EC aspects

The end of 1992 saw the removal of non-tariff barriers and the consequent harmonization of food hygiene controls. Food hygiene is increasingly an EC issue and it is no longer possible to view it in an exclusively national context. It can be seen from measures dealt with earlier that specific products are being covered by EC proposals and existing measures which already apply to intra-Community trade are being modified and extended to cover all production, including domestic trade and imports from non-Member States. In addition, new horizontal legislation, laying general across-the-board hygiene and inspection requirements will require revision of existing UK law on general hygiene requirements.

A vertical Directive, such as Directive 92/5/EEC, which lays down rules for the production and placing on the market of meat products has major implications in that it will apply a single European standard to thousands of UK meat-product plants previously subject only to national rules.

A horizontal measure such as Directive 89/397/EEC on the Official Control of Foodstuffs sets out the criteria for inspection of all Community food establishments.

For the future, new domestic standards for imports to the UK from EC member states can only be applied if an exception to the free movement of goods principle can be justified. Article 36 of the EC Treaty can allow action where protection of health is justified, but it is the obligation of Member

States not to go beyond what is strictly necessary for health protection. Measures must be shown to be proportionate to the end to be achieved and not a disguised restriction on trade. Article 36 does not come into play where there are in existence agreed Community rules, what comes into play in these circumstances is any 'safeguard' article in the particular EC legislation. The safeguard article may permit a Member State to act in an emergency for public health reasons. However, even this is subject to notification to the EC Commission and to the Commission's subsequent confirmation that the action is acceptable. To be acceptable, any action must be the minimum necessary to achieve the safeguard.

In conjunction with Directive 89/397/EEC (Official Control of Foodstuffs) the Commission intends to avoid the use of the term 'food inspector', but suggests the more appropriate term of 'food-control official'. No detailed Community requirements are to be drawn up for the training of food-control officials, but the Commission feels it essential to define areas in which personnel responsible for official food control must have received training to an appropriate professional level. A proposal for a Council Directive (COM(91)526) sets out additional measures necessary relating to the Official Control of Foodstuffs. They include a proposal for a team of Community officials to exercise control acts in the Member States. This team would:

(i) help coordinate the uniform application of community law;
(ii) assist in resolving Member States' problems emanating from the free movement of foodstuffs in the Community;
(iii) promote the optimal use of the Commission's 'rapid alert' system when emergencies or food hazards occur.

18

Microbiological specifications

Microbiological specifications for foods were considered to be desirable when it became clear, in the late fifties, that factories preparing commodities for distribution to more than one country were under pressure to conform to specifications that differed from one country to another; in some instances, no specifications were stipulated. Medical and non-medical microbiologists, food scientists, food administrators and, later, representatives from industry formed a committee to consider drawing up guidelines for the permitted microbial content of certain foods in international trade. The first foods selected for discussion were those most likely to cause trouble – cooked frozen seafoods (shrimps, prawns, crab and lobster), frozen egg products, frozen pre-cooked whole meals and frozen raw meat, both in bulk and comminuted.

It was decided from the outset (its first meeting was in 1962) that legalized standards were not the objective of this committee, but that its task was to obtain knowledge and reach agreement regarding reasonable levels of bacteria in foods subject to good manufacturing practice (GMP) and to good commercial practice (GCP). No conclusions could be reached until the results of examination of large numbers of food samples from various parts of the world had been studied. The methods of examination and the rationale of sampling were later considered in detail. The founding organization became known as the International Commission on Microbiological Specifications for Foods (ICMSF), under the auspices of the International Association of Microbiological Societies (IAMS) (now known as the International Union of Microbiological Societies), and has continued to meet annually since 1962. As well as its role in response to the need for internationally acceptable and authoritative decisions on microbiological limits commensurate with public health and safety, activities have expanded into the field of food technology and the hygiene of production. References to the five books published by the Commission are given in Appendix B. Books 1–5 are described briefly at the end of this chapter.

Other international organizations such as the WHO, *Codex Alimentarius*, the International Organization for Standardization (ISO) and the Nordic Committee on Food Analysis, based at Statens Livsmedelsverk, Uppsala,

Sweden, also meet regularly to discuss and produce standard methodology for the examination of various products and to consider microbiological specifications.

The two principal requirements for food are (a) freedom from microbial populations which threaten the health of the consumer and (b) reduction to a minimum of spoilage organisms, as loss of quality reduces the nutritive value and shelf life and renders the food uneconomic and even inedible. The two aspects of food control, safety and quality, cannot or should not be regarded as separate entities, because some bacteria are active in both fields.

The Commission's main concerns are:

(i) objective information about the microbiological characteristics of each food product under examination and the consideration of a range of results from many factories and countries;
(ii) defined sampling plans with due regard to changes in food formulation, production, processing and bacteriological content, as arbitrary sampling may give inconclusive or misleading results;
(iii) standardized methods which are simple to perform and take into account changes in technology and the formulation and processing of foods;
(iv) to study the ecology of organisms in relation to food production;
(v) the application of the concept of HACCP (Hazard Analysis Critical Control Point) to the hygiene of food manufacture;
(vi) the compilation of data related to the behaviour of pathogens in food.

The Commission's first aim was to find and, where possible, to enumerate those organisms deemed to be hazardous to health. In addition, the quantitative estimation of the approximate numbers of common contaminating organisms per gram of food is an informative indicator of the general hygienic condition of the food. The enumeration of *Escherichia coli* in food may be used to demonstrate its presence as a pathogen in its own right and, as in testing water, as representative of the *Enterobacteriaceae*, members of which can cause illness. Enumeration of other aerobic and anaerobic bacteria may be used to indicate the adequacy of storage conditions.

Sampling

The number of samples selected for examination from a batch should be sufficient to indicate the extent of undesirable contamination likely to threaten health or lead to spoilage. In practice only a small number of units from the batch can be examined, because of limited laboratory capacity and the expenses incurred. When the preliminary sampling of foods has yielded results which are regarded as dubious, careful attention should be paid to them with regard to numbers of samples, frequency of sampling and the interpretation of results. The ICMSF devised a scheme for sampling which incorporates 2-class and 3-class plans.

The 2-class plan provides presence or absence criteria for intestinal pathogens, such as salmonellae from 'n' sample units. It was also proposed that if the number of samples containing a pathogen ('c') was zero in the 2-class plan, for example n=10, c=0, samples might be bulked to reduce the length of time and amount of media required for tests. However, for epidemiological purposes, examination of individual samples may be necessary to find particular serotypes involved in outbreaks of food poisoning and the intensity of sampling may also be influenced by the degree of hazard arising from particular agents of food-borne disease. Factors to be considered are frequency of incidents, clinical severity, duration of illness, the extent of distribution of the pathogen in the implicated or suspected food and the previous history of a food as a vehicle of infection or intoxication.

The Commission suggested grouping the food-borne pathogens into three categories according to the degree of hazard: (1) 'severe' – for example *Clostridium botulinum*, which is uncommon but often fatal, and the enteric (typhoid and paratyphoid) organisms, *Vibrio cholerae* and *C. perfringens* type C; (2) 'moderate with a potential for spread' – for example, some serotypes of salmonellae, *V. parahaemolyticus, E. coli* and *Campylobacter*; (3) 'moderate with limited spread' – for example, *Bacillus cereus, C. perfringens* type A and staphylococci, all of which are dependent on large numbers for toxin production in food or the intestine.

The sampling plans are then divided into cases, numbered according to degree of hazard, case 1, for example, being 'no direct health hazard' and cases 13, 14 and 15 'severe direct hazard', depending on conditions of use.

The 3-class plan is based on enumeration of the general bacterial flora and takes into account minimum and maximum levels. In the ICMSF notation on sampling plans, n = the number of sample units, m = the acceptable level and M = the rejection level. Values between m and M are undesirable, but subject to a number ('c' – reached by applying results and experience of previous tests), which can raise the acceptable level to meet the capabilities of the food industries concerned.

A 3-class plan is recommended by FAO/WHO Expert Consultations and Working Groups for pre-cooked shrimps and prawns; it is based on the examination of five samples per product lot:

Mesophilic aerobic bacteria	n=5, c=2, m=100 000, M=1 million.
Staphylococcus aureus	n=5, c=2, m=500, M=5 000.
Salmonella	n=5, c=0, m=0.

Food inspection

Inspection, whether of food premises (including factories, abattoirs, shops and catering services) or of food itself as it arrives at the port of entry, is an important part of the control system for all foods. There may be systematic sampling carried out by food inspectors, as part of the environmental health programme, and in close collaboration with laboratories responsible for food examination. The results of microbiological examination can be used to support inspection services and to improve the condition of food throughout production and service; investigations may extend overseas following poor results from imported food. It is necessary to ensure that the workload imposed upon the laboratory does not surpass its capability and that emphasis is placed on foods known to give rise to health hazards. Where there is routine sampling, the choice of food will vary according to the local need, and the occurrence of cases and outbreaks of food-borne illness in the area or nationally. For example, there may be an agreed investigation for the *Salmonella* content of chickens or sausages coming into or produced in a district, or the staphylococcal content of ham or cream cakes. If conflicts arise between results from official laboratories and those used on a private basis by multiple stores or importers, it may be necessary to involve a third laboratory to act as arbiter under carefully controlled conditions.

There are unavoidable differences in technique from person to person and between the sources and method of composition of media ingredients. Because of these variations, mandatory standards on microbiological content are considered to be inadvisable and for some foods impracticable. Foods should not be needlessly destroyed; when there are excessive counts it may be possible to use them for purposes other than retail sale, for example, in canning or other processes requiring heat treatment. After 1992 when the single European market came into force standards will become uniform. Foods cannot be stopped at the borders of Member States for examination so that reliance will be placed on examination and conformance with standards in the exporting state (see Chapter 17).

For all stages of manufacture the GMP and HACCP concepts are applicable. The HACCP system for quality control provides a means of improving inspection services. It is described as a cost-effective approach ensuring the safety of processes and products. It requires the identification of sites throughout production lines which are most likely to lead to contamination of the finished product, and consists of three parts:

(1) The identification of hazards, the assessment of their severity and the risk of their occurrence. Within this context, hazard means the unacceptable growth, survival of contaminating microorganisms known to challenge safety or with spoilage potential and ability to produce unacceptable metabolic products which may persist in the food.
(2) Observation of critical control points (CCP), locations, practices, procedures or processes whereby contamination with unwanted organisms could be prevented or controlled.
(3) The institution of methods to monitor the CCPs to ensure that they are regulated or to carry out remedial action.

The following data must be made available:

(1) Descriptions of the physico-chemical characteristics of the perishable products, those factors that influence the growth, death and survival of microorganisms. They include storage conditions, pH, water activity, preservation, packaging and gas atmosphere and also the anticipated consumer use and misuse.

(2) The flow sheets showing all stages in the manufacture of products and details of each process such as drying and pasteurization, the holding temperature in the storage tanks, and the prediction of what could happen at each point, the frequency and methods of cleaning of equipment and the materials required for cleaning.
(3) Knowledge and understanding of the microbial ecology of the food products under consideration and, where possible, the results of challenge tests with microorganisms, inoculation studies and the effect on survival and growth of the various procedures necessary before the final product is reached.
(4) The identity of the CCPs in the whole production system and records of monitoring results. Also observations on the regulation of equipment to provide trend analysis for food inspectors.

Frequent multidisciplinary consultations are required. The system is intended for quality control, rather than safety, but when quality is well controlled the safety of products, if not assured, is markedly improved.

Laboratory methods

So far as possible methods of examination of food samples should be standardized and a limited number of methods chosen for their simplicity and reproducibility for application in any country. Plate counts should be interpreted with understanding and with caution because of the difficulty of obtaining reproducible results. Media and equipment also need to be standardized. Automated systems will relieve the drudgery of repetitive counts and they are likely to improve the reproducibility. The clumping of organisms in food will increase the variability of most probable number (MPN) counts of *E. coli* and of plate counts. The adherence of food particles and organisms to the lumen of pipettes and to other glassware cannot be eradicated entirely by homogenization and dilution. Water is the best medium for the even distribution of organisms. In spite of these drawbacks in the interpretation of results from counting methods, it is observed that the higher the general count as well as that of organisms with potential for food poisoning, the greater the risk.

C. perfringens, *B. cereus* and *Staph. aureus* amongst the food-poisoning agents are commonly found in food samples in small numbers. When isolated in large numbers, millions per gram of sample, they are indicative of both food poisoning and spoilage. Factors leading to growth in food are similar for most organisms.

If legislation is required for specifications, it might be more appropriately applied to process control rather than to microbiological standardization of the end product. Laboratory methods would then be concerned with tests for the destruction of enzymes or the change in colour of dyes, indicating reduced oxygen potential from growth of organisms, where such tests are applicable. When a heat process is involved, the destruction of the enzymes phosphatase in milk and α-amylase in egg (liquid/whole melange), for example, can be used to monitor the process. Dye tests are more arbitrary, but economical of time, effort and expense. The methylene blue reductase test is applicable to milk, cream and ice-cream although it is no longer included in the statutory tests for milk (1989, see Chapter 17), and a resazurin test may be used for the oxidation-reduction (redox) potential as a measure of the growth of organisms in foods. The quantitative evaluation of metabolites formed by organisms growing in food may be developed.

Terminology

Various terms are used for the microbiological analysis of foods:

Microbiological criteria are derived from the microbiological evaluation of foods using prescribed methods of examination (ICMSF and WHO) and in accordance with accepted sampling plans; they may be applied for the acceptance or rejection of a food. As already described, the values include the measurement of the number of microorganisms per gram of food and the presence or absence of agents of gastrointestinal disease. Microbiological limits are formulated from criteria recommended by an authoritative body for particular purposes or regions, but not incorporated into law.

Microbiological purchasing specifications (MPS) are agreed between manufacturers and purchasing agents. Factories responsible for the production of food commodities for a number of countries would like to work to specifications agreed by all the purchasing countries. This was the format discussed by the ICMSF at its first meeting; the specifications which were eventually agreed were the results of years of discussion and laboratory studies between representatives from many countries.

Microbiological standards are formulated from microbiological criteria, they are incorporated into laws or regulations controlling food produced, processed and stored in or imported into the area of jurisdiction of a regulatory agency. The ICMSF chose to apply specifications rather than legislative standards to foods. The microbiological consistency of processed foods such as cooked frozen food can be maintained and assessed with a fair degree of accuracy, which is helpful for inspection services. Raw foods are far less definitive with regard to their microbiological condition; they have passed through many stages of production and been subjected to transport conditions where it would be difficult, if not impossible, to apply the HACCP concept.

The factors of importance for the application of microbiological criteria are summarized as follows: the criteria should be technically attainable by GMP and the microbiological condition of the raw materials must be considered as well as the effect of processing. The possibility and consequence of contamination and growth of organisms during handling, transport and storage should be recognized. The category of the consumers at risk is important; the young and the aged especially need safe food. The cost to benefit ratio associated with criteria necessitating surveillance and a monitoring system should be studied in relation to the hygienic advantages. Meaningful criteria for worldwide application require close coordination and international agreement on sampling and methodology.

A small number of methods should be chosen, relatively easy and rapid to perform, available internationally and readily interpreted. The criteria should be reviewed after approximately 3 years, and abandoned or retained according to whether or not they are serving a useful purpose; it may be necessary to make them more effective.

The advantages of microbiological specifications are the assurance of good standards of hygiene in production and service and thus the prevention of food poisoning and food wastage; attention will be focussed on the condition of food in competitive production. The disadvantages include the expense of the surveillance and monitoring of food samples and the possible rejection of large quantities of food unjustifiably, because results have been viewed in too much detail with regard to numerical limits. They may also be used for exploitation purposes between manufacturers. The growing international trade in increasing numbers of food commodities suggests that guidelines and specifications are unavoidable.

The ICMSF has published five books (see Appendix B, Bibliography). The first is a general introduction – *Microorganisms In Foods*, 1. *Their Significance and Methods of Enumeration*.

The second, *Microorganisms in Foods*, *2. Sampling for Microbiological Analysis: Principles and Specific Applications*, is now in its second edition. The opening chapter considers 'Meaningful microbiological criteria', and the volume is divided into two parts – the first deals with principles, the various concepts of sampling, choice of plan for pathogens and general investigation, and the collection and handling of samples. The second part considers specific proposals for sampling and sampling plans for many commodities of food and drink.

The third book, *Microbial Ecology of Foods*, has two parts: Volume 1, *Factors affecting the Life and Death of Microorganisms* and Volume 2, *Food Commodities* in detail; book three is a useful corollary to book two and is under revision.

The fourth book, *Application of the Hazard Analysis Critical Control Point (HACCP) System to Ensure Microbiological Safety and Quality* is in two parts: Part 1, *Reasons for Action and Principles of Control* and Part 2, *Critical Control Points of Operation and their Control and Monitoring*, it offers a rational and effective approach to the control of microbiological hazards at all stages in the food chain. Examples are given of the application of the HAACP concept to factory practice.

The fifth book, *Characteristics of Microbial Pathogens* provides a compilation of factors influencing the growth and death of intestinal and other pathogens in foods.

19

Education

The need for information

Food-borne disease occurs because many people employed in all aspects of the food industry – in processing, preparation, animal husbandry, water supply – are uninformed, negligent, or economically unable to carry out safe practices. Those involved in any capacity with foodstuffs for human or animal consumption should be concerned with the prevention of food poisoning and other food-borne disease. Even the engineers who design plant and equipment and the architects responsible for planning kitchens, manufacturing establishments, abattoirs, markets and farms should be aware of their part in controlling the spread of infection. Many people need to be trained in food-hygiene measures, but the basic facts are the same for all disciplines, although the method of approach may differ from one group to another.

There are different backgrounds, levels of perception and purpose and educators must adapt information to the needs of the particular occupational group. Experience has shown that it is better to give a wide view of the subject rather than to limit the information given, and the interest and learning ability of non-technical groups should not be underestimated.

People needing education and training include government officials, so that they may know the basic facts for prevention; food hygienists – including physicians and other medical personnel; veterinarians and sanitarians; food microbiologists; managers in charge of food-processing personnel and the plant workers; food service managers and workers; transportation officials and workers; field personnel; farmers and all those engaged in animal husbandry; abattoir workers; fishermen and those who harvest shellfish and, not least, the general public.

Medical, veterinary and agricultural government officials need to keep pace with the changing phases of food technology and food science. The growing world population and consequent necessity for more food brings with them difficulties in ensuring food hygiene. New preservatives and additives and the behaviour of certain pathogenic organisms influence the microbiological content of food.

Information should be given on the various causes of food poisoning, the bacterial, viral, fungal and parasitic agents, their sources and means of spread, and the main control factors. Man and animals require protection against the oral intake of pathogenic organisms and their toxins. With the importation of foods – for man and for animals – from many countries, the hazards increase and careful laboratory vigilance is necessary to assess significance of contamination. Salmonellae, for example, may be scattered throughout batches of feed for animals to be distributed all over the country.

Intestinal pathogens such as *Salmonella*, *Campylobacter* and *Escherichia coli* enter abattoirs and food processing plants in or on live animals and birds and they are transferred to carcasses and cuts of meat and poultry. Fish and shellfish may harbour *Vibrio parahaemolyticus*, other vibrios and viral particles. Cereals such as rice and flour are usually contaminated with *Bacillus cereus*; *Clostridium perfringens* is almost always present on and in meat and poultry. *Listeria monocytogenes* is widespread in the environment and may be found in a wide range of raw and cooked products. All these contaminated foods will reach kitchens of food service establishments, institutions and homes. The chances of cross-contamination from raw to cooked food are likely by many means: hands, surfaces, kitchen tools and other equipment; clear teaching is necessary to cut down the risks. Faults in processing such as canning, fermentation and curing are contributory factors.

Major hazards arise through lack of attention to correct time and temperature exposures for cooking and storage. Storage of cooked food at ambient temperature for long periods of time will encourage massive multiplication of bacteria in the food, resulting in spoilage or food poisoning if the bacteria are pathogenic.

Outbreaks and notification

There is legislation to ensure that outbreaks and cases of food poisoning are notified to the Department of Health (DH). Yet many incidents, probably thousands of cases, are not reported to the doctor so that information does not reach the Public Health Laboratory Service (PHLS) or the DH. Sometimes a large number of single incidents regarded as sporadic cases may be part of one large outbreak from a contaminated foodstuff distributed over a wide area and unrecognized as a source of danger.

Annual figures for food poisoning and salmonella infections in England and Wales are compiled by the Communicable Disease Surveillance Centre.

Over the 3 years 1989 to 1991, 54 918 incidents involving 76 357 cases due to the main bacterial agents of food poisoning (excluding *Campylobacter*) (see Tables 3.2 and 3.3) were reported. These figures include general and family outbreaks and sporadic cases where the causal organism was identified; there were more incidents where it was not recognized, and there were probably many more unreported cases. In many episodes of food poisoning, no specimens are sent for laboratory examination or, if sent, no aetiological agent can be found.

The outbreaks most commonly notified are those which occur in locations where food is prepared for large numbers of people, such as hospitals, schools, factories, geriatric institutions, hostels, halls of residence, nurseries and children's homes. Faults in the daily preparation of food for the same groups of people and leading to illness cannot fail to be observed. The same faults causing food poisoning in persons making chance visits to hotels and restaurants, or in travellers, may pass unnoticed or unrecorded.

Meat and poultry in their various forms are the foods most commonly implicated and they are responsible for about 50–70% of the notified general and family outbreaks each year. Organisms of the salmonella group have hitherto been regarded as the predominant causal agents but notification of other incidents where the causal agent is thought to be *Campylobacter jejuni* continues to increase and the number of individual cases reported now surpasses that due to salmonellae, although the number of incidents where a known food is incriminated is still small; the main vehicles of infection for campylobacters are untreated milk and undercooked poultry. *C. perfringens* outbreaks are the next most common.

An appreciable reduction in the incidence of salmonella food poisoning could be achieved by reducing the animal sources. To control staphylococcal food poisoning there must be strict attention to techniques used for the manipulation and storage of cooked foods such as meat and poultry and prepared sweets, custards, trifles and cream products. Food poisoning caused by *C. perfringens* could be prevented with more care in rapid cooling and cold storage of bulks of meat and poultry. In all instances conscientious cold storage of cooked food and environmental design for cleanliness are important. However, it should be remembered that *L. monocytogenes* can grow, albeit slowly, at refrigeration temperatures. Provision of good conditions is the responsibility of management, with whom, as well as with food handlers, rests the safety of the food provided and sold. Widespread and persistent instruction is necessary to correct faults and avoid hazards leading to food contamination. As yet there is no evidence to suggest that the incidence of food poisoning is declining in England and Wales.

The food industry

In food-service establishments, managers must be trained in the principles of food hygiene so that they can in turn train and supervise the workers responsible for processing, preparation, storage and service of food. Those who maintain the cleanliness of the environment in which foods are in constant preparation also have an important function and should know the principles according to which they are working.

Hospitals

Catering personnel in hospitals have a particularly responsible position because they are feeding sick people whose immunity may be low. Whereas healthy people are usually able to withstand small numbers of intestinal pathogens in food, those who are ill, especially the elderly and young children, will succumb readily to infection and intoxication; fatalities from food-borne infection are most common amongst these groups.

Each regional hospital board has an officer responsible for training schemes. Courses are arranged for hospital officers either at their own centres or in technical colleges. Several regions give supervisory courses for catering staff including food safety and quality. The teaching of hygiene in hospitals is considered in the Department of Health Circular HM (64) 34 (see Appendix B; Bibliography)

A Health Service Catering Manual is published by the Catering and Dietetic Branch of the DH (see Appendix B; Bibliography). In 1974 a booklet entitled *Hygiene* was published by the then DHSS (now DH) based on a Code of Practice designed for the Birmingham Regional Hospital Board; a new edition appeared in 1986.

Restaurants, hotels, factories and food handlers in retail shops

Factory managers and food handlers are responsible for the safety and quality of the various daily commodities used by the public. Faults in production can lead to widespread food poisoning on a national and even international level. Managers and workers in industry processing meat, poultry, dairy products, cereals and vegetables sold fresh, canned, frozen or dehydrated need to be aware of the condition of the basic ingredients as well as that of the finished product.

The Hazard Analysis of Critical Control Points (HACCP) (see Chapter 12), described in detail by the International Commission on Microbiological Specifications for Food (ICMSF) in its fourth book, *Application of the Hazard Analysis Critical Control Point (HACCP) System to Ensure Microbiological Safety and Quality*, focuses attention on the critical processes of any food operation, which allow contamination, survival and growth of microorganisms. Routine monitoring at these points requires skill in sampling and checking temperatures, pH, a_w and disinfectant concentration and in the interpretation of chart recordings. The effectiveness of the monitoring process must be evaluated.

There should be an awareness of the importance of communication between factory managers and staff, and between management and medical and environmental health officers. In the event of outbreaks of disease, the speed at which reports are made is vital for immediate observations, sampling and examinations to halt the spread of infection and to find the source.

Meat and fish producers

Farmers and all those involved with animal husbandry are responsible for the health of herds and flocks. Intensive rearing, particularly of poultry and pigs, carries special problems with regard to environmental contamination from feeds and excreters on the premises. Slaughtermen and those who dress carcasses need to know and practise techniques that will reduce the contamination of carcasses by intestinal contents.

Fishermen may not be aware of the need for potable water for cleaning fish. Cold storage of their catch and transportation in suitable containers are important factors. The structure of ponds for fish-rearing must aim to minimize the occurrence of *C. botulinum* in mud. Shellfish harvesters should be aware of the hazards of polluted areas and the benefits of cleansing systems.

Training

University and independent colleges train students in domestic science, including food hygiene, and special courses up to degree level are available for students wanting responsible work in the catering industry as well as in hospitals. Correspondence courses are also available through colleges of further education and the Hotel, Catering and Institutional Management Association, whose annual examinations include questions on food hygiene and nutrition.

Managers, supervisors and other key personnel from commercial eating establishments, retail shops, small cafés, snack bars and operators of portable food vans can be invited to participate in lecture demonstrations in school kitchens and other institutions, with three or four talks given at weekly or shorter intervals. Films, slides, charts and pictorial descriptions are available, and short notes can be provided to help disseminate the information. Teaching in large catering establishments and multiple department stores may be allowed within the shop or store. In establishments where it is difficult for staff to be spared, talks can be given to small groups at their work station. Local environmental health officers can advise and assist with training in food hygiene. Under the provisions of the Food Safety Act, 1990, legislation was considered on compulsory education for food handlers. A consultation document was issued. It has been decided that a basic requirement would be introduced whereby food businesses must train their staff in a way proportionate to their tasks (see p. 286).

The public

Home-makers, consumers' associations and individuals who prepare food for parties and other special occasions may not fully understand the precise hazards involved. The mass media are required to reach them, and full use should be made of television, radio, newspapers, magazines, leaflets and books to disseminate food hygiene messages – which could save lives. Exhibitions accessible to the public can be set up in strategic places such as town halls and mother-and-baby clinics and health centres.

A catering trade working party published a report in 1951 in which it was stated that 'no large scale and lasting improvements in the hygienic conditions of catering establishments can be brought about unless informed public opinion demands it. We would stress the word "informed" because many people, while appreciating generally the value of cleanliness, have little knowledge of the real risks attendant upon particular faulty practices. The dissemination of knowledge of the principles underlying food hygiene should, therefore, be carried out by practical means'. There is much improvement in the standard of kitchens since that was written, but faults leading to food poisoning are still not always fully understood, even by responsible people. In 1989, in response to the growing concern on food safety, the Government issued a simple leaflet *Food Safety – a guide from HM Government* aimed at helping the general public understand the hazards associated with the incorrect handling of food. Many millions of the leaflets were distributed through supermarkets, chemist shops, doctors' surgeries and clinics. Some of the major supermarkets have sponsored a food safety advisory centre, which provides, by means of a free foodline telephone service, advice to the general public on any problem relating to food. They have also prepared a series of useful leaflets and booklets. (See Appendix B: Bibliography).

Children

Schools have a responsibility to teach children, however young, about infection and how bacteria spread. Hygienic habits are not instinctive, so that training in food hygiene should be included in the curriculum of domestic science and home management (home economics). Elementary bacteriology, its application to the contamination of food from a variety of sources and its prevention can be conveniently taught and demonstrated practically during cookery lessons. Supplementary education can be supplied to schools by the local environmental health departments. The environmental health officers (EHOs) are active in this field. Some local authorities employ their own advisory staff in the Education Advisory Service. Some sections of the food industry, for example the National Dairy Council, have produced educational packages including games and quizzes to assist with food-hygiene education in schools.

Practical examples for schoolchildren could include: bacterial cultures of swabs from hands, or impressions of fingers on culture media before and after touching food and washing; cultures from samples of foods and swabs of surfaces and utensils; also demonstration of the effects of heat and of cold on the growth of bacteria in food.

When public demonstrations of food hygiene are available, school parties can be taken round the exhibits and given an explanatory running commentary. Competitions could be held between schools to design posters showing aspects of food hygiene, and the winning entry be displayed at an exhibition and featured in the local Press.

The school meals service provides regular midday meals and many schools have introduced a cafeteria service with a wide choice of hot and cold food; catering advisers are employed throughout the country to supervise the school catering service. Although most schools have their own kitchens, for those that do not, central kitchens may produce more than 1000 meals each day for several schools as well as providing meals-on-wheels for elderly and sick people, such preparation may be conventional or using the cook–chill system.

In-service training courses for school catering staffs are held at colleges of further education, covering hygiene and nutrition. Students may study for the City and Guilds qualifications. Courses leading to certificates in the hygienic handling of food are held for organizers, cooks and other members of staff. The county of North Yorkshire has been active in training courses and has an excellent record of freedom from food poisoning. Domestic science/home economics teachers in schools and hospital kitchen supervisors request talks and hold weekend training courses. The World Health Organization (WHO) is also active in food-hygiene education and has sponsored many publications aimed at different sections of the community. *Safe Food Handling* by M. Jacob, produced in 1989, is directed at those involved with training workers in food-service establishments while *Food, Environment and Health*, published in 1990, is a guide for primary school teachers for the implementation of health-education programmes, including the safety of food, and is aimed particularly at the less developed countries.

Provision of information

Information on food hygiene is available from many sources, but much of the responsibility for disseminating facts is assumed nationally by EHOs, medical officers for environmental health (MOsEH), consultants in communicable disease control (CCDCs), health education officers, and other public health officers including, in hospitals, the infection control nurse. In many European countries and in the USA the veterinary officer takes a prominent part also. The EHO is the link between the MOEH, the CCDC and the food trader, restaurant and canteen supervisor and other food handlers in the area; the routine inspection work within the district keeps the EHO constantly in touch with a wide variety of food workers. The success of any programme for the establishment and improvement of conditions and techniques amongst food handlers depends largely on his or her efforts.

EHOs whose training has been financed by local authorities are also employed by food and catering companies. They help in the education of food handlers and also in the maintenance of high standards of hygiene in food production and preparation.

The DH provides expert advice and its staff lecture on food hygiene and related subjects. The Ministry of Agriculture, Fisheries and Food (MAFF) provides an advisory and laboratory service for the veterinary profession and also a limited advisory service for the food industry. MAFF are also concerned with food-hygiene education in schools. Jointly with the DH they commissioned the Central Office of Information to produce a schools food-hygiene programme for the 5–12-year-old group entitled *Food Hygiene with Hy-Genie*, which includes a video plus written material.

Unilever are also currently developing a project on working with food in primary schools that will give ideas and activities for primary schools including teacher's notes and activity sheets.

The PHLS in England and Wales is active in the field of education on matters relating to the prevention of bacterial and viral diseases. Amongst the many functions of the PHLS, which has laboratories scattered throughout the country, are included the examination of foodstuffs associated with outbreaks of food poisoning and the investigation of the sources and paths of spread of the bacterial agents responsible. The PHLS Communicable Disease Surveillance Centre collates data on infectious disease and compiles annual reports on the incidence of food poisoning including salmonella infection in England and Wales. Such reports provide up-to-date information on agents, sources and foods currently implicated in outbreaks for those involved in education.

PHLS laboratories regularly examine both home-produced and imported foods and carry out surveillance studies and research on foods known to be public health hazards and suspected of spreading bacterial agents in manufacturing establishments, shops and kitchens. The zoonotic illnesses due to bacteria shared between humans and animals, causing both to suffer, are investigated jointly with the veterinary profession. The data obtained from these studies are made available in a wide range of scientific and medical journals as well as through lectures, discussion groups and occasionally radio and television interviews and documentary programmes.

Agricultural and food research council laboratories such as those of the Food Research Institute at Reading and Norwich, both involved with many kinds of food and methods of examination, the Poultry Research Station, the food research laboratories of the manufacturing, preservation and bakery industries, the MAFF fishery laboratories and other independent laboratories are continually checking and investigating means to prevent food spoilage, as well as the contamination and build-up of pathogenic organisms in food.

Independent organizations active in the field of food hygiene include the Health Education Council, St. John Ambulance Brigade, the Red Cross, Women's Institutes and the Women's Royal Voluntary Service.

Three organizations provide nationally recognized food hygiene courses: the Institution of Environmental Health Officers (IEHO), the Royal Institute of Public Health and Hygiene (RIPHH) and the Royal Society of Health (RSH). The IEHO programme includes (a) a basic course in food hygiene, considered to be the minimum level of hygiene education required by food handlers involved in the preparation of high-risk foods; (b) an intermediate certificate course designed to bridge the gap between the basic and advanced courses to enable the less academically inclined persons to progress successfully to the higher course; and (c) an advanced food hygiene certificate intended for managers and supervisory staff. The RIPHH offers a Primary Certificate in Hygiene for Food Handlers, a Certificate in Food Hygiene and the Handling of Food, a Diploma in Food Hygiene and also Diplomas in specialist food-hygiene subjects designed for managers in the industries concerned (bakery, canning, dairy and meat and poultry), and the RSH give a Certificate in the Hygiene of Food Retailing and Catering. Persons who successfully complete the IEHO Advanced Course or the RIPHH Diploma or obtain a credit award in the RIPHH or RSH certificate courses may themselves organize and lecture on the Basic Course in Food Hygiene under the control of the local environmental health departments.

For many years the King Edward VII Hospital Fund for London supported an excellent Hospital Catering Advisory Service, which provided a centre with model kitchens, equipment and facilities for lectures. They instituted trials of new feeding systems for hospitals. Reports on some of these projects covering aspects such as the use of frozen cooked meals, preparation of single items either in central kitchens or by the food manufacturer, and methods of thawing and reheating, including hot air circulation and microwave ovens, may be consulted by arrangement with the Librarian at the King's Fund Centre, 126 Albert Street, London NW1.

The Central Office of Information, various industries and overseas teaching laboratories such as the Centers for Disease Control, Atlanta and universities in the USA produce films, slides and bulletins. Many commercial firms engaged in the manufacture of foods, equipment and detergents provide exhibits for educational purposes.

A detailed list of films and slide presentations on Food Safety has been updated by David Bates. It provides information on international sources of audiovisual materials and is available from The British Life Assurance Trust Centre for Health and Medical Education, London (1987).

Media

Television and broadcast programmes for both adults and schools have wide appeal. The teletext services, Ceefax and Oracle and the Prestel system could be used to disseminate food-hygiene information, particularly at peak food-poisoning times such as at Christmas and during hot weather. So that newspaper reports of outbreaks of food poisoning will be as accurate as possible, the Press should be given full information on the faults and reasons which lead to outbreaks, the methods of spread of infection and contamination and preventive measures.

Teaching methods

There are many different ways of describing the essential facts about food hygiene.

Lectures

The best lecturer is an authority on the subject, but experts are not always available and it may be necessary to provide authoritative material for an available good speaker. Supportive visual aids can include slides, film strips, films, video films, tape recordings, flannelgraphs, charts of facts and outbreaks and photographs of good and bad establishments and practices.

Overhead projection facilities are useful for the presentation of instant material, diagrams and notes written during the lecture or pre-prepared. Several layers of transparent film may be overlaid, to build up pictures illustrating successive events; for example, in a food-poisoning outbreak the chain of infection and the means by which the links in the chain can be broken.

Time should be allowed for discussion, which may be stimulated by a panel of leaders each describing their own approach and with audience participation encouraged. Contributions from both sides can be summarized by the panel's chairman, whose role is also to balance the contributions and to prevent monopolization by any particular speaker, and to control the time devoted to each.

Conferences

Day or weekend conferences in pleasant surroundings are popular and provide opportunities for detailed and extensive demonstrations, for learning new facts and for exchanging views both formally and informally, supplemented by symposia which may be held within the conference structure. The conference leader will summarize the proceedings.

Workshop

Practical work is carried out in addition to theory sessions, and a workshop is of particular value for cooks and other kitchen workers. Swabs of the nose, hands, environment and of foods will demonstrate the extent of bacterial contamination and the beneficial or otherwise effects of washing and of disinfectants such as hypochlorite. They may include sessions for role-playing when unrehearsed participants simulate people involved with food poisoning or other situations. Prepared video recordings of the events leading to food-poisoning incidents can be used in a similar way. Five to 10 minute dramas can help to highlight mistakes which lead to food poisoning. The case method uses a real or simulated situation in more detail, for analysis, discussion and solution.

Home study

Individual study texts including questions are provided for people who find it difficult to reach training areas. The correspondence is carried out according to the students' level of interest and available time.

Programmed instruction

Short introductory printed material with examples of realistic situations is provided. The trainees respond in notebooks or through teaching machines, where automatic course monitors provide immediate confirmation of the accuracy of the answers; if not, the various steps must be repeated. Tape-slide presentations are useful when a qualified teacher is not available.

Audiovisual aids

Such aids include displays, large and small, with charts and pictures, slides of bacterial cultures and notes on methods of examination for food and water, as well as posters giving pictorial examples of outbreaks of food poisoning with short explanatory notes; the faults leading to the outbreaks are listed together with the lessons to be learned. Such materials should be portable; charts may be carried in special portfolios. Alternatively, sets of slides, filmstrips, video and tape recordings may be made, thereby also making them available for loan or sale.

Sets of petri dish cultures may be prepared to illustrate hand, nose, hair, utensil and food cultures, and can be made semi-permanent by the careful application of a liquid preparation of formica. Bacterial cultures may be prepared on various agar media and the petri dishes shown by means of an illuminated viewing box. Plate cultures can demonstrate bacterial colonies from fingers rubbed lightly on the surface of the media. Comparative culture plates will show the effect of touching food and the efficacy or otherwise of washing the hands, the growth from clean and dirty handkerchiefs, kitchen cloths, towels and surfaces. Hairs and flies may be placed on agar plates also. Colonies of bacteria may be grown from imperfectly washed utensils. Portions of food may be cultured to illustrate bacteriologically clean and lightly or heavily contaminated food and the effect of refrigeration on the prevention of bacterial growth.

Cine and video films are useful audiovisual aids and many have been prepared in the UK and the USA by both government and commercial agencies. They tell the story of food-poisoning outbreaks giving the source of the agents concerned and their ways of spread. Faults in technique, eventually giving rise to the growth of the food-poisoning agents in the foods presented for consumption, are emphasized.

Demonstrations

Lecture demonstrations can take place in canteen kitchens. Equipment may be used and cleaned *in situ*. The storage, use and cleaning of cooking and other equipment can be demonstrated together with the ways of spread of organisms entering kitchens in foods or on people. Fluorescent dye in vaseline can be used to demonstrate the spread of infection via hands to the handles of vessels and taps, and back to hands after washing. The splashing of poured liquids over a wide area may be demonstrated by means of organisms that produce brightly coloured colonies. The splash from the bowl of flushed toilets and the droplet spread can be illustrated in a similar way. The necessity to clean and disinfect cleaning materials such as cloths, mops, brushes and buckets will be evident from culture plates before and after treatment.

The flannelgraph (see Fig. 19.1) giving cut-out picture sequences is an old method of teaching but nonetheless effective. Imaginative thinking can no doubt devise other methods of approach to the teaching of the facts which are simple, but knowledge of which is so urgently needed in the practice of food hygiene.

Table 19.1 groups the agents of food poisoning according to source, methods of public health control and laboratory diagnosis and typing.

Means to control one organism may do little to stop the activities of another. For example, emphasis on personal hygiene alone may reduce the incidence of staphylococcal food poisoning, but it will do little to prevent salmonellosis and nothing to stop *C. perfringens* food poisoning. On the whole, people are unaware of the presence of salmonellae in raw poultry and meat, and thus in the environmental surroundings of these products. The onus of producing salmonella-free foods lies with those involved at the start of the chain of production. The risks can be reduced in the kitchen and recommendations to help those in the retail and consumer trade to do this are essential; nevertheless they expect to be given safe food in the first place.

Precautions against *C. perfringens* and *B. cereus* food poisoning will succeed only if it is understood that the spores of both organisms survive cooking and that unless hot food is quickly cooled and refrigerated there will be rapid growth of vegetative cells. Perhaps the most effective of all measures against food poisoning is the prevention of bacterial growth after cooking by attention to cooling times and cold storage. The control of temperature during manufacture, storage, preparation and service in the factory, canteen and home is essential to keep bacterial numbers low.

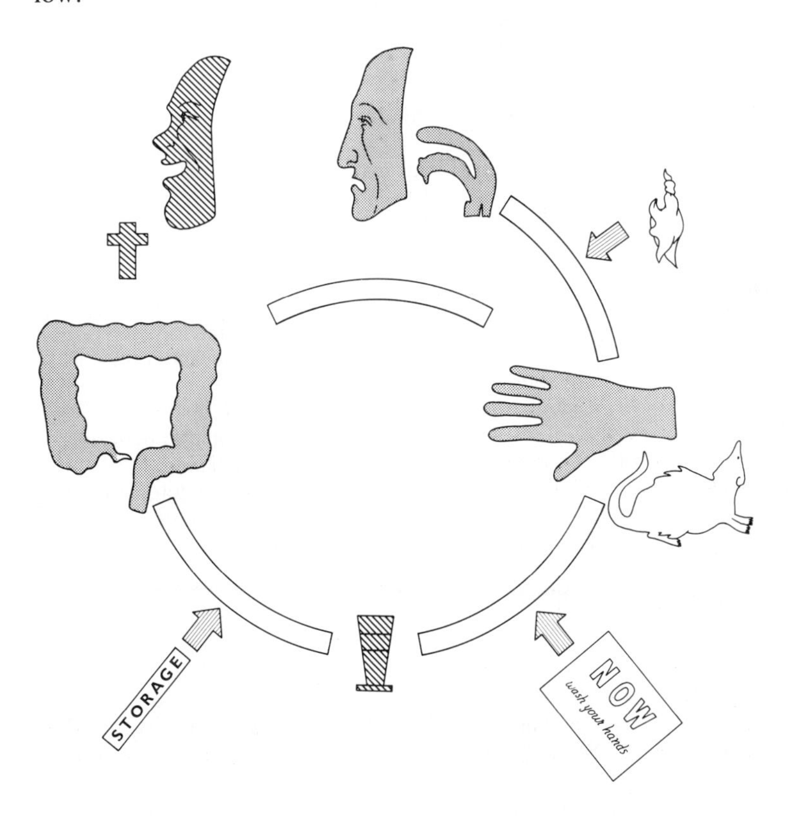

Fig. 19.1 Flannelgraph showing chain of infection

Table 19.1 Sources and control of food-poisoning bacteria

Source	Public Health control	Laboratory control
Salmonella		
Animal stool, coat, hooves, paws	Rearing methods Feeding stuffs Farm hygiene Slaughterhouse hygiene	Diagnostic media for stool samples, swabs, and food Bacteriological counts on foods and feeds Biochemical tests Serological and bacteriophage typing
Foodstuffs (animal origin) meat and poultry feedingstuffs for animals, egg products, raw milk	Hygiene of production Treatment to render safe Storage	
Environment of food preparation	Cleanliness of equipment, utensils and surfaces	
Water for drinking and pre-preparation of food	Treatment by filtration and chlorination	
Human stool, hand	Care of handling foods Avoidance of cross-contamination from raw to cooked food Personal hygiene	
Staphylococcus aureus		
Human nose, throat, hand, skin, lesions	Care in handling foods Storage of cooked foods Personal hygiene and habits	Diagnostic media for swabs and food Bacteriological counts on food
Animal cow, goat	Care of mastitis	Coagulase test Bacteriophage and serological typing
Foodstuffs (dairy) milk, cheese, cream	Hygiene of milk production Heat treatment of milk intended for drinking and for cream and cheese	Enterotoxin detection by immunological techniques
Clostridium perfringens		
Foodstuffs meat and poultry, dehydrated foods	Cooking and cooling techniques Storage of cooked foods	Diagnostic media for stool samples and food Bacteriological counts on food

Table 19.1 (cont.) Sources and control of food-poisoning bacteria

Source	Public Health control	Laboratory control
Environment of food preparation (food and dust)	Cleanliness of equipment and surfaces	*C. perfringens* counts on stools
Human stool		Serological typing
Animal stools and dust		Enterotoxin testing
Clostridium botulinum		
Soil and mud		Toxin identification (neutralization tests in mice)
Fish	Processing and cooking	
Foodstuffs fish, meat, and vegetables		Diagnostic media
***Bacillus cereus* and other *Bacillus* spp.**		
Foodstuffs (cereals) dust, soil	Storage after cooking Cleanliness of environment	Diagnostic media Bacteriological counts on food Serological typing Toxin detection
Vibrio parahaemolyticus		
Seafoods	Warning against eating raw fish and other seafoods Avoidance of cross-contamination from raw to cooked seafood	Diagnostic media Bacteriological counts on food Serological typing
Campylobacter jejuni		
Animals	Pasteurization	Diagnostic media
Water	Chlorination	Serological typing
Foodstuffs milk, poultry	Hygiene of production	
Listeria monocytogenes		
Foodstuffs raw and cooked, plant and animal	Hygiene of production Treatment to render safe Storage	Diagnostic media Bacteriological counts on food Serological and bacteriophage typing
Environment of food preparation	Cleanliness of equipment and surfaces	Swab cultures
Yersinia enterocolitica		
Foodstuffs raw and cooked	Hygiene of production Treatment to render safe Storage	Diagnostic media Bacteriological counts on food Serological typing

Table 19.1 (cont.) Sources and control of food-poisoning bacteria

Source	Public Health control	Laboratory control
Environment of food preparation	Cleanliness of equipment and surfaces	
Other organisms, e.g. streptococci		
Human	General care of food	Diagnostic media
Animal	Storage	Bacteriological counts on food
Foodstuffs		Serological typing

Teaching requires a knowledge of the statistical facts about food poisoning and also familiarity with the sources, habits and general behaviour of the various organisms responsible for food-borne illness.

Appendix A gives suggested outlines for lectures to various groups of people. Appendix B gives a list of references to supplement the information given in this book.

Travelling and camping

It is relatively straightforward to apply good food-hygiene practices when preparing and consuming food in the home or in catering premises; that is, where there is a ready supply of clean, potable water (both hot and cold), and electricity or gas for cooking and refrigeration equipment. It is not as easy to maintain good hygiene when travelling or camping. The following notes are aimed at helping to prevent food poisoning when food preparation facilities are less than ideal.

Prevention of food poisoning abroad

Most people travelling in some parts of Europe and in Asia, the Far East, South America and Africa experience gastroenteritis. Some accept it as inevitable and include in their luggage prophylactic and treatment pills and medicines of various kinds.

It follows meals eaten on intercontinental ships, trains and planes, in hotels, boarding houses, restaurants, cafés and private houses, and during camping and other self-catering holidays after eating food bought in local shops and markets and from drinking unchlorinated water.

The patterns of illness follow those described in the previous chapters and all types of bacterial food poisoning are encountered, as well as illness due to viruses and parasites.

There is much so-called diarrhoea or 'gippy-tummy' which is said to be due to a change of food, overeating, overdrinking, heat, rapid change of temperature, iced drinks and so on. Unfortunately, the facilities for a full and competent investigation of these illnesses are rarely available. Bacteriological laboratories are not carried routinely, even in the largest and most important shipping lines, nor, of course, on trains.

The traveller on land accepts his predicament, knowing that he will soon recover; when medical aid is summoned, the doctor recognizes the well-known symptoms and seeks to relieve the condition, often without taking time to study the cause.

While allowing for the fact that loose stools are likely to occur sometimes during travel overseas, due to strange foods, and cooking habits, and to overdoses of aperients following constipation, the vast majority of incidents are likely to be microbial in origin. The principles underlying the provision of safe food are the same by whatever means one chooses to travel and in whatever country one chooses to be. The countries with hot climates will have greater problems than those in temperate zones; refrigerators are necessary where it is consistently hot and in cooler countries accustomed to central heating. Certain food habits may expose both inhabitants of a country and travellers to special hazards. For example, the eating of raw fermented fish in the Arctic and Japan has given rise to botulism, and also in Japan to much food poisoning caused by *Vibrio parahaemolyticus*. This latter type of food poisoning is now reported from other countries and the danger of cross-contamination from raw to cooked seafoods stressed. Care in the processing of canned foods, a moderate distaste for raw fish and careful education on safe methods of home preservation – which include heat treatment of acid fruits, but rarely meat, poultry, game or vegetables – have no doubt protected the population of the British Isles from botulism; the few incidents reported have been mostly from imported foods. This fatal food poisoning occurs more frequently in countries where household curing and preservation of foods is common. The prevalence of *C. perfringens* food poisoning in England and Wales may be related to the habit of pre-cooking and warming meat dishes, a practice which became common in the war years due to shortage of meat.

Many illnesses which occur during overseas travel can be prevented by simple precautions, in particular with water and food.

Rules for travellers

(1) Avoid drinking and cleaning teeth in cold water unless it has been boiled or personally chlorinated by the addition of proprietary brands of tablets or liquids containing sodium hypochlorite, or pumped through a filter such as the Berkefeld filter. Wines, bottled mineral water and soft drinks are usually safe, especially if carbonated. Locally produced drinks, including some in bottles, may not have been sterilized and may have been prepared with local and polluted water supplies. Hot tea and coffee are safe standbys.

(2) Avoid the addition of ice to drinks as the ice may be prepared with polluted water.

(3) Foods to be avoided unless their safety is assured include:

(i) ice-cream; the method of production can be subject to many faults in hygiene unless legislation covers the procedure for manufacture, as in the United Kingdom;

(ii) milk, unless given a controlled heat treatment process such as pasteurization, sterilization or ultra-heat treatment, or when boiled. Avoid any form of fermented milk;

(iii) soft and semi-hard cheese, particularly goats' milk cheese. In many continental countries, soft cheeses are prepared from pasteurized or boiled milk and are safe; this may or may not be stated on the packet. In the Middle or Far East, and in some Mediterranean countries, such cheese is usually prepared from raw or inadequately heated milk and may, therefore, be bacteriologically dangerous. It should be noted that soft cheeses have been involved in listeriosis from the organism *L. monocytogenes*; pregnant women have been warned against the consumption of the softer kinds of cheese;

(iv) cooked or semi-preserved cold meats, sausages and meat sandwiches, which may be subjected to much handling and poor storage in warm countries, with the exception of those products known to have a high-salt content. Ready-to-eat snacks sold from market stalls and street vendors may be prepared and stored in an unhygienic manner;

(v) meatballs, which are popular in some countries and may be prepared with meat cooked ahead of requirements and thus be a suitable medium for the growth of *C. perfringens*;

(vi) curried foods prepared some hours before required and kept warm pending consumption;

(vii) shellfish and other seafoods, particularly from warm countries;

(viii) 'bar tasties', particularly those including prawns and other seafoods;

(ix) cooked chickens, and other meats such as kebabs, hamburgers, hot dogs, sold hot, unless the food is seen to be freshly cooked in a reputable shop;

(x) salads and dessert fruits, unless carefully washed in water containing hypochlorite;

(xi) dishes prepared with raw or lightly cooked egg.

It may be reasoned that similar precautions should be taken in the home country and this is true, but when crowds in holiday countries must be fed quickly, and atmospheric temperatures are soaring, the hazards of mass feeding are magnified.

Campers and travellers who buy food for self catering can take care of their own purchases and preparation of food. Meals eaten out at night should be chosen with regard to freshly cooked meat, for example grills and fresh roasts and the avoidance of raw ingredients.

Those who live in tents and caravans need care with tea towels and dish cloths, which should be washed thoroughly and stored or wrung out in disinfectant after use; disposable paper is preferable to cloths. A similar precaution applies to face cloths and towels. Children are particularly susceptible to gastroenteritis, and to protect against infection from hands, each child should rinse the hands in disinfectant before and after meals. This practice should apply also to adults preparing food.

The floors of caravans should be washed daily and a disinfectant such as hypochlorite (bleach) may be used.

These precautions should not be considered unnecessary as the alternative might be at best a short attack of gastroenteritis or at the worst typhoid or paratyphoid fever or dysentery during the holiday or after returning home. Holidays are expensive and days lost by illness are misspent.

The World Health Organization acts as the directing and coordinating authority on international health work. WHO fosters close working relationships between health administrations, airlines, shipping companies and other bodies associated with international traffic to ensure coordination for health protection in international traffic. (World Health Organization (1974), *International Health Regulations (1969)*, 2nd annotated edition, Geneva.)

A number of publications have been produced by the WHO to improve hygiene in relation to travel and tourism, such as the *Guide to Sanitation in Tourist Establishments* (1976), *Aviation Catering* (1976), *Guide to Hygiene and Sanitation in Aviation* (1977) (J. Bailey). In the UK, guidance is given to travellers in leaflet SA 35 issued regularly by the DH – *Protect Your Health Abroad*. Other references are given in Appendix B.

Canvas and hutted camps

Certain aspects of general food hygiene advice need emphasizing in canvas and hutted camps where large numbers of young people are accommodated in often primitive conditions. These notes are especially applicable to camps in the UK.

(1) *Water supply*. In urban areas, mains water supplies are taken for granted and can be used with confidence. In the country, where a mains water supply is not available, great care must be taken. In some areas there may be a local piped supply from an artesian well or other source; the advice of the local health department should be taken before it is used for drinking or even for the preparation of food and dish washing (unless boiled). Water from rivers or streams should never be used for drinking or for food preparation.
(2) *Milk supply*. In towns, most milk is pasteurized or otherwise heat treated and therefore safe, unless it is mishandled after delivery. In the country, raw milk is often the only available supply, and it may be contaminated with harmful organisms. Every effort should be made to obtain a supply of pasteurized, UHT or sterilized milk. If none is available, all milk used cold must be boiled, preferably immediately on delivery. If raw milk is used for cooking, it must be regarded as contaminated and kept strictly away from other foods.
(3) *Food storage and preparation*. As refrigeration is seldom available, the separation of raw and cooked foods is important. Raw meat, sausages and

poultry, in particular, must be kept in a separate place, together with utensils and implements used for handling them. Meat and poultry should be purchased fresh regularly and cooked as near to the mealtime as possible, and served immediately afterwards. Reheating should be discouraged and unused left-over gravy discarded. Latecomers should be given safe food, such as freshly cooked eggs or bacon, bread and cheese, or a freshly opened canned product. Cold foods such as sliced ham, meat pies or brawn must be obtained from a reliable source and eaten as quickly as possible after purchase.

If it is necessary to serve cold meat the day after it has been cooked, small joints should be used, preferably not more than 2.7 kg (6 lb) in weight. They should be well roasted, and after draining away the gravy, cooled quickly in the open air or in an outside safe as a protection against marauders, and kept covered until sliced just before use. If the gravy is not needed immediately, it should be discarded.

Custard, blancmange, and similar milk dishes should be prepared immediately before use and not kept afterwards.

If traditional methods of camp cooking are employed, e.g. overnight slow cooking in a hay box, such foods should be reheated to boiling before serving. Foods containing cereals should not be cooked in this way, because of the hazard of the very heat-resistant enterotoxin produced by *Bacillus cereus*.

(4) *Hand washing*. Hand washing is as important in camp as elsewhere. An ample supply of clean water, soap and towels should be provided, particularly in the latrine and food preparation areas. Campers should be encouraged to use personal towels because of the hazard of cross-infection from those provided for communal use.

(5) *Dish washing*. Crockery and cutlery should be scraped to remove food debris, washed clean, rinsed in hot water and allowed to drain dry. Utensils and other implements used for raw meat, poultry or eggs should be cleaned and scalded with boiling water after use.

(6) *Local health contacts*. It is a wise precaution to contact the local environmental health department before camp. Advice can be given about water and milk supplies and action in the event of food poisoning (or other large-scale illness). Many health authorities advise that samples of all meals should be kept for at least 24 hours, so that if gastroenteritis occurs, relevant food samples are available for examination; specimen containers may also be supplied for samples of faeces or vomit. If there is any suggestion of food poisoning, a local doctor should be informed and he will pass on this information to the health department. On no account should a camp be closed and the campers dispersed, without the knowledge of the health department. It is their responsibility not only to report and investigate outbreaks, but also to trace and clear contaminated food, and people who may be excreting harmful organisms.

20

Food hygiene in the tropics

Although the basic factors or faults leading to food-borne disease, infection and intoxication, are the same in the tropics as in temperate zones, there are conditions which enhance the dangers. The high temperatures, sometimes high humidity, general lack of refrigeration, local habits, impure water, poor sanitary facilities and profusion of intestinal pathogens and parasites combine to establish diarrhoeal cases in large numbers. The majority of microorganisms responsible are worldwide in occurrence, but they are encouraged by the warm climate and by national habits. Most of this chapter is based on experience in India, but the situations described will apply to many countries with hot climates, impure water, poor sanitation and similar habits.

The general lack of facilities leads to malpractices which have unhygienic implications for the spread of infection. Although most village houses have toilets of the bore-hole type, some have septic tanks, primitive WCs or the open ground is used; in rural areas children may squat over drains. The hands are used for cleaning the perianal region with plenty of water, where available. Otherwise clay or earth is used and the hands washed later possibly in the village pond. Traditionally the left hand is used for toilet purposes and the right hand for eating, the food is gathered up from plates by hand. The hands are habitually washed before and after meals. Handkerchiefs are rarely seen, and the nose is blown on to the ground outside or into the nearest sink; similarly for the disposal of sputum, a contributory factor to the high incidence of tuberculosis.

Despite these shortcomings, bathing is considered important, and is carried out by splashing water over the body, as bathing in a pool of water is regarded by the Indian people as unhygienic.

Major hazards with regard to the diarrhoeal diseases are polluted water and lack of understanding about the importance of the time/temperature factor in the holding of food between cooking and eating.

Water

Water is probably the main source of infection, both by direct consumption and by contamination of food and the environment of food preparation areas. Large cities may have chlorinated supplies in most districts, but the smaller towns and rural areas are dependent on crude supplies.

In a busy industrial city of approximately one million inhabitants in the Punjab, results for *Escherichia coli* counts on water samples collected from many parts varied between less than 1/100ml (satisfactory) and more than 1100/100ml of water (unsatisfactory). In rural areas where untreated water is used, counts of faecal coliform bacteria (which include *E. coli*) may be in excess of 10 000/100ml. The pollution is consistent, but rises and falls according to the season, and is highest at times of heavy rain.

On the whole, the deep tube wells give good results unless the water is channelled through overhead tanks to give pressure for domestic purposes. An accumulation of silt in the tanks provides food for microorganisms, including coliform bacilli. In some houses the kitchen tap water passing through a tank is polluted, whereas the garden water, straight from the pipeline coming from the tube well, is almost pure. When the summer temperature rises to 43.3–48.8°C (110–120°F), there is increased consumption of water either alone or in fruit drinks.

Polluted water is used not only for drinks in various forms, but also in food preparation and cleaning operations. Dishes, utensils, cooking vessels and premises are washed in cold water with little or no soap or detergent. There are many small eating places sited in areas served with impure water. Water and sewage pipes become old, rusty and leaking; lack of initiative and funds delay replacement and repair. Incidents of hepatitis, typhoid fever and cholera are reported particularly from areas served by polluted water. The incidence of these diseases may not necessarily be reduced by the introduction of tubewells, especially if people continue to use polluted surface water for domestic purposes.

Market stalls are packed with vegetables and fruit, which are also piled on the ground and, to delay wilting in the extreme heat, they are sprinkled with water from stand pipes and hand pumps. Strips of coconut, displayed on stalls or in jars of water, slices of radish, cucumber and banana are eaten with salt from stalls. Oranges are crushed for fresh drinks or juice, and little attention given to cleaning out the pith that accumulates inside the equipment. Thus, water and hands are sources of contamination and infection.

In addition, flood water from land and river pollutes ground vegetables during the monsoon season of rain. Damp conditions followed by hot sun are conducive to microbial growth. Fungal growth on cut wheat is rapid under such conditions and the production of mycotoxins is inevitable.

Remote rural areas have their own foci of infection such as the village ponds which may be the only source of water and are thus used by animals and man for bathing, as well as for man's domestic purposes, laundry and even for drinking. The incidence of infection from protozoa, helminths, nematodes and other parasites must be high in such areas.

Water from village wells may be pure or polluted according to the care in construction of the walls and upper surround of the well. The more affluent villagers possess hand pumps for use by one family only. Most of the poor people share a common water pump constructed with a stand pipe sunk about 50 feet into the ground. The surrounding cement should be sound and hold the pipe firmly in position with the rim sufficiently high to protect the pump from flood and drainage water, including rain water and buffalo urine in the courtyard. If the cement surround is too low or cracked, sewage will leak into the well. The danger of pollution is enhanced by rain. Epidemics of cholera and other intestinal pathogens occur in villages dependent on well water prone to pollution in the rainy season. Conversely a parallel has been noted between areas of water shortage and diarrhoeal disease, usually a seasonal effect associated with dense fly populations. The availability of water is affected by the proximity of wells to villages. Similarly the feasibility of boiling water is affected by the availability of firewood, which may not always be easy to obtain. A further disincentive to the boiling of water is the time taken for water to boil on an open fire and the slow cooling rate at high ambient temperatures, which result in unboiled water being given to infants to drink in some countries. As the necessity for water increases in the hot weather, so does the consumption of ice, often made with impure water.

Ice

Ice is supplied from factories served by local municipal water supplies regardless of pollution, and is sold for the small-scale preparation of ice-cream and to vendors of cold drinks. It is doubtful whether special precautions are taken to clean the trays used for ice blocks or if the surfaces and equipment used for cutting the blocks into pieces are properly cleaned; the blocks are wrapped in cloth and crushed with a hammer. However unclean the surfaces and equipment are, the main source of contamination is the water supply, under the misapprehension that it is clean and harmless. Ice-cream, fruit juice, coloured water drinks and water alone are consumed in large quantities by everyone, whether rich or poor. Milkshakes are also popular and taken with ice. Crushed or grated ice made into ice balls and saturated with coloured sweet syrups is popular with school children and frequently sold from stalls outside schools. There is no check on the safety of the product.

Milk

In the large cities. Delhi for example, milk is pasteurized by modern equipment. In Ludhiana in the Punjab, there are two systems; pasteurized milk is delivered in sealed plastic bags, and raw milk from buffaloes and cows is collected in churns from dairies outside the city and carried on bicycles for delivery. Milk, even when pasteurized, is habitually boiled on arrival, but this practice of boiling cannot be guaranteed in the villages, where the dairy facilities are often in the open and far from clean. Milk may be adulterated with water, but little attention is given to the source of the water and microbial pollution.

Unhygienic practices have been observed in a study of milk collections and storage in rural West Africa, where the cows are milked in the open field and the milk collected in unwashed calabashes (gourd bowls). Dirt and dead flies are removed from the milk by skimming with a gourd spoon, and the milk is then filtered through a muslin cloth. The calabashes used for collecting the milk from the cows are left upturned in the field or covered with a raffia mat. The milk is taken to the village and sold either as fresh milk on the day of collection or as sour milk on the following day. *E. coli* was isolated in large numbers from freshly collected milk. The numbers were reduced by souring but were still unacceptable. These milks were fed without boiling to infants; impure water may be added.

Food

In Indian and similar cultures much attention is paid to the chemical adulteration of food, and little to microbiological contamination. Dilution of milk with water and the addition of prohibited colouring matter to drinks and sweets are not uncommon, so that sand may be found in flour and iron filings in tea. Fats reputed to be vegetable in origin may contain animal constituents, thus contravening the rules of the strict vegetarian. For such deliberate faults there are convictions under the law; many remain pending in the law courts.

Sampling and methods of examination in India follow those laid down by the Prevention of Food Adulteration Act, 1954, but there is no tradition of or legal requirement for routine microbiological examination of retail foods and there are no directives for sampling or means of guiding or punishing offenders. Nevertheless, there are microbiological specifications for a few foods such as milk, baby foods and tomato purée, and quality control for food factories includes the examination of pickles, meat in cans and ketchup in bottles; a few university laboratories are under contract to manufacturers. The routine sampling of food for microbiological examination is mostly confined to army laboratories and to the export trade, such as for fishery products and frogs' legs. Corporation and State laboratories, although largely concerned with chemical analysis, examine water, dehydrated milk and other baby food for microbiological control.

The Indian Standards Institution publishes standardized methods for the examination of milk, ice-cream, animal feeds and other products and also gives recommended limits for microbiological content. Methods are also published for the isolation and identification of intestinal pathogens such as salmonellae, shigellae, clostridia and vibrios, and for colony and coliform counts.

The regular examination of food from market stalls, except to check chemical adulteration and during the routine inspection of eating establishments, is rarely considered to be part of the inspector's work because of lack of sampling directives and enforceable standards. The food inspector has yet to be convinced of the value of surveillance and persuasion to improve hygiene, rather than reliance on statutory condemnation of food and premises. The contamination of food for infants is a major concern, since acute and chronic diarrhoea is a predominant cause of infant mortality. Most village mothers breast feed their babies, but there is little guidance on the time and frequency of feeding. As a supplementary measure diluted 'top' milk is given. Unfortunately, the bottles and teats are rarely sanitized satisfactorily and left-over milk may be kept for the next feed. Unclean bottles are thought to contribute to high mortality rates. Diarrhoea is four times more common in babies on 'top' feed than in those who continue with breast milk; also, the appetite is diminished and there is weight loss. Eight-month-old children are allowed to crawl around the courtyard where they are likely to pick up many microorganisms, particularly parasites. After weaning, chapatis may be the only food items available and pieces dropped on the floor will be picked up and eaten; if softened in tea, the chapati will be easier to consume. Food for adults is mostly made daily; left-over food may be in covered pots and is usually reheated before it is required. Eating and cooking utensils and vessels are scoured with ash or mud, washed with cold water and dried in the sun.

In the homes of professional people, working wives usually prepare meals which are consumed throughout the day and the food is not necessarily thoroughly re-heated before being eaten. The various dishes made for guests and parties require early preparation and refrigeration is not assured.

The cooking of food for large gatherings at weddings and other festivities commences in the morning, often in the open, shaded from the sun; storage throughout the day will be at atmospheric temperature. Bacterial growth may be enhanced when cooked food is stored covered near the cooking range. There is an added risk for the bride's party at Hindu weddings, because the bridegroom and his relatives and friends eat first and the bride's party 1, 2 or more hours later. It is not unknown for the bride and her friends to suffer food poisoning while the bridegroom and his party remain well. There may be many hundreds of guests at large weddings and vast quantities of food prepared. The amount of food eaten is a major factor in food-poisoning incidents and those with small appetites may escape with minor symptoms.

The milk casein/sugar/syrup sweets are a feature of India and other eastern countries. They are prepared on a large scale for shops and for special occasions such as festivals and weddings; they may be stored in dark rooms unfit for habitation but a harbour for rats, mice, cockroaches and flies. Staphylococcal enterotoxin outbreaks from these products are the most frequently reported cause of food poisoning in India. There is little follow-up of outbreaks and few records of the investigation of premises and personnel involved in the process of preparation. It is assumed that the main hazard is contamination of the milk concentrate from hands, other skin surfaces and nose, particularly in hot weather when the food handlers are perspiring. Once the mix is contaminated with staphylococci, growth and toxin production will continue until reduced water content prohibits multiplication; even so the static organisms may continue to produce enterotoxin slowly. Subsequent treatment will not destroy the organisms or the toxin in 'barfee', for example. Some casein preparations such as 'gulab jamun' are dipped in hot oil and soaked in sugar syrup; 'rasgulla' are similar, but not fried. Even gulab jamun sweets have been involved in outbreaks of staphylococcal food poisoning with cases distributed over wide areas. Shops in the cities display the attractive and palatable sweet meats in glass cases at atmospheric temperature; further growth on or in the finished product is unlikely or slow, because of the reduced a_w and sugar content.

Salty fried preparations such as 'samosa' are dependent for safety on the time allowed in hot fat. The cooked filling of meat and vegetable or vegetable alone is wrapped in a layer of wheat flour pastry before frying. The filling is mixed in vats and may be contaminated with spoilage organisms and pathogens; flora should be killed by the heat of frying, except some spores. Toxins may or may not be destroyed. Although suspected, there is no confirmed evidence that these products cause food poisoning; microbiological examination has shown small and occasionally large numbers of Gram-positive bacilli by direct smear, but few organisms have been grown in culture.

'Pakoras' are deep fried crispy gram flour products, which often contain spinach or sprigs of cauliflower and onion. They are small, fried well and unlikely to be contaminated. The safety of bread 'pakoras' is more suspect, as the sandwich filling mixed in large vats is layered by hand on to bread slices to make a sandwich, which is deep fried. Again the safety of the process must depend on the exposure time in hot oil.

There is a misapprehension that curries are safe whatever the storage time and temperature after cooking. Although the basic ingredients are cooked in hot oil, spores will be present in the spices which are added at a later stage, and some will survive subsequent heat treatment. Ground spices may contain large numbers of spores resistant to drying and to the antibacterial properties that high concentrations of some spices may possess. The heat shock received in cooking enables many spores to germinate into vegetative cells, which multiply when the temperature of the cooked food drops to a convenient level. Large and small volumes of curried meat and vegetables kept warm in tanks or wells for conferences and in vessels at home will support the growth of *Clostridium perfringens*. People may come for meals at irregular times and latecomers will ingest large numbers of *C. perfringens* with their meal. Some of these organisms survive passage through the stomach, colonize the intestine, sporulate and release the toxin responsible for diarrhoea and pain.

Staphylococci and salmonellae have been isolated from spices, but they are likely to be destroyed by minimal heat. Aflatoxin has been identified in certain spices prepared under conditions allowing mould growth at some stage. Much attention is given in India to quantitative and qualitative assay methods for mycotoxins, which are known to be major hazards in cereals. Warm climatic conditions, especially when rain is followed by hot sun, encourage mould growth in cut wheat and maize.

The cereals, wheat, maize, pulses and rice make up the bulk of the Indian diet; the majority of people are vegetarians and the cereals provide protein. The proportion of each cereal eaten varies according to the diet in different parts of the country; wheat flour in 'chapatis', 'puris' and 'paratha' together with pulses (gram or dahl) make up the staple diet in the Punjab, while in the south of India, rice and rice flour are eaten flavoured with coconut. In Calcutta, more bread is consumed than in the rest of the country, perhaps because of the British influence in the days of occupation.

In the Gambia, millet gruels and boiled rice are often used for infant feeding. The food may be prepared at breakfast and lunchtime and subsequently fed to the child on demand throughout the day. It is not uncommon for food left over from the adults' evening meal to be offered to the child at breakfast on the following morning.

Millet gruels are prepared by pounding millet flour with water followed by simmering in a metal bowl on an open fire. Cooking is minimal, as prolonged cooking causes the gruel to become gelatinous and unpalatable, the water used in the preparation may be heavily contaminated with faecal bacteria and the millet with *Bacillus cereus*. Heat penetration of the gruel is poor and a proportion of these bacteria has been found to survive cooking. A further hazard arises when the cooked product is transferred to a gourd bowl which has a rough fibrous inner surface that is difficult to clean and may be washed with polluted water.

Naturally fermented maize gruels and porridges are also used as weaning foods in some African countries, for example, Ghana. The low pH prevents the growth of *Enterobacteriaceae*, which provides a safety factor.

Dehydrated products are safe, except for the possible introduction of microorganisms from hands and insects; the low a_w inhibits growth but as soon as water is added a growth medium is available.

Powdered baby milks constitute a risk if not prepared correctly. Preparation with unboiled polluted water is dangerous, and even if boiled water is used, the milk is transferred to a dirty bowl and left at ambient temperature for several hours.

The various flours contain spores, particularly those of *B. cereus* and also those of anaerobes such as *C. perfringens*, so that there is the possibility of spore outgrowth in rehydrated mixes, for example multiplication of the vegetative cells and toxin production from *B. cereus* in cooked rice stored at atmospheric temperature. *B. cereus* has been isolated in large numbers from baked cakes of moist consistency in the centre and sold wrapped in cellophane. Even *C. perfringens* and *Staph. aureus* have been isolated from a cake stored in the refrigerator and eaten from time to time. Two persons were affected, after consuming the cake, with symptoms of staphylococcal enterotoxin food poisoning in one person and *C. perfringens* food poisoning in the other. Further investigation is necessary on the microbial, including spore, content of flour and the time/temperature processes in the bakery.

The fermented product 'dahi' is similar to yoghurt except for the organisms used for fermentation. *Lactobacillus bulgaricus* and *Streptococcus thermophilus* are used for yoghurt, and *Lactobacillus bulgaricus* and various lactic streptococci for 'dahi' which is prepared daily in the home. In the Punjab it is eaten with curries, either plain or mixed with chopped onion, cucumber, tomato or other vegetable. In the south it is more often eaten with sugar as a sweet dish. During fermentation the pH drops rapidly and thus the growth of pathogens, such as staphylococci and salmonellae, will be inhibited. When washed chopped vegetables are added the pH may rise and allow the growth of contaminants from the raw materials. Occasional high coliform counts have been found in samples; care must be taken in preparation and storage.

'Lassi' is the whey from 'dahi' and buttermilk is the whey residue from butter. They are similar products with a pH of approximately 6.0; salt may be added to lassi. Both lassi and whey drinks are nutritious and popular, and are sold bottled commercially. They also are good media for bacterial growth and should be sold and stored refrigerated.

A popular mixture of various fruits including soft bananas, bruised apples and guavas blended together wth an extract of tamarind and sprinkled with salt, pepper and spices is served on small plates and picked up with birch twigs. 'Gazrela', also, which is a blend of shredded carrot, sugar and ghee with dried fruit and nuts, may be seen in large pans in the bazaar for sale without safety checks.

Education and training

Some food handlers have little education and are unable to read or write. The preparation of food is done by hand without utensils. The washing of hands may be infrequent, soap may or may not be available and the water is likely to be cold. Nails will be neglected and dirty without use of a nailbrush, which even if provided will quickly disappear. In the more sophisticated kitchens, some surfaces are covered with stainless steel, but cutting surfaces will be wooden. Enormous counts of Gram-negative organisms have been obtained from thin cutting boards used both for raw and cooked foods in mess kitchens; wooden sinks are still found.

Hospital kitchens provide special diets for patients and cook for staff members and guests, the relatives buy food and drink for the patients from nearby shops and stalls. Students, nurses and junior doctors eat in their own canteens. In some instances the kitchens are inadequate for the needs of those they aim to serve, and there is no overall supervision with regard to hygiene.

The habit of eating out is becoming more prevalent as incomes increase; many restaurants are without proper facilities for storage of food supplies and cooked foods. Celebration parties are particularly at risk, because of the early preparation of food and the lack of cooling and refrigeration facilities. There is no inspection, except in a few enlightened areas.

In large cities, Catering and Food Craft Institutes hold courses for students and the community, mainly housewives. Subjects include the principles of nutrition, simple food microbiology and food hygiene. There are practical classes on cookery and the fermentation processes for pickles and wines, for example, so that there are opportunities for teaching hygiene in practice. The Punjab Agricultural University has a Department of Home Economics.

Certain large Institutes of Public Health and Hygiene and of Food Technology, State Public Health Laboratories and veterinary complexes with facilities for Veterinary Public Health courses are active in teaching graduates, technicians and occasionally sanitary inspectors, but information about food hygiene rarely if ever reaches the general public.

The World Health Organization together with local university or other institutional personnel occasionally organize courses and workshops on food microbiology and food hygiene for national attendance from wide areas, yet there seems to be little evidence that the information is passed on by those attending such courses. Housewives, of whatever class, have little knowledge of the faults in kitchen practice that lead to food-borne disease. The larger schools with senior classes sometimes include home science in their curriculum, encompassing cooking, nutrition and hygiene, but the majority of schools would not include such training.

Combined courses on food hygiene, health education, health administration and microbiology should be planned by Institutes of Hygiene and Public Health, including Veterinary Public Health and State Public Health Laboratories. Catering and food industries could take part in planning the courses and giving lectures to microbiologists, graduates and technicians concerned with food microbiology, sanitary inspectors, food handlers, students in schools and universities and particularly the general public.

Factors such as the necessity to wash hands before and after handling raw foodstuffs as well as after visiting the toilet, for the scrupulous cleanliness of surfaces, utensils and vessels, and for the provision of facilities for rapid cooling and cold storage may not be recognized. The use of modern appliances, pressure cookers, ovens and grills at home as well as in the kitchens of eating establishments is growing, but an understanding of their proper use may not be clear.

An additional problem is that the correct methods for freezing, thawing and storage of food are not always understood and bulks of cooked food may be placed in cupboards.

Because of fluctuations in and irregular supply of electricity, refrigerator temperatures need frequent checking. The current may be off for several hours at a time, so that the foods stored in the refrigerator should be checked carefully for signs of spoilage including mould growth. Frozen food will thaw and judgement must be made about its acceptance for consumption or refreezing.

It may not be fully appreciated that vegetables, fruit and eggs should be washed only immediately before use, vegetables and fruit rot more quickly if stored wet and eggs lose their protective skin so that the shells become more permeable. People may not realize that unrefrigerated food should not be left packaged in plastic wrapping because the condensed moisture inside the bag will encourage spoilage and rot particularly in warm climates.

Again because of the high ambient temperature, special care is needed not to prolong the time between cooking and eating, between cooking and refrigeration and between refrigeration and eating. Only the quantity sufficient for the required meal should be removed from refrigerated food.

Teaching must include factors about conditions in the rural areas because they are such that the usual rules of hygiene cannot be applied. Certain high-caste communities remove shoes and change clothes before entering the kitchen or eating place; yet the floors may be smeared with fresh dung from cows or buffalos. Cooking takes place with primitive apparatus – brick or mud kilns or over an open fire, using natural fuel, coke or wood. Fuel can be made by hand from cow dung, the pats are dried on the outer walls of dwellings. Calor or gobar gas may be available; gobar gas is collected from the fermentation of dung.

The food will be mainly cereals, pulses and vegetables grown locally. Both preparation and eating are done by hand without utensils except for wooden stirring rods.

Food is prepared for one day at a time, and while prolonged boiling in water (or more usually, oil) will ensure safety for the consumption of freshly cooked food, this is not effective if food is stored for longer than 2–3 hours. Because there is no refrigeration milk is almost invariably boiled, even the pasteurized milk in plastic packs which has more recently come to urban areas. Fermented milk products are safeguarded by their low pH.

In some parts of the world, people do not have tables, so food is prepared on the floor inside or outside the home, exposed to animals. The garbage pit may be a short distance away and it is covered when full. Also, rubbish may be thrown into the courtyard or road. Water is drawn by hand pump or bucket, from deep or shallow wells, or even from the local ponds used for both human and animal purposes.

Infants and young children suffer from gastroenteritis more than adults, as their food may be left standing at atmospheric temperature, and the children play with the surrounding earth. Excreta will be dropped in fields or primitive types of latrine. Amoebae and giardia from water and parasites such as nematode worms from animals and man infest the surroundings and are recycled over and over again.

Basic habits in various countries predispose towards particular diseases. For example, the consumption of raw or uncooked meat and fish may give rise to taeniasis, trichinosis, toxoplasmosis and salmonellosis. *Diphyllobothrium latum* may originate from undissolved fat. Consumption of wild animal meat (bear or boar) can give rise to trichinosis. Mosquito-borne diseases, such as malaria, dengue and other viral infections will be prevalent together with a high incidence of tuberculosis, due to dark, cramped and overcrowded living conditions. Malnutrition is an important factor; it is common and leads to increased susceptibility to disease.

Perhaps in its full context the teaching of food hygiene in the tropics should include plans for construction of wells, cooking equipment, sanitary arrangements and garbage disposal. Without well-designed facilities it is difficult to apply the hygienic principles, which ultimately affect the safety of food and drink and the health of the people.

Appendix A

Suggested lecture material

The following notes and headings for talks on food hygiene have been useful for various groups studying the safety and keeping quality of foodstuffs. The course of talks with suggestions for charts was compiled by Miss C.F. Scott, and the late Miss I.J. Martin, formerly of the School Meals Service.

Types of food poisoning (Session 1)

The term food poisoning describes a disturbance of the gastrointestinal tract with diarrhoea, nausea and vomiting with or without fever. It follows consumption of food contaminated with certain bacteria, viruses or occasionally chemicals.

The time interval between eating the food and onset of symptoms (incubation period) can be as short as 1 hour to 24–48 hours or even longer depending on the organism (agent) responsible and also the number of organisms or amount of toxin ingested; the toxins are poisonous substances produced by certain bacteria as they grow in food or in the intestine. The age and condition of the individual consumer are factors of importance with regard to the time of incubation, the symptoms and their severity. Sharp attacks of vomiting within 1 hour or less of eating food may be due to bacterial toxins or to chemical substances, such as zinc, copper and tin, which may be dissolved from containers by acid foods and alkaloids or even pesticides and herbicides.

Description of bacteria and other microorganisms

Microorganisms, as the name suggests, are minute living cells, varying in size, shape and mobility, and they are visible only through a microscope. Viruses are among the smallest organisms and require special means, the electron microscope, to make them visible; unlike bacteria they can only proliferate in living cells. Bacteria can be found everywhere, most are harmless and some are useful to man, but a small proportion are dangerous and responsible for disease in man and animals and other living things.

Chart Causes of food poisoning

(1) Foods	Plants, fungi, some shellfish	
(2) Allergies		
(3) Microorganisms		
	BACTERIA	
	Small dose*	*Campylobacter* spp Dysentery bacilli
	Large dose† (usually after growth in food)	*Salmonella* spp *Staph. aureus* *C. perfringens* *C. botulinum* *B. cereus* *V. parahaemolyticus* *E. coli*
	VIRUSES	
	Small dose*	
(4) Chemicals		
	Zinc Copper Tin Alkaloids Pesticides	
(5) Parasites		
	Trichinella *Taenia*	

* Small dose = few organisms only
† Large dose = thousands to millions of organisms

Given suitable conditions for growth, such as, nutriment, temperature and time, bacteria will divide into two every 10–30 minutes, the time of division may be shorter if conditions are favourable. One organism can develop into many millions within 10 hours, and colonies of bacteria are visible on agar media.

The fungi grow more slowly and their hyphal filaments and also sporing bodies are readily visible on foodstuffs.

Allergic reactions may be mistaken for food poisoning; they occur in certain persons sensitive to protein constituents of particular foods, such as shellfish and milk. Parasites in animals may be present in meat, for example *Trichinella* (see chart).

Conditions which affect growth (Session 2)

Bacteria increase by doubling in number every 10–30 minutes depending on the suitability of conditions; growth and multiplication require the following factors:

(1) *Food*, the vehicle of infection, most foods eaten by man, support and encourage the growth of bacteria: meat and meat products, poultry, eggs and egg foods, milk and milk products, seafood. In the laboratory special media, either liquid or set with agar, are required; they may be formulated to encourage and select the growth of certain organisms, such as those causing disease, while suppressing the growth of others.
(2) *Optimum temperatures* for bacterial growth are 20–45°C (68–113°F), outside this range bacteria that infect man and animals grow slowly or not at all. The most favourable temperature for food poisoning is 37°C (98.4°F) similar to that of the human body. Most spoilage organisms grow at lower temperatures. Bacterial spores survive temperatures that kill vegetative cells, some survive boiling; spores can remain dormant in dust and soil indefinitely. The toxins produced in food by bacteria vary in their susceptibility to heat, some are destroyed readily while others may survive boiling or even autoclaving.
(3) *Time*, when food and temperature conditions are good for growth, time is needed for division of the cells and for maximum rates of multiplication.
(4) *Moisture*, water is essential for the survival of living cells although most bacteria can remain alive indefinitely when dry as in powdered food, particularly in the sporing form.
(5) *Atmosphere*, most bacteria require air to grow actively, and the growth of many is enhanced by carbon dioxide. Some bacteria, the anaerobes, will grow only in the absence of oxygen. Between the extremes there are various atmospheric conditions for the optimum requirements of different bacteria.

It should be noted that bacterial agents of food poisoning may grow in foods to large numbers without changing the character of the food by appearance, odour or taste. The spoilage organisms are mostly proteolytic and give rise to 'off' odours and flavours.

Conditions which discourage growth

(1) *Salt* (as a preservative) and *sugar* (syrups and honey) in high concentrations will repress growth.
(2) *Heat*, bacteria are destroyed by temperatures above 63°C (145.4°F) if held for a sufficient length of time in cooking processes, due to irreversible protein changes (coagulation of whole egg), and in the pasteurization of milk; spores may not be killed by these treatments.
(3) *Cold*, extreme cold will not kill all bacteria, but most will cease to multiply under refrigerated storage and there will be no growth in the deep freeze.

(4) *Dehydration*, the moisture content can be reduced to levels below which microorganisms cannot grow although they may survive in a static condition.
(5) *Fat*, lard, butter and margarine will not permit growth although a few organisms may survive.
(6) *Acid*, as acidulants in fermented foods, pickles and sauces acts as preservative.

Summary

Factors which encourage bacterial growth are warmth, high moisture low salt, (with the exception of staphylococci, which have a somewhat greater salt tolerance than other pathogenic bacteria), low acid, low fat and moderate sugar. Growth is repressed by extremes of heat, cold and dehydration and by high concentrations of salt, sugar, acid and fat.

Growth	No growth
Warm	Too hot or too cold
Moist	Dry
Nutritious – meat, milk, egg	Non-nutritious – high sugar, high salt, high fat
Neutral pH	Acid
Time	No time

Food poisoning bacteria (Session 3)

The main types of bacteria causing food poisoning are:

(1) *Salmonella* – infection by living bacteria in food.
(2) *Staphylococcus aureus* – toxin from growth of bacteria in food.
(3) *Clostridium perfringens* – toxin released in intestine from living bacteria swallowed in food.
(4) *Clostridium botulinum* – toxin from growth of bacteria in food.
(5) *Bacillus cereus* – toxin from growth of bacteria in food.
(6) *Vibrio parahaemolyticus* – infection by living bacteria in food.
(7) *Escherichia coli* – infection by living bacteria in food.
(8) *Campylobacter jejuni* – infection by living bacteria in food.
(9) *Aeromonas, Yersinia* and *Listeria* occasionally.

Salmonella

This is the commonest cause of food poisoning in the British Isles and elsewhere and the most serious.
Source. Carried in the human and animal intestine, and excreted in stools. Likely to come into the kitchen on raw foods of animal origin; for example, meat, poultry, pet meat, sausages, egg products, or to be brought in by human and animal excreta and fertilizers. Insects, birds, vermin and domestic pets may play a part.

Cause of illness. The living bacteria in cooked or uncooked food. The organisms usually need to multiply to large numbers to cause illness.
Illness. Begins 6–36 hours after eating contaminated food. Symptoms include fever, headache, abdominal pain, diarrhoea, and vomiting. The illness lasts from 1–7/8 days, and can be fatal in the elderly, very young or sick people. The typhoid bacillus is a member of this group, carried by humans only and found in stools, sewage, water and food.

Kitchen control
(1) Kitchen hygiene: (a) separation of raw from cooked foods, particularly meat and poultry, using different surfaces and equipment (and personnel, if possible), to prevent cross-contamination by boards, benches, cutting and mincing machines, cloths and kitchen tools; hands can also pass organisms from food to food (raw to cooked); (b) thorough cleaning of all surfaces, equipment and tools.
(2) Care of personal hygiene by washing hands before and after handling food – especially raw meat and poultry.
(3) Cold or hot storage of food to prevent multiplication of bacteria.

Staphylococcus aureus

Staphylococci are another common cause of food poisoning.
Source. Comes mainly from the skin of human carriers, from the nose, hands, throat, boils, carbuncles, whitlows, styes, septic lesions, burns and scratches. Also comes from the milk of cows and goats; the organisms and toxin can be carried over to cream and cheese made from raw milk. *Staph. aureus* is found on poultry also.
Cause of illness. A toxin produced by the bacteria as they grow in food handled after cooking.
Illness. Begins 2–6 hours after eating contaminated food. Symptoms include acute vomiting, pain in abdomen, diarrhoea and sometimes collapse; there is no fever. The illness lasts not more than 24 hours, and is seldom fatal.

Kitchen control
(1) Care of personal hygiene, washing and drying of hands; it is impossible to sterilize the hands, therefore cooked food should not be touched.
(2) Cold storage of food to prevent multiplication of bacteria.
(3) Kitchen hygiene to prevent bacteria becoming numerous in equipment, boards, cutting and mincing machines, cloths, Savoy bags, and other tools; clean thoroughly after use.

Clostridium perfringens

C. perfringens food poisoning occurs frequently in the UK and elsewhere.
Source. Found in the human and animal intestine, soil, dust, flies and other insects. Raw meat and poultry and some dried products are frequently

contaminated with C. *perfringens* spores, which survive normal cooking temperatures. The organism grows without oxygen.
Cause of illness. Toxin released in the intestine by ingested bacilli. Large numbers of bacilli grow from spores which have survived in food. Important foods are cooked meat and poultry dishes allowed to cool slowly and stored warm.
Illness. Begins 8–22 hours after eating contaminated food. Symptoms include abdominal pain and diarrhoea, but rarely vomiting. The illness lasts from 1–2 days; it may be fatal in elderly and sick people.

Kitchen control
(1) Consideration of cooking methods and size of joints.
(2) Rapid cooling of cooked food to prevent multiplication of bacteria.
(3) Cold or hot storage of cooked food to prevent multiplication of bacteria.
(4) Separation of raw from cooked food to prevent cross-contamination.
(5) Kitchen hygiene; thorough cleaning of all surfaces and equipment, boards, cutting machines, cloths, and other tools after use. Regular removal of soil and dust from vegetable store and preparation area and dust from cereals in dry goods store.

Clostridium botulinum

C. botulinum is an uncommon but serious and often fatal cause of food poisoning.
Source. Found in soil, meat and fish in some areas; the spores survive cooking and other processes. The organism grows without oxygen, and may be found in spoiled cans of imperfectly preserved food.
Cause of illness. A toxin produced by the bacilli as they grow in food.
Illness. Begins 12–96 hours after eating contaminated food. There is fatigue, headache and dizziness. There may be diarrhoea at first but not later. The nervous system is attacked and vision and speech are disturbed. Death often occurs within 8 days, unless antitoxin is given soon after onset of illness.

Kitchen control
(1) Avoid home preservation of meat, poultry, game and fish except by freezing.
(2) Avoid eating raw and fermented fish.
(3) Discourage smoking and curing of fish in certain areas of the world where *C. botulinum* is common.
(4) Careful inspection of cans and contents.
(5) Care with curing solutions, concentrations of salt and nitrite/nitrate.

Bacillus cereus

B. cereus and other organisms of the *Bacillus* group can be a cause of food poisoning.
Source. Found in soil, dust, cereals, spices, vegetables, dairy products and

many other foods; some spores may survive cooking. The organism grows best with oxygen.
Cause of illness. Toxins produced by the bacilli as they grow in food.

Illness
(1) *Vomiting-type*. Begins 1–6 hours after eating contaminated food, usually a cooked rice dish. Symptoms include nausea and vomiting, and sometimes diarrhoea a little later. The illness lasts not more than 24 hours.
(2) *Diarrhoeal-type*. Begins 8–16 hours after eating contaminated food – a wide variety of foods have been incriminated. Symptoms include diarrhoea and abdominal pain, but rarely vomiting. The illness lasts not more than 24 hours.

Other *Bacillus* species

Members of the *B. subtilis–licheniformis* group are the species which have mainly been involved. Symptoms are either diarrhoea 8–12 hours after eating contaminated food or rapid onset of vomiting (less than 1 hour). Incriminated foods include minced meat, chicken, rice and various meat and pastry products (pasties, sausage rolls).

Kitchen control
(1) Avoid cooking large bulks of rice and other cereals for storage, cook in smaller quantities.
(2) After cooking keep all foods containing cereal grains or flour hot (above 63°C, 145.4°F) or cool quickly and refrigerate (4°C, 39.2°F).
(3) Kitchen hygiene; thorough cleaning of all surfaces, equipment, and tools. Regular removal of cereal dust from storage and preparation areas.

Vibrio parahaemolyticus

V. parahaemolyticus food poisoning is frequently reported in Japan; and occasionally recorded in other countries.
Source. Found in sea creatures and coastal waters (usually warm).
Cause of illness. The living bacteria in raw and cooked sea foods such as fish, prawns, crabs and other shellfish.
Illness. Begins approximately 15 hours after eating contaminated food. Symptoms include acute diarrhoea, abdominal pain, and vomiting with fever. The illness lasts for 2–5 days or more.

Kitchen control
(1) Care that cooking methods destroy the organisms.
(2) Separation of raw and cooked food to prevent cross-contamination; by hands, utensils, equipment and surfaces.
(3) Rapid cooling to prevent growth of vibrios.
(4) Cold storage of food to prevent growth.
(5) Kitchen hygiene; thorough cleaning of all surfaces, equipment and tools after use.

Escherichia coli

Certain serotypes of *E. coli* cause gastroenteritis sometimes associated with traveller's diarrhoea.
Source. Carried in the human and animal intestine and excreted in stools. Likely to come into the kitchen in the human and animal excreter and in raw foods of animal origin; for example, meat, poultry and pet meat. Insects, birds, vermin and domestic pets help to spread the organisms.
Cause of illness. The living bacteria in food, and toxins also.
Illness. Begins 18–48 hours after eating contaminated food (or drinking polluted water). Symptoms include pain and diarrhoea and sometimes pyrexia and vomiting. The illness lasts from 1–5 days.

Kitchen control
(1) Care of personal hygiene by washing hands before and after handling food – especially raw meat and poultry.
(2) Care of water used for washing salad and vegetables.
(3) Separation of raw from cooked food to prevent cross-contamination.
(4) Kitchen hygiene; thorough cleaning of all surfaces, equipment and tools after use.
(5) Cold or hot storage of food to prevent multiplication of bacteria.

Campylobacter

C. jejuni gastroenteritis is reported mostly as sporadic cases, but outbreaks do occur.
Source. Wide variety of animals. Infection can be spread from person to person and from animal to person. Likely to come into the kitchen in the human and animal excreter and in raw foods of animal origin, particularly poultry. Raw milk has been responsible for a number of outbreaks. Untreated natural water is also a source.
Cause of illness. The living bacteria in food. It is thought that illness can be initiated by a low dose of the organisms.
Illness. The incubation period is 3–5 days, with a range of 1.5–10 days. Symptoms include diarrhoea preceded by fever and sometimes malaise, headache, dizziness, backache, myalgia, abdominal pain and rigors. The illness lasts from 2–3 days; relapses can occur, but milder than the first attack.

Kitchen control
(1) Kitchen hygiene: (a) separation of raw and cooked foods, using different surfaces, equipment and personnel to prevent cross-contamination; (b) thorough cleaning of all surfaces, equipment and tools.
(2) Care of personal hygiene by washing hands before and after handling food – especially raw meat and poultry.
(3) Thorough cooking of food.
(4) Avoid the use of raw milk.

Other organisms, such as certain streptococci, *Proteus* and those in the Providence group, are sometimes suspected of causing food poisoning when reaching abnormal numbers in food. Control measures are similar to those described for *Salmonella* and *E. coli*.

How bacteria reach food – contamination (Session 4)

Chart Cycle of infection and contamination

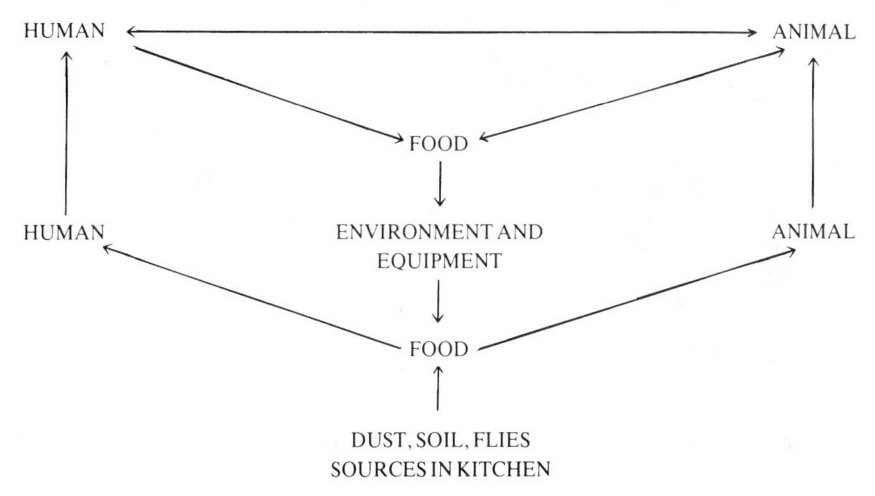

Food handlers and other living things

Foodstuffs, e.g. meat, poultry and sea foods

Surfaces and equipment

Means by which bacteria reach food – contamination. The foods that encourage the growth of bacteria include raw and cooked meat and poultry, foods with meat as a base, soups, stocks, gravies, made-up meat dishes; also eggs and egg products, milk and milk products.

Raw foodstuffs. Disease-producing organisms can be carried in the animal intestine, with or without harm to the animal, and they may be spread to the carcass during slaughter and dressing. Thus meat and poultry can reach the kitchen already contaminated with salmonellae, clostridia and campylobacters. Contact between carcasses and between meats and other foods will spread bacteria; food handlers may pick up food-poisoning organisms from the foods they handle.

Food handlers. Careless handling during transport, manufacture preparation and service may add and spread bacteria to foods. Hands can transfer food-poisoning germs from raw to cooked foods and to utensils. Personal bacteria from the nose, mouth, skin, stool and hands can contaminate food.

Environment. Unclean kitchen surfaces, equipment and utensils can harbour bacteria and passage contamination to other foods, especially from raw to cooked food. Flies, vermin, cats, dogs, birds and other creatures can carry bacteria to food. Dust, soil and dried excreta with organisms and spores can reach food.

Prevention of food poisoning

(1) Prevent spread of contamination in the kitchen:
- (i) separation of raw and cooked food to avoid cross-contamination
- (ii) care on part of food handler;
- (iii) cleanliness of kitchen environment.

(2) Prevent bacteria already in food from growing and spreading:
- (i) cold storage facilities must be adequate and efficiently maintained
- (ii) avoid long storage in warmth – close time gaps between preparation and serving;
- (iii) provide facilities for rapid cooling.

} In the retail store and in the kitchen

(3) Cook to destroy most bacteria (but not spores), with particular attention to:
- (a) careful preparation and adequate cooking; if possible, serve at once. If not:
- (b) keep hot; or
- (c) cool rapidly and store cold;
- (d) avoid recontamination.

Measures used to control bacterial growth in foods

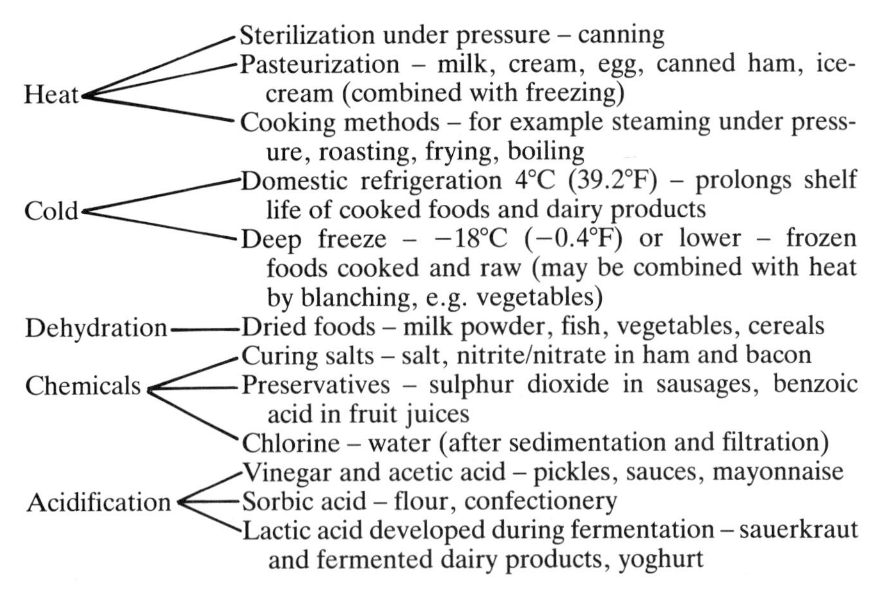

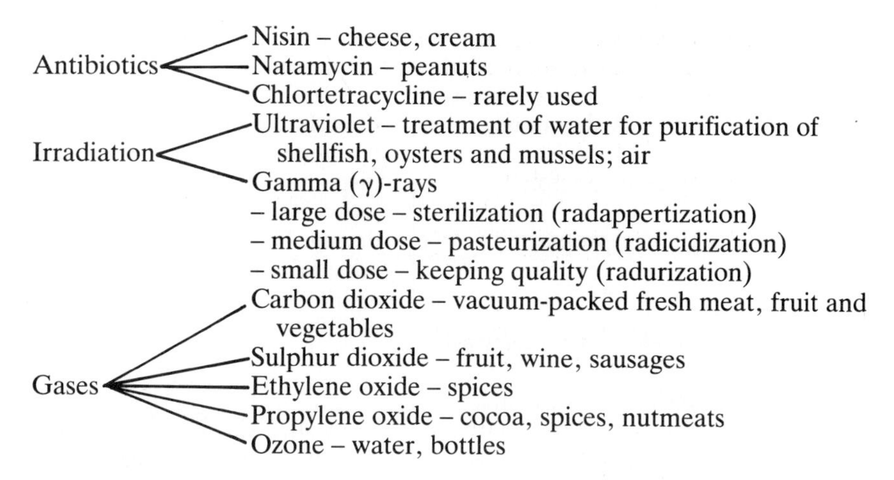

Foods and liquids that are safe

Water, filtered and chlorinated, and bottled mineral waters.
Milk, pasteurized, ultra-heat treated or sterilized.
Ice-cream, mix pasteurized, cooled rapidly and kept cold until frozen; manufacture controlled by regulations.
Liquid egg, pasteurized (frozen process controlled by regulations).
Canned food, with exceptions such as large cans of ham, which are not sterile and must be kept cold.
Other foods, such as bread (except for mould and growth of organisms of the *Bacillus* group), flour (except for attack by pests such as weevils), jams, syrups and honey, pickles and acid fruits, fats.

Personal hygiene (Session 5)

It is the responsibility of the food handler to take scrupulous care that personal bacteria are not added to food. Bacteria from the following sources can be passed to food by the hands:
(i) secretions from the nose, throat, and skin; dust, dandruff and loose hairs from the scalp;
(ii) excreta from the bowel;
(iii) other people's secretions and excreta;
(iv) liquor from raw meat and poultry and from other foods, powdered foods;
(v) utensils and equipment;
(vi) cloths – tea towels, dishcloths, meat cloths.

Personal hygiene can break the chain of infection in the following ways:

Hands
Hands should be kept in good condition, free from roughness:

(1) *Washing*. Thorough washing with hot water and soap is essential after using the WC, immediately before, during and after food preparation,

after handling raw meat or poultry, and after using a handkerchief. Hand washing should not be done in kitchen sinks.

Hand-wash basins should be:

(a) in or adjacent to a toilet

(b) in the kitchen preparation area

} they should be supplied with hot and cold water, a soap dispenser, nail brush and an individual method of hand drying. Hand cream with added disinfectant should be available.

Hands are NOT sterile after washing.

(2) *Nails.* Keep short and clean.
(3) *Cuts and abrasions.* Cover with a non-porous dressing or Newskin and a fingerstall or glove if necessary. Persons with septic lesions should not work with food until the infection is completely healed.
(4) *Fingers.* Do not lick when tasting or serving food or separating sheets of paper.

Soaps and hand creams.
The hands must be kept in good condition. Cracks, blemishes, and broken nails will all harbour bacteria, and a good hand cream containing a bactericidal substance should be used after washing with soap from a dispenser.

Good habits
Take care not to touch the nose, hair or face; avoid sneezing and coughing near food; maintain a high standard of bodily cleanliness.

Smoking
Smoking should be prohibited in food premises.

Handkerchiefs
Handkerchiefs carry infection from nasal secretions. They can contaminate hands and clothing and, indirectly, foods. Paper tissues are useful and easily disposable.

Hair
Hair should be well covered with cloth, net or paper caps.

Clothing
Clean overalls and aprons are essential; changed regularly and frequently.

Health
Illness, particularly diarrhoea, vomiting, throat and skin infections, should be reported. Severe septic conditions should also be reported.

Purchasing and storage (Session 6)

Purchasing
The standards of hygiene maintained by the supplier should be noted, such as cold and chilled storage, separation of raw and cooked foods,

handling of unwrapped raw and cooked foods, cleanliness of premises and equipment and type of equipment. The amount of food purchased should not be greater than can be stored in the available deep freeze cabinet or refrigerator (4°C, 39.2°F).

Perishable foods
Raw and cooked meat and poultry, milk, cream and fish should be bought in quantities sufficient for 1 day only unless there is ample refrigerated space. Meat and fish should not be refrigerated for more than 3 days.

Non-perishable foods
Dry goods, preserves and canned foods should be bought in reasonable quantities. Avoid overstocking; there is danger from vermin and deterioration where storage facilities are poor. These products should be stored in a dry, well-ventilated storeroom, and care should be taken to rotate stocks.

Storage of perishable foods
The following foods must be refrigerated: raw and cooked meat and poultry, dairy products, fats, fish, jellies, trifles and cooked rice.

Raw meat and poultry must be refrigerated on delivery. Store raw and cooked meat and poultry separately to prevent cross-contamination.

Refrigerator
The temperature should be 1–4°C (33.8–39.2°F), checked with a thermometer. Bacteria are not destroyed, but those that are significant to man do not grow readily.

(1) Packing
  (i) do not overcrowd the shelves and prevent circulation of cold air;
  (ii) use the coldest part for the most susceptible foods;
  (iii) cover food to prevent loss of moisture and transfer of smells;
  (iv) ensure that all containers are dry and clean; milk bottles should be wiped.
(2) Food must be cooled before refrigeration, otherwise the humidity and temperature will rise in the refrigerator.
(3) Open the door as little as possible and close it quickly.
(4) Defrost and clean weekly.
(5) Regular servicing is essential.

Storage of non-perishable food – storeroom and larder

(1) Keep cool, well ventilated and dry.
(2) Proof against vermin and flies.
(3) Surfaces of shelves, walls and floors should be easy to clean.
(4) Arrange packs in an orderly manner:
  (i) food 450 mm (18 inches) from floor unless in mobile metal bins;
  (ii) stock rotated – dated on delivery.

Deep freeze unit
The temperature, –18°C (–0.4°F), is much lower than the refrigerator. Use

for long-term storage; food must be well protected. Defrost periodically; regular servicing is essential.

Read and follow instructions on packs of frozen foods. Completely thawed food should not be refrozen so that good quality is maintained. Rotate stocks.

Vegetables
Keep in cool dry storage away from direct sunlight, preferably screened from the kitchen to keep soil bacteria away from food preparation and cooking areas. Store in bins or on pallets raised from the floor.

Rapid cooling larders
Rooms with extractor fans and adequate intake of air through filter pads to give moving air currents should be available. Impervious materials should be used for all surfaces.

Food preparation – cooking and serving (Session 7)

Remember

(1) Conditions which favour growth of bacteria must be avoided, especially the length of time food is left at temperatures suitable for bacterial growth.
(2) Foods which are sources of food-poisoning bacteria and those which encourage growth need special care; for example, raw and cooked meat, foods with meat as a base, made-up meat and fish dishes, sandwiches, raw and cooked poultry, sweets with lightly cooked or raw eggs (ducks' or hens'), milk products and fresh and imitation cream. Keep raw pet foods out of commercial kitchens; in the home, store apart from household foods and use separate utensils. Cook well.
(3) Particular care is needed in the summer months (June to October in the northern hemisphere); bacteria grow readily in warm weather.

Kitchen practice

Personal hygiene
Bacteria from all sources can be passed to food by hands. Wash frequently before and during food preparation, particularly after handling raw poultry and meat, after use of the WC and after using a handkerchief. In spite of washing do not touch cooked food.

Cooking techniques
Heat penetrates slowly and is lost slowly in traditional cooking methods; with microwaves heat transfer is rapid, but not necessarily uniform, depending on the design of oven. Infra-red rays are used for browning only. Cooking does not always destroy bacterial spores or even bacteria, particularly in rolled and stuffed joints, poultry, large meat pies and sausages.

(1) Avoid partial cooking.
(2) Cool cooked food rapidly and refrigerate within 1.5 hours. Cooling can be speeded up by:

  (i) provision of rapid cooling larders;
  (ii) breaking up bulk and placing in shallow containers in a moving current of air;
  (iii) limiting the size of joints to not more than 2.7 kg (6lb)

(3) Avoid reheating food. If it is essential the food must be boiled or recooked thoroughly in all parts.
(4) Keep hot food hot and serve quickly.
(5)

Safe cooking methods	*Less safe cooking methods*
Pressure cooking	Boiling
Grilling	Stewing
Frying	Braising
Roasting of small or thin joints	Roasting of large bulky and rolled joints
Microwave cooking	

Foods requiring special attention

Poultry. Poultry frequently harbour food-poisoning organisms, including *Salmonella* and *Campylobacter*, on the skin, within the carcass and in offal. Surfaces, utensils and hands should be carefully cleaned after processing and particularly between raw and cooked stages. Cloths should not be used to wipe carcasses inside or outside or to cover poultry and meat. Frozen meat and poultry should be properly thawed before being cooked.

Sausages. Sausages, raw meat scraps and minced meat may be contaminated with salmonellae, they should be prepared carefully and cooked well. If sausages are pricked the fork or other utensil should be cleaned in hot water at once.

Egg powders. Meringue powder and dehydrated egg products should be handled with care and used only for well-cooked confectionery and other foods. Prepared cake mixes should not include powdered egg.

Shell eggs. Eggs, both hens' and ducks' should be cooked thoroughly and not used for lightly-cooked foods; all used receptacles and utensils should be cleaned and scalded.

Meat pies. Casseroles, cottage, shepherd's and steak and kidney pies, pasties and similar dishes should be prepared freshly from raw meat. If cooked meat is used, steaming under pressure is the best way to ensure the destruction of heat-resistant organisms including spores. If leftovers are 'warmed up' they must be cooked thoroughly to boiling point to destroy post-cooking contaminants and toxins which may have been formed.

Rolled meats. Rolled joints can be responsible for *C. perfringens* food poisoning, because spores from the outside are rolled into the centre where they are likely to survive. Such rolls of meat should be cooked thoroughly and eaten freshly cooked. Slow cooling and warm storage will encourage the growth of *C. perfringens* in the centre where conditions will be anaerobic.

Cold cooked meats including tongue and ham. Sliced meats and tongues are subject to handling and thus to contamination by staphylococci, which will grow during storage times pending service at dinners and parties.

Chart Food-borne infection in the kitchen and methods of control

	Hazard	*Action in kitchen*
FOOD DELIVERY	Raw meat and poultry	Immediate cold storage Care in handling
PREPARATION	Hands	Wash thoroughly and frequently
	Surfaces, containers, equipment, kitchen tools, boards and cloths	(1) Clean thoroughly with hot water and detergent (2) Heat and chemical disinfection; e.g. hypochlorite solutions (combined with compatible detergent or after washing)
COOKING PROCESSES	*Less safe methods* Boiling Stewing Braising	*Safer methods* Pressure cooking Roasting Grilling Frying
	Large bulks of food	Divide into small quantities
COOLING AFTER COOKING	Long slow cooling	Rapid cooling within 1.5 hours. Refrigerate
SERVING	Long storage in warmth	Serve freshly cooked. Keep hot
	Contamination from hands	Do not handle cooked food
	Contamination from equipment	Suitable equipment kept clean
COOKED MEAT		Serve hot and keep hot, above 63°C (145.4°F), or cool quickly, serve cold; keep cold, below 5°C (41°F)
REHEATING	Survival of bacteria in centre of meats and other dishes	Heat right through to boiling point. Serve immediately
KEEP TIME SHORT:	(i) between cooking and eating; (ii) between cooking and cold storage; (iii) between cold storage, serving and cold storage of remainder	

Sandwiches. Sliced meats are frequently used for sandwich fillings. They should be prepared as near the time required as possible, and if prepared the night before they should be refrigerated wrapped or in a container.
Rice. Cooked rice should not be stored overnight unless cooled and refrigerated. Care must be taken with bulks of rice required in eating places specializing in curries and other dishes intended both for sit-down and take-away meals.
Salads and fruits. Salad, vegetables, including watercress and lettuce, and dessert fruits which cannot be peeled, should be well washed preferably in water containing hypochlorite.
Seafoods. Oysters, mussels and other shellfish should be obtained only from reputable sources. After thawing, frozen cooked prawns and shrimps should be stored cold until required. Care is needed to prevent cross-contamination from raw to cooked food.

Kitchen equipment

Ease of cleaning is an important factor in selecting all surfaces, equipment and utensils.

(1) Keep surfaces, equipment and utensils clean and in good repair; they should not be old or worn.
(2) Slicing machines, mincing machines and can openers require frequent and thorough cleaning; they must be easy to dismantle and reassemble. In-plant cleaning may be necessary for fixed parts of equipment.
(3) Use separate boards for raw meat, cooked meat and vegetables. Choose appropriate materials for ease of cleaning; for example, synthetic and/or natural rubber hardened with plastic fillers, high molecular weight, medium-density polyethylene, or phenolic fibre laminates.

For cleaning use hot water and an anionic or non-ionic detergent combined with or followed by a disinfecting agent such as hypochlorite. Preferably use disposable paper. Mops and cloths should be washed thoroughly, disinfected by heat or hypochlorite daily and hung to dry – if possible in the sun.

Serving

(1) Avoid exposure of susceptible foods in a warm atmosphere. Keep cold food cold, below 5°C (41°F).
(2) Avoid warm storage of cooked food. Keep hot food hot, above 63°C (145.4°F).
(3) Keep displayed food cold and under cover.
(4) Do not reheat food from cold in a warm holding apparatus (hot cupboard, *bain marie*). Place hot food only in such equipment.
(5) Minimize handling of cooked foods; use suitable kitchen tools.
(6) Use new clean paper for wrapping and covering food.
(7) Keep animals and insects out of the kitchen.

Washing-up (Session 8)

Efficient washing-up is necessary to clean and remove bacteria from all dining room and kitchen equipment. The essential provisions are:

(1) Good layout of washing-up area.
(2) Correct temperature of wash and rinse water.
(3) A good detergent suited to the type of water.
(4) Orderly methods of work in rinsing, stacking, racking and storage.

Preparation
The aim is to remove as much food waste as possible before washing.

(1) Cutlery should be rinsed.
(2) Tableware should be scraped and rinsed.
(3) Tableware should be stacked according to kind for hand-washing, or racked according to kind for machine washing. Cutlery should be racked with bowls and prongs uppermost.

Methods

Dish-washing machine
The large machines operate automatically and provide:

(1) A pre-rinse to soften and remove food particles. When using small machines this process must be done by hand.
(2) A detergent wash, temperature 60°C (140°F); an automatic detergent dispenser can be fitted.
(3) A rinse at higher temperature 82–87°C (180–190°F). Remove clean dry ware from racks and store.

Two- or three-sink method. Suitable for domestic and large-scale use.
(1) Rinse, scrape, or wipe off, with paper, food particles.
(2) Wash in hot water, 46–50°C (115–122°F) with measured detergent or detergent/disinfectant.
(3) Rinse in racks in hot water, 77–82°C (170–180°F) (maintained). Both wash and rinse waters should be changed as soon as they become soiled or lose temperature.
(4) Rack for drying before storage. The use of rubber gloves will permit the hands to be immersed in hotter water and will prevent them becoming dry and cracked. Gloves must be dried inside and out and replaced regularly.

One-sink or bowl method
Inefficient since crockery and cutlery may still be contaminated with bacteria. Two bowls for wash and rinse are better.

Drying
When the hot rinse is at the right temperature, dishes in racks will air-dry in 30–40 seconds.

Cloths
Dish cloths and tea towels harbour bacteria and require daily washing and disinfection, preferably by heat. Hang to dry. Unclean cloths can contaminate hands, equipment and cutlery.

Paper
Disposable paper should be used in place of dish cloths and tea towels.

Storage
Covered storage should be provided for tableware.

Waste disposal (Session 9)

In kitchen
Food scraps on floors and surfaces encourage bacterial growth and attract vermin and insects.

Waste can be collected:
(1) In pedal-operated bins with liners which can be emptied regularly.
(2) In paper or plastic bags on pedal-operated stands. Bags can be sealed and put into dustbins, incinerated, or collected by the local authority refuse collection service.

Outside kitchen
Provide sufficient waste bins or paper or plastic sacks to prevent over-spilling.

Bins with well-fitting lids should be placed in the shade on a stand 250 –300 mm (10–12 inches) high above a concrete area with drainage, which can be hosed down. Wet-strength paper or plastic bags with lids should be wall-mounted to give good ground clearance for hosing down.

Paper waste is usually kept separately.

Cans should be rinsed, both ends removed and beaten flat.

Bins (other than for pig food and compost) should be kept as dry as possible by wrapping wet waste.

Waste disposal unit
The unit is attached to the sink waste pipe. The ground waste food passes into the waste pipe and drain. Cans and clothing materials cannot be disposed of in this way.

Vermin and fly control

Rats, mice, and insects
Rats, mice, flies, cockroaches and ants are the most common pests.

If premises do not provide food and shelter, infestation is unlikely.

For extermination, seek expert advice from the local authority.

Flies, including bluebottles
Feet and hairs spread bacteria acquired from excreta and other waste food.

Control

(1) Avoid breeding grounds such as uncovered refuse bins. Bins (other than for pig food and compost) should be kept as dry as possible.
(2) Spray refuse areas in summer to destroy flies and prevent breeding.
(3) Prevent access to kitchen and food by fly-proof windows, doors and ventilators. Cover food.
(4) Electrical devices kill flies without the hazards associated with aerosol sprays.

Cockroaches

Active at night. Attracted to warm places, such as heating pipes. Seal off crevices which provide hiding places.

Close-fitting lids prevent access to food.

Treat area with suitable insecticide.

Sprays and powders

Care must be taken to prevent pesticides reaching food, preparation surfaces and equipment.

Premises (Session 10)

The layout of a kitchen must be planned with the principles of hygiene in mind, with regard to the sources of food-poisoning bacteria (human, raw foods, environment), the importance of hot and cold storage and the prevention of cross-contamination. So far as possible there should be separate work areas for raw and cooked food; for large-scale catering this is essential and should include separate equipment and personnel in the different areas.

Floors

Durable, non-slip surfaces should be impervious to moisture and easy to clean. Equipment preferably is raised or mobile to allow floor to be cleaned underneath.

Walls

Smooth, impervious walls should reduce condensation and be easy to clean. Junctions of wall and floor should be coved for easy cleaning, and equipment should be fixed away from the wall to allow for cleaning.

Ceiling

The surface that encloses the inside of the roof should be smooth, anticondensation, and easy to clean. Acoustic tiles may be used.

Lighting

Illumination must be good, both natural and artificial, particularly over work and preparation areas, sinks and cooking equipment. Shadows should be avoided.

Ventilation

Good natural and mechanical ventilation is necessary to prevent a rise in temperature and humidity (discomfort and rapid growth of bacteria).

Sanitary convenience
Toilets must not open directly into food-preparation rooms. Foot-operated flushes are desirable.

Wash-basins should be available:
(1) In or adjacent to the toilet.
(2) In kitchen preparation areas. They should be supplied with:
  (i) hot and cold water; foot operation is preferable;
  (ii) soap dispenser, which must be kept in an hygienic condition;
  (iii) nail brush with plastic or nylon back and bristle, although not all plastics will withstand hot water disinfection;
  (iv) an individual method of hand drying, such as paper towels, continuous roll towels or hot-air dryers. Roller towels (old type) should not be used because there is danger of cross-infection;
  (v) antiseptic hand cream should be available.

Cloakroom
Adequate hanging space for outer clothing must be provided in a separate room for drying also.

Food preparation surfaces
Wood is unsuitable because it is too difficult to clean. The best materials are stainless steel or a laminate; these are smooth, impermeable surfaces, which are easy to clean, thus the number of bacteria in the environment will be reduced.

Chopping boards
Wood is still used but it constitutes a hazard. In recent years several new cutting surfaces have been developed, and their use in this country has become increasingly popular. The new proprietary boards or pads are usually made of (a) synthetic and/or natural rubber hardened with plastic fillers, (b) high molecular weight, medium density polypropylene, or (c) phenolic fibre laminates. Thorough cleaning is necessary after use, followed by a disinfectant or 'sanitizing' agent, such as hypochlorite. Separate surfaces are required for raw meat, cooked meat and vegetables.

Hot water
There should be a plentiful supply of hot water.

Sinks
Stainless steel with sink and drainer in one piece is recommended. There should be separate sinks for vegetable preparation.

Refrigerator and/or cold room
Facilities for cold storage are essential for safety and they must be of adequate size. There should be adjustable metal shelves in the cold room.

Fixed equipment
The design must allow easy cleaning, and avoid ledges, nooks, and crannies; fix away from walls.

Facilities
Rapid cooling is most essential.

Cleaning materials
Mops, brushes and cloths should be cleaned and heat-disinfected regularly and stored dry to prevent multiplication of bacteria in the wet materials. Where possible single-use disposable cloths should be used. Articles for cleaning should not be stored in food rooms, but in cleaners' cupboards or small rooms with draining and drying racks and a supply of hot and cold water. Buckets should be cleaned, disinfected, drained and stored inverted.

Summary of preventive care

(1) Do not touch with the hands: cooked, prepared food – particularly meats and poultry.
(2) Keep time short between: cooking and eating, cooking and refrigeration, refrigeration and eating.
(3) Watch environment: clean well and disinfect – by heat where possible, separate raw and cooked foods.

Appendix B

Bibliography and further reading

Catering

Airline Caterers Technical Coordinating Committee (1990). *Airline Catering. Code of good catering practice*. London : ACTCC (Heathrow Airport).
Catering Research Unit, The University, Leeds (1975). *A Manual on Cook-Freeze Catering*. Luton: Local Government Training Board.
Charles, R.H.G. (1983). *Mass Catering*. WHO Regional Publications. European Series No. 15. Copenhagen: World Health Organization.
Department of Health (1989). *Chilled and Frozen. Guidelines on Cook-Chill and Cook-Freeze Catering Systems*. London : HMSO.
Department of Health and Social Security (1986). *Health Service Catering Manual, Hygiene*. London : HMSO.
Department of Health and Social Security, Scottish Home and Health Department, Northern Ireland Ministry of Health and Social Services, Welsh Office (1972). *Clean Catering*. London: HMSO.
Majewski, C. (1990). Sous-vide – New technology catering? *Environmental Health* **98**, 100–102.
Schafheitle, J.M. (1991). The *Sous-vide* system for preparing chilled meals. *British Food Journal* **92**, 5, 23–7.
Sous-Vide Advisory Committee (1991). *Code of Practice for sous-vide catering systems*. (Voss Training services – distribution).
Wilkinson, P.J., Dart, S.P. and Hadlington, C.J. (1991). Cook-chill, cook-freeze, cook-hold, *sous-vide*: risks for hospital patients? *Journal of Hospital Infection* **19** (Suppl A), 222–9.

Disinfection

Ayliffe, G.A.J., Coates, D. and Hoffman, P.N. (1984). *Chemical Disinfection in Hospitals*. London: Public Health Laboratory Service.
Block, S.S. (ed.) (1983). *Disinfection, Sterilization and Preservation*. 3rd edition. Philadelphia: Lea and Febiger.
Department of Health and Social Security (1980). *Sterilizers*. Health

Technical Memorandum No. 10. London: HMSO.

Hoffman, P.N., Cooke, E.M., McCarville, M.R. and Emmerson, A.M. (1985). Microorganisms isolated from skin under wedding rings worn by hospital staff. *British Medical Journal* **290**, 206–7.

Maurer, I.M. (1985). *Hospital Hygiene*. 3rd edition. London: Edward Arnold.

Perkins, J.J. (1969). *Principles and Methods of Sterilization in Health Sciences*. 2nd edition. Springfield, Illinois: Charles C. Thomas.

Pawa, R.R. and Hobbs, B.C. (1980). Control of infection in a neonatal nursery. *Indian Journal of Paediatrics* **47**, 375–80.

Russell, A.D., Hugo, W.B. and Ayliffe, G.A.J. (eds) (1992). *Principles and Practice of Disinfection, Preservation and Sterilization*. 2nd edition. Oxford: Blackwell Scientific Publications.

Education

Bates, D. (1987). *Food Safety. An international source list of audiovisual materials*. London: British Life Assurance Trust Centre for Health and Medical Education.

Food Safety Advisory Centre (1991). *Food Safety. Your questions answered*. London: Food Safety Advisory Centre.

Food Safety Advisory Centre leaflets:
The Good Food Safety Guide
Safe Food
A Consumers Guide to Safe Handling of Food in Shops
All you need to know about nutrition
A Consumers guide to Genetic Engineering
The facts about BSE
Food Irradiation

Hayes, S. (1991). *Caring for our food*. London: National Dairy Council.

Jacob, M. (1989). *Safe food handling. A training guide for managers of food service establishments*. Geneva: World Health Organization.

Whyte, B.H. (1984). Medical Microbiology. In: *Information Sources in the Medical Sciences*. Morton, L.T. and Godbolt, S. (eds). 3rd edition. London: Butterworths pp. 250–80.

Williams, T., Moon, A. and Williams, M. (1990). *Food, Environment and Health*. Geneva: World Health Organization.

Epidemiology

Barker, D.J.P. (1982). *Practical Epidemiology*. 3rd edition. Edinburgh: Churchill Livingstone.

Last, J.M. (ed.) (1983). *A Dictionary of Epidemiology*. International Epidemiological Association. Oxford: Oxford University Press.

Lowe, C.R. and Kostrzewski, J. (eds). (1973). *Epidemiology. A Guide to Teaching Methods*. International Epidemiological Association. Edinburgh: Churchill Livingstone.

Foodborne diseases

Anderson E.S. and Hobbs B.C. (1973). Studies of the strain of *Salmonella typhi* responsible for the Aberdeen typhoid outbreak. *Israeli Journal of Medical Science* **9**, 162–74.

Appleton, H., Palmer, S.R. and Gilbert, R.J. (1981). Foodborne gastro-enteritis of unknown aetiology: a virus infection? *British Medical Journal* **281**, 1801–2.

Bartholomew, B.A., Berry, P.R., Rodhouse J.C., Gilbert R.J. and Murray, C.K. (1989). Scombrotoxic fish poisoning in Britain: features of over 250 suspected incidents from 1976 to 1986. *Epidemiology and Infection* **99**, 775–82.

British Association for the Advancement of Science (1978). *Salmonella; The Food Poisoner*. Report by Study Group 1975–1977.

British Egg Marketing Board (Symposium) (1971). *Poultry Disease and World Economy*. Gordon, R.F. and Freeman, B.M. (eds). Symposium No. 7. Edinburgh: Longmans.

Brouwer, R., Mertens, M.J.A., Siem, T.H. and Katchaki, J. (1979). An explosive outbreak of campylobacter enteritis in soldiers. *Antonie van Leeuwenhoek* **45**, 517–19.

Cowden, J.M., O'Mahony, M., Bartlett, C.L.R., Rana, B., Smyth, R., Lynch, D., Tillett, H., Ward. L., Roberts, D., Gilbert, R.J., Baird-Parker, A.C. and Kilsby, D.C. (1989). A national outbreak of *Salmonella typhimurium* DT 124 caused by contaminated salami sticks. *Epidemiology and Infection* **103**, 219–25.

Doyle, M.F. (ed) (1989). *Foodborne Bacterial Pathogens*. New York: Marcel Dekker, Inc.

Fleming, D.W., Cochi, S.L., MacDonald, K.L., Brondum, J., Hayes, P.S., Plikaytis, B.D., Holmes, M.B., Audurier, A., Broome, C.V. and Reingold, A.L. (1985). Pasteurized milk as a vehicle of infection in an outbreak of listeriosis. *New England Journal of Medicine* **312**, 404–7.

Gilbert, R.J. (1983). Food-borne infections and intoxications – recent trends and prospects for the future. In: *Food Microbiology: Advances and Prospects*, Roberts, T.A. and Skinner, F.A. (eds). Society for Applied Bacteriology Symposium. Series No. 11. London: Academic Press pp. 47–66.

Gill, O.N., Sockett, P.N., Bartlett, C.L.R., Vaile, M.S.B., Rowe, B., Gilbert, R.J., Dulake, C., Murrell, H. and Salmaso, S. (1983). Outbreak of *Salmonella napoli* infection caused by contaminated chocolate bars. *Lancet* **i**, 574–7.

Health and Welfare Canada. *Canada Diseases Weekly Report*. Ottawa: Bureau of Epidemiology, Laboratory Centre for Disease Control.

Health and Welfare Canada (1991). *Food-borne Disease in Canada. A 10-year Summary* 1975–1984. Ottawa: Health Protection Branch.

Kirov, S.M., Wellock, R. and Goldsmid, J.M. (1984). *Aeromonas* species as enteric pathogens. *Australian Microbiologist* **5**, 210–14.

Kramer, J.M., Turnbull, P.C.B., Munshi, G. and Gilbert, R.J. (1982). Identification and characterization of *Bacillus cereus* and other *Bacillus* species associated with foods and food poisoning. In: *Isolation and*

Identification Methods for Food Poisoning Organisms. Corry, J.E.L., Roberts, D. and Skinner, F.A. (eds). Society for Applied Bacteriology Technical Series No. 17. London: Academic Press pp. 261–86.

Lancet Review (1991). *Foodborne Illness*. London: Edward Arnold.

McLauchlin, J., Hall, S.M., Velani, S.K. and Gilbert, R.J. (1991). Human listeriosis and pâté: a possible association. *British Medical Journal* **303**, 773–5.

Morbidity and Mortality Weekly Report (1985). Listeriosis outbreak associated with Mexican-style cheese – California. *Morbidity and Mortality Weekly Report* **34**, No. 24, 357–9.

O'Mahony, M., Cowden, J., Smyth, B., Lynch, D., Hall, M., Rowe, B., Teare, E.L., Tettmar, R.E., Rampling, A.M., Coles, M., Gilbert, R.J., Kingcott, E. and Bartlett, C.L.R. (1990). An outbreak of *Salmonella saint-paul* infection associated with beansprouts. *Epidemiology and Infection* **104**, 229–35.

O'Mahony, M., Mitchell, E., Gilbert, R.J., Hutchinson, D.N., Begg, N.T., Rodhouse, J.C. and Morris, J.E. (1990). An outbreak of foodborne botulism associated with contaminated hazelnut yoghurt. *Epidemiology and Infection* **104**, 389–95.

Parker, M.T. and Collier, L.H. (1990) (eds). *Principles of Bacteriology, Virology and Immunity*. 8th edition. London: Edward Arnold.

Pearson, A.D., Bartlett, C.L.R., Page, G., Jones, J.M.W., Lander, K.P., Lior, H. and Jones, D.M. (1983). A milk-borne outbreak in a school community – a joint medical veterinary investigation. In: *Campylobacter II*. Proceedings of the Second International Workshop on Campylobacter Infections, Brussels. London: Public Health Laboratory Service pp. 97–8.

Riemann, H. and Bryan, F.L. (1979). *Food-borne Infections and Intoxications*. 2nd edition. New York: Academic Press.

Roberts, D. and Roberts, C. (eds) (1992). Cholera Update. *PHLS Microbiology Digest* **9**, 13–44.

Rodhouse, J.C., Haugh, C.A., Roberts, D. and Gilbert, R.J. (1990). Red kidney bean poisoning in the UK: an analysis of 50 suspected incidents between 1976 and 1989. *Epidemiology and Infection* **105**, 485–91.

Ryser, E.T. and Marth, E.H. (1991). *Listeria, Listeriosis and Food Safety*. New York: Marcel Dekker, Inc.

Skirrow, M.B. (1989). Campylobacter perspectives. *PHLS Microbiology Digest* **6**, 113–7.

Stringer, M.F. (1985). *Clostridium perfringens* Type A food poisoning. In: *Clostridia and gastrointestinal disease*. Borriello, S.P. (ed). Boca Raton, Florida: CRC Press, pp 117–43.

Turnbull, P.C.B., Lee, J.V., Miliotis, M.D., van de Walle, S., Koornhof, H., Jeffrey, L. and Bryant, T.N. (1984). Enterotoxin production in relation to taxonomic grouping and source of isolation of *Aeromonas* species. *Journal of Clinical Microbiology* **19**: 175–80.

WHO (1988). *Foodborne Listeriosis*. Report of a WHO Informal Working Group. WHO/EHE/FOS/88.5. Geneva: World Health Organization.

WHO (1985). *Programme for Control of Diarrhoeal Diseases*. Fourth Programme Report, 1983–1984. WHO/CDD/85.13.

Williams, L.P. and Hobbs, B.C. (1975). Enterobacteriaceae Infections. In: *Diseases Transmitted from Animals to Man*. Hubbert, W.T., McCulloch, W.F. and Schnurrenberger, P.R. (eds). 6th edition. Springfield, Illinois: Charles C. Thomas, Publisher pp. 33–109.

Wilson, P.G., Davies, J.R., Hoskins, T.W., Lander, K.P., Lior, H., Jones, D.M. and Pearson, A.D. (1983). Epidemiology of an outbreak of milk-borne enteritis in a residential school. In: *Campylobacter II*. Proceedings of the Second International Workshop on Campylobacter Infections, Brussels, 143. London: Public Health Laboratory Service p. 143.

Woolaway, M.C., Bartlett, C.L.R., Wieneke, A.A., Gilbert, R.J., Murrell, H.C. and Aureli, P. (1986). International outbreak of staphylococcal food poisoning caused by contaminated lasagne. *Journal of Hygiene* **96**, 67–73.

Food hygiene

Christie, A.B. and Christie, M.C. (1971). *Food Hygiene and Food hazards*. London: Faber and Faber.

Department of Health and Social Security (1964). *Food Hygiene in Hospitals*. Ministry of Health Circular HM(64)34.

Hobbs, B.C. and Gilbert, R.J. (1975). Food Hygiene and Sanitation. In: *Current Topics in Applied Microbiology*. Tauro, P. and Varghese, T.M. (eds). International Bioscience Monographs 2. Hissar, Madras: International Bioscience Publishers.

Hobbs, B.C. and McLintock, J.S. (1979). *Hygienic Food Handling*. 3rd edition. London: The St. John Ambulance Association.

Linton, A.H. (ed) (1983). *Guidelines on prevention and control of salmonellosis*. VPH/83.42. Geneva: World Health Organization.

Ministry of Agriculture, Fisheries and Food (1981). *Home Preservation of Fruit and Vegetables*. Bulletin 21. London: HMSO.

Roberts, D. (1982). Factors contributing to outbreaks of food poisoning in England and Wales 1970–1979. *Journal of Hygiene* **89**, 491–8.

Sprenger, R.A. (1985). *Hygiene for Management*. 5th edition. Rotherham: Highfield Publications.

WHO Expert Committee Convened in Cooperation with FAO (1974). *Fish and Shellfish Hygiene. Technical Report Series*, No. 550. Geneva: World Health Organization.

WHO Expert Committee with the Participation of FAO (1976). *Microbiological Aspects of Food Hygiene. Technical Report Series*, No. 598. Geneva: World Health Organization.

WHO (1977). *Food Hygiene in Catering Establishments. Legislation and Model Regulations*. Offset Publication No. 34. Geneva: World Health Organization.

WHO (1985). *Report of Round Table Conference on Meat Hygiene in Developing Countries*. VPH/85.60. Geneva: World Health Organization.

WHO (1985). *Report of the Working Group of the WHO Veterinary Public Health Programme on Prevention and Control of Salmonellosis (and*

other Zoonotic Diarrhoea Diseases). VPH/85.61. Geneva: World Health Organization.
Wood, P.C. (1976). *Guide to Shellfish Hygiene*. WHO Offset Publication, No. 31. Geneva: World Health Organization.

Food microbiology / food contamination

Austwick, P.K.C. (1975). Mycotoxins. In: *Chemicals in Food and Environment. British Medical Bulletin* **31**, 222–9.
Brown, M.H. (ed) (1982). *Meat Microbiology*. London: Applied Science Publishers.
Edel, W. and Kampelmacher, E.H. (1969). *Salmonella* isolation in nine European laboratories using a standardized technique. *Bulletin of the World Health Organization* **41**, 297–306.
Edel, W. and Kampelmacher, E.H. (1974). Comparative studies on *Salmonella* isolations from feeds in ten laboratories. *Bulletin of the World Health Organization* **50**, 421–6.
Frazier, W.C. and Westhoff, D.C. (1978). *Food Microbiology*. 3rd edition. New York: McGraw-Hill Book Co., Inc.
Gilbert, R.J. (1982). The microbiology of some foods imported into England through the Port of London and Heathrow (London) Airport. In: *Control of the Microbial Contamination of Foods and Feeds in International Trade: Microbial Standards and Specifications*. ed. Kurata, H. and Hesseltine, C.W. (eds). Tokyo: Saikon Publishing Co. Ltd, pp 105–119.
Greenwood, M.H., Roberts, D., and Burden, P. (1991). The occurrence of *Listeria* spp. in milk and dairy products: a national survey in England and Wales. *International Journal of Food Microbiology* **12**, 197–206.
Harrigan, W.F. and McCance, M.E. (1976). *Laboratory Methods in Food and Dairy Microbiology*. New York: Academic Press.
Hayes, P.R. (1985). *Food Microbiology and Hygiene*. London: Elsevier Applied Science Publishers.
Hersom, A.C. and Hulland, E.D. (1980). *Canned Foods*. 7th edition. Edinburgh: Churchill Livingstone.
Hobbs, B.C. (1976), Microbiological hazards of international trade. In: *Microbiology in Agriculture, Fisheries and Food*. Skinner, F.A. and Carr, J.G. (eds). Society for Applied Bacteriology Symposium Series No. 4. London: Academic Press pp. 161–80.
Hobbs, B.C. (1982). Observations on public health microbiology including experience in India. *Food Technology in Australia* **34**, 501–7.
International Commission on Microbiological Specifications for Foods (1978). *Microorganisms in Foods, 1. Their significance and methods of enumeration*. 2nd edition. Canada: University of Toronto Press.
International Commission on Microbiological Specifications for Foods (1980). *Microbial Ecology of Foods, 3. Volume I. Factors Affecting Growth and Death of Microorganisms. Volume II. Food Commodities*. London: Academic Press (2nd edition in preparation).

Lawrie, R.A. (1985). *Meat Science*. 4th edition. Oxford: Pergamon Press Ltd.
Mathur, R. and Reddy, V. (1983). Bacterial contamination of infant foods. *Indian Journal of Medical Research* **77**, 342–6.
McLauchlin, J. and Gilbert, R.J. (1990). *Listeria* in Food. *PHLS Microbiology Digest* **7**, 54–5.
Parry, T.J. and Pawsey, R.K. (1973). *Principles of Microbiology for Students of Food Technology*. London: Hutchinson Educational Ltd.
Public Health Laboratory Service Working Group, Skovgaard, N. and Nielsen, B.B. (1972). Salmonella in pigs and animal feeding stuffs in England and Wales and in Denmark. *Journal of Hygiene* **70**, 127–40.
Roberts, T.A. and Skinner, F.A. (eds) (1983). *Food Microbiology. Advances and Prospects*. Society for Applied Bacteriology Symposium Series No. 11. London: Academic Press.
Roberts, D., Gilbert, R.J., Nicholson, R., Christopher, P., Roe, S. and Dailley, R. (1989). The microbiology of airline meals. *Environmental Health* **97**, 56–62.
Velani, S.K. and Roberts, D. (1991). *Listeria monocytogenes* and other *Listeria* spp. in pre-packed mixed salads and individual salad ingredients. *PHLS Microbiology Digest* **8**, 21–2.

Food Safety

Association for the Study of Infectious Disease (Symposium) (1974). Food. Is it Safe? Medlock, J.M. (ed). *Postgraduate Medical Journal* **50**, No. 588.
Campden Food and Drink Research Association (1987). *Microbiological and Environmental Health Problems Relevant to the Food and Catering Industries*. Proceedings of a joint CFDRA/PHLS Symposium, Stratford-upon-Avon, January 1987. Chipping Campden: CFDRA.
Campden Food and Drink Research Association (1990). *Microbiological and Environmental Health Issues Relevant to the Food and Catering Industries*. Proceedings of a joint CFDRA/PHLS Symposium, Stratford-upon-Avon, February 1990. Chipping Campden: CFDRA.
Committee on the Microbiological Safety of Food (Chairman, Sir Mark Richmond). (1990). *The Microbiological Safety of Food*. Report, Part 1, London: HMSO.
Committee on the Microbiological Safety of Food (Chairman, Sir Mark Richmond). (1991). *The Microbiological safety of food*. Report, Part II, London: HMSO.
FAO/WHO (1980). *Safe Food for All*. Proceedings of the 1st World Congress, Foodborne Infections and Intoxications. Institute of Veterinary Medicine, Berlin: FAO/WHO Collaborating Centre for Research and Training in Food Hygiene and Zoonoses.
FAO/WHO (1986). *Safe Food for All*. Proceedings of the 2nd World Congress, Foodborne Infections and Intoxications. Institute of Veterinary Medicine, Berlin: FAO/WHO Collaborating Centre for Research and

Training in Food Hygiene and Zoonoses.
FAO/WHO (1992). *Safe Food for All*. Proceedings of the 3rd World Congress, Foodborne Infections and Intoxications. Institute of Veterinary Medicine, Berlin: FAO/WHO Collaborating Centre for Research and Training in Food Hygiene and Zoonoses.
Harrigan W.F. and Park, R.W.A. (1991). *Making Safe Food*. A management guide for microbiological quality. London: Academic Press.
Hobbs, B.C. and Christian. J.H.B. (eds) (1973). *The Microbiological Safety of Food*. London: Academic Press.
Johnson, R. (ed) (1981). *Food Safety Services*. Public Health in Europe 14. Copenhagen: World Health Organization.
Jowitt, R. (ed.) (1980). *Hygienic Design and Operation of Food Plant*. Chichester: Ellis Horwood Ltd.
WHO (1988). *Salmonellosis control: the role of animal and product hygiene*. Report of a WHO Expert Committee. Technical Report Series 774. Geneva: World Health Organization.
WHO (1988). *Food Irradiation. A technique for preserving and improving the safety of our food*. Geneva: World Health Organization.

HACCP

Bryan F.L. (1992). *Hazard Analysis Critical Control Point Evaluations*. A guide to identify hazards and assessing risk associated with food preparation and storage. Geneva: World Health Organization.
Campden Food and Drink Research Association (1987). *Guidance to the establishment of Hazard Analysis Critical Control Points (HACCP)*. Technical Manual No. 19. Chipping Campden, Gloucs: CFDRA.
International Commission on Microbiological Specifications for Foods (1988). *Microorganisms in Food. Book 4: Application of the Hazard Analysis Critical Control Point (HACCP) system to ensure microbiological safety and quality*. Oxford: Blackwell Scientific Publications.
International Association of Milk, Food and Environmental Sanitarians (1991). Procedures to implement the hazard analysis critical control point system. Ames, Iowa: IAMFES Inc.
Mayes, T. and Kilsby, D.C. (1989). The use of HAZOP hazard analysis to identify critical control points for the microbiological safety of food. *Food Quality and Preference*. **1**, 53–7.
WHO (1992). *Report of the WHO/ICMSF Meeting on Hazard Analysis Critical Control Point System in Food Hygiene*. VPH/82.37. Geneva: World Health Organization.

Legislation and Codes of Practice

Anderton, A., Howard, J.P. and Scott, D.W. (1987) *Microbiological Control in Enteral Feeding*. A Guidance Document. London: Parenteral and Enteral Nutrition Group, The British Dietetic Association.
Association of Public Analysts; Dairy Trade Federation; Institution of Environmental Health Officers; Milk Marketing Board of England

and Wales; National Farmers Union; Public Health Laboratory Service (1989). *Guidelines for the Sampling and Testing of Pasteurized Milks for Enforcement Purposes*. London: HMSO.

Brown, K.L. and Oscroft, C.A. (1989). *Guidelines for the hygienic manufacture, distribution and retail sale of sprouted seeds with particular reference to mung beans*. Technical Manual No. 25. Chipping Campden: Campden Food and Drink Research Association.

Creamery Proprietors' Association (1988). *Guidelines for good hygienic practice in the manufacture of soft and fresh cheeses*. London: Creamery Proprietors' Association.

Committee of Inquiry into the Future Development of the Public Health Function (Report). (1988). *Public Health in England*. London: HMSO.

Department of Health. (1990). *Guidelines on the Food Hygiene (Amendment) Regulations 1990*. London: HMSO

Department of Health. (1992). *Guidelines for the Catering Industry on the Food Hygiene (Amendment) Regulations 1990 and 1991*. London: HMSO.

Department of Health and Social Security, Ministry of Agriculture, Fisheries and Food, and Welsh Office. *Food Hygiene Codes of Practice*:

No.3 (1960). *Hygiene in the Retail Fish Trade.*

No.4 (1960). *The Hygienic Transport and Handling of Fish Trade.*

No.5 (1961). *Poultry Dressing and Packing.*

No.6 (1966). *Hygiene in the Bakery Trade and Industry.*

No.7 (1967). *Hygiene in the Operation of Coin Operated Food Vending Machines.*

No.8 (1969). *Hygiene in the Meat Trades.*

No.9 (1972). *Hygiene in Microwave Cooking.*

No.10(1981). *The Canning of Low Acid Foods.*

London: HMSO.

DHSS Advisory Memorandum on the Processing of Pasteurised Large Canned Hams. (Covering letter to Medical Officers of Health, Reference A/C291/7, dated 14 January 1972).

Milk Marketing Board. (1989). *Guidelines for Good Hygienic Practice for the manufacture of soft and fresh cheeses in small and farm-based production units*. Thames Ditton, Surrey: Milk Marketing Board.

Ministry of Agriculture Fisheries and Food, Department of Health. Food Safety Act 1990. Code of Practice No 1. *Responsibility for Enforcement of the Food Safety Act 1990*. London: HMSO.

Ministry of Agriculture Fisheries and Food, Department of Health. Food Safety Act 1990. Code of Practice No 9: *Food Hygiene Inspections*. London: HMSO.

Ministry of Agriculture Fisheries and Food, Department of Health. Food Safety Act 1990. Code of Practice No 10: *Enforcement of the Temperature Control Requirements of Food Hygiene Regulations*. London: HMSO

Ministry of Agriculture Fisheries and Food, Department of Health. Food Safety Act 1990. Code of Practice No 11. *Enforcement of the Food Premises (Registration) Regulations*. London: HMSO.

Review of Law on Infectious Disease Control. Consultation Document. (1989). Department of Health.

Technical Committee DAC/17 (now AFC/17). (1990). Draft British Standard Code of Practice for the Pasteurization of Milk on Farms and in Small Dairies. Document 90/50231. Cleaning and disinfection of dairy equipment. 15 February 1990.

Reports

Department of Health and Social Security and Ministry of Agriculture, Fisheries and Food (1969). *Joint Committee on the Use of Antibiotics in Animal Husbandry and Verterinary Medicine; Report*. Cmnd. 4190. London: HMSO.

Department of Health and Social Security (1986). *The Report of the Committee of Inquiry into an Outbreak of Food Poisoning at Stanley Royd Hospital*. Cmnd. 9716. London: HMSO.

Scottish Home and Health Department (1964). *The Aberdeen Typhoid Outbreak 1964: Report of the Departmental Committee of Enquiry*. (Chairman: Sir David Milne) Cmnd. 2542. Edinburgh: HMSO.

Sampling/microbiological specifications

Board, R.G. and Lovelock. D.W. (eds) (1973). *Sampling – Microbiological Monitoring of Environments*. Society for Applied Bacteriology Technical Series No. 7. London: Academic Press.

Christian, J.H.B. (1983). *Microbiological Criteria for Foods*. Summary of Recommendations of FAO/WHO Expert Consultations and Working Groups 1975–1981. VPH/83.54. Geneva: World Health Organization.

FAO/WHO Expert Consultation (1975). *Microbiological Specifications for Foods*. GC/Microbiol/75/Report 1. Rome: Food and Agriculture Organization of the United Nations.

FAO/WHO Second Joint Expert Consultation (1977). *Microbiological Specifications for Foods*. EC/Microbiol/77/Report 2. Rome: Food and Agriculture Organization of the United Nations.

International Commission on Microbiological Specifications for Foods (1986). *Microorganisms in Foods, 2. Sampling for microbiological analysis: principles and specific application*. 2nd edition. Oxford: Blackwell Scientific Publications.

National Research Council (US) Food Protection Committee. Subcommittee on Microbiological Criteria (1985). *An evaluation of the role of microbiological criteria for foods and food ingredients*. Washington: National Academy Press.

WHO Study Group on Food-borne Disease (1974). *Methods of Sampling and Examination of Surveillance Programmes*. Technical Report Series, No. 543. Geneva: World Health Organization.

Surveillance/food-poisoning statistics

Bryan, F.L. (1973). *Guide for Investigating Foodborne Disease Outbreaks and Analyzing Surveillance Data*. Atlanta, Georgia. Department for Health, Education and Welfare, Center for Disease Control.

Communicable Disease Centers (1983). *Foodborne Disease Surveillance*. Annual Summary 1981. USA: Department of Health and Human Services.

Department of Health and Social Security (1982). *Food Poisoning. The Investigation and Control of Food Poisoning in England and Wales*. Memo. 188/Med. London: HMSO.

Health and Welfare, Canada (1991). *Foodborne and Waterbourne Diseases in Canada 1985–86*. Ottawa: Health Protection Branch.

Public Health Laboratory Service Salmonella Subcommittee (1990). *Notes on the Control of Human Sources of Gastro-Intestinal Infections and Bacterial Intoxications in the United Kingdom*. Communicable Disease Report Supplement 1. London: PHLS.

Sockett, P.N. (1991). Food poisoning outbreaks with manufactured foods in England and Wales: 1980–1989. *Communicable Disease Report* **1**, Review No. 10, R105–9.

US Department of Health and Human Services/Public Health Service. *Morbidity and Mortality Weekly Report*. Atlanta, Georgia: Communicable Disease Centers.

WHO (1991). *Surveillance programme for the control of foodborne infections and intoxications in Europe*. Fourth Report 1983/84. Institute of Veterinary Medicine, Berlin: WHO/FAO Collaborating Centre for Research and Training in Food Hygiene and Zoomoses.

WHO Surveillance Programme for Control of Foodborne Infections and Intoxications in Europe. *Newsletter*. Berlin: FAO/WHO Collaborating Centre for Food Hygiene and Research and Training in Zoonoses. (2–3 issues per annum).

Travel

Royal Society of Edinburgh and the Royal College of Physicians, Edinburgh (Symposium) (1982). *Travel: Disease and Other Hazards. Proceedings of the Royal Society of Edinburgh* **82B**, 1–144.

Walker, E. and Williams, G. (1983). ABC of Healthy Travel. Preventing illness while abroad. *British Medical Journal* **286**, 960.

WHO Report on a Working Group (1977). *Aviation Catering*. Copenhagen: World Health Organization.

Miscellaneous

Drasar, B. and Hill, M.J. (1974). *Human Intestinal Flora*. London: Academic Press.

Steele, J.H. (ed). (1979). *Handbook Series in Zoonoses. Volume 1*. (Volume 2, 1980); Boca Raton, Florida: CRC Press.

White, W.R. and Passmore, S.M. (eds) (1985). *Microbial Aspects of Water Treatment*. Society for Applied Bacteriology Symposium Series No. 14. *(Supplement to Journal of Applied Bacteriology* **59**, 1S–245S.)

Index